D0218393

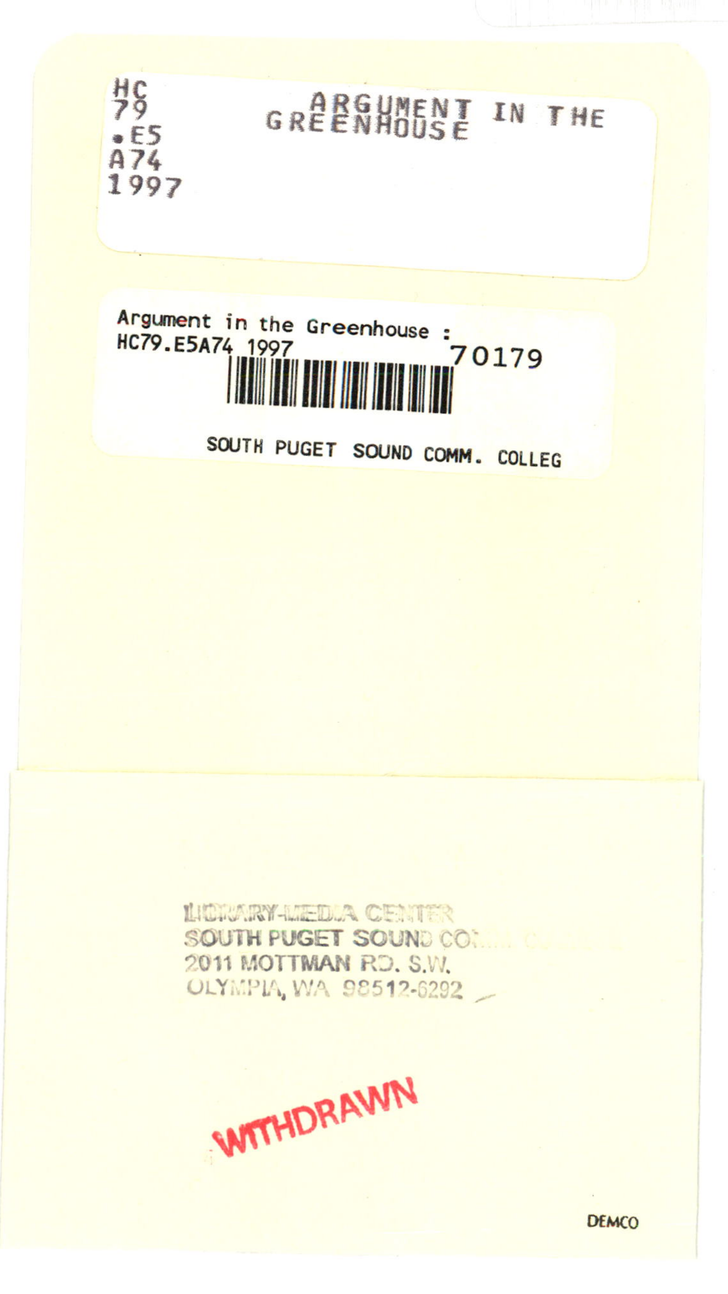
HC
79
.E5
A74
1997
ARGUMENT IN THE
GREENHOUSE
Argument in the Greenhouse :
HC79.E5A74 1997
70179
SOUTH PUGET SOUND COMM. COLLEG
SOUTH PUGET SOUND
2011 MOTTMAN RD. S.W.
OLYMPIA, WA 98512-6292
WITHDRAWN
DEMCO

ARGUMENT IN THE GREENHOUSE

How can greenhouse gases be controlled and reduced? Will it be in time? Who will pay?

Argument in the Greenhouse adds a significant new contribution to the climate change/global warming debate. Incorporating the key political and legal considerations into 'real world' applied economic analysis, the authors provide a unique focus on the wider political economy of the problem.

All the key issues of controlling climate change (costs, timing and degree of stabilisation, ecological tax reform, role of developing countries, and evolution of international agreements) are placed firmly within the current legal and political economy context, with state-of-the-art economic theory, econometrics and economics modelling introduced to analyse different policy proposals and address this applied policy problem.

Covering both the developing and developed world, the role of uncertainty, inaccuracy and learning in defining policy is extensively discussed and its implications explained. Detailed policy analysis of near term options is carried out with reference to the ongoing process of negotiating international controls.

Argument in the Greenhouse looks at the potential for action to prevent global warming in the next 5–10 years and in the longer term, and identifies the important factors for fostering effective agreements on emissions reductions. The authors suggest realistic policies which are likely to receive actual support at both international and domestic levels.

Students and policy makers alike, within the field of environmental policy or economics, will find this book an invaluable new contribution to the literature.

Nick Mabey is a Research Fellow at the Centre for Economic Forecasting, London Business School; **Stephen Hall** is Professor of Economics, Imperial College; **Clare Smith** is a Consultant on energy and environment issues; **Sujata Gupta** is a Research Fellow at the Tata Institute, India.

GLOBAL ENVIRONMENTAL CHANGE SERIES
Edited by Michael Redclift, Wye College, University of London, Martin Parry, University of Oxford, Timothy O'Riordan, University of East Anglia, Robin Grove-White, University of Lancaster and Brian Robson, University of Manchester.

The *Global Environmental Change Series*, published in association with the ESRC Global Environmental Change Programme, emphasises the way that human aspirations, choices and everyday behaviour influence changes in the global environment. In the aftermath of UNCED and Agenda 21, this series helps crystallise the contribution of social science thinking to global change and explores the impact of global changes on the development of social sciences.

Also available in the series:

ENVIRONMENTALISM AND THE MASS MEDIA
The North–South divide
Graham Chapman, Keval Kumar, Caroline Fraser and Ivor Gaber

ENVIRONMENTAL CHANGE IN SOUTH-EAST ASIA
People, politics and sustainable development
Edited by Michael Parnwell and Raymond Bryant

POLITICS OF CLIMATE CHANGE
A European perspective
Edited by Timothy O' Riordan and Jill Jäger

THE ENVIRONMENT AND INTERNATIONAL RELATIONS
Edited by John Vogler and Mark Imber

GLOBAL WARMING AND ENERGY DEMAND
Edited by Terry Barker, Paul Enkins and Nick Johnstone

SOCIAL THEORY AND THE GLOBAL ENVIRONMENT
Edited by Michael Redclift and Ted Benton

ARGUMENT IN THE GREENHOUSE

The international economics of controlling global warming

Nick Mabey, Stephen Hall, Clare Smith and Sujata Gupta

Global Environmental Change Programme

London and New York

First published 1997
by Routledge
11 New Fetter Lane, London, EC4P 4EE

Simultaneously published in the USA and Canada
by Routledge
29 West 35th Street, New York, NY 10001

Typeset in Garamond by
J&L Composition Ltd, Filey, North Yorkshire

Printed and bound in Great Britain
Redwood Books, Trowbridge, Wiltshire

British Library Cataloguing in Publication Data
A catalogue record for this book is available from
the British Library

Library of Congress Cataloging in Publication Data
Argument in the Greenhouse: the international economics
of controlling global warming/Nick Mabey [*et al.*].
(Global environment change series)
Includes bibliographical references and index.
1. Environmental economics. 2. Global warming.
3. Greenhouse effect, Atmospheric. 4. International
economic relations.
I. Mabey, Nick. II. Series.
HC79.E5A74 1997
333.7–dc20 98–8982

ISBN 0–415–14908–8 (hbk)
ISBN 0–415–14909–6 (pbk)

CONTENTS

FIGURES

TABLES

ACKNOWLEDGEMENTS

This book was based on research funded by the ESRC's Global Environmental Change Programme (Grant No. L320253010), and the European Commission's JOULE II research initiative (Grant No. JOU2-CT92-0259). Many researchers at London Business School have helped in this research over the years, but particular thanks must go to David Kemball-Cook and Laurence Boone for their work on modelling energy demand, James Nixon for his advice on macroeconomic modelling issues and Ann Sheppard for developing and maintaining the solution software for the model. A final dedication must go to Roberta Bivins, who had to live with the final stages of this project, and was always willing to correct typescripts if necessary.

Nick Mabey
London Business School
March 1996

Part I

THE SCIENCE AND POLITICAL ECONOMY OF CLIMATE CHANGE

1

AN INTRODUCTION TO CLIMATE CHANGE

ANTHROPOGENIC CLIMATE CHANGE: THE PROBLEM

In the last decade there has been growing concern that major changes in the global climate will be caused by a build-up of heat-trapping pollutants in the atmosphere. Though the magnitude, timing and socio-economic impacts of these changes are at present highly uncertain, the mechanism linking climate and polluting gases unambiguously exists because the current mild temperatures on the earth are maintained by a naturally existing layer of these 'greenhouse gases'.

Over the earth's long history the concentration of the main heat-trapping gas in the atmosphere, carbon dioxide (CO_2), has markedly decreased as plants incorporated it into their structures as carbon; these plants were then themselves geologically fixed under sedimentary rocks as fossil fuel deposits. This process has markedly cooled the earth, and led to significant changes in fauna and flora over the long time spans involved. Mankind is currently reversing this effect by uncovering the deposits of fossilised carbon, and burning them to produce energy. However, unlike the natural processes that formed the deposits, we are releasing greenhouse gases at a rate unprecedented in geologic time; in fact since the beginning of the industrial revolution the concentration of warming gases has increased by ≈ 30 per cent, and with the expansion of the world economy this rate is continually increasing. Not only will plants and animals find it hard to adapt to the potentially rapid climate change this behaviour implies, but human systems – cities, agriculture and communities – have also grown up around the assumption of a stable, if variable, climate and sea-level. The overturning of these assumptions is likely to have serious impacts on both human well-being and the viability of the remaining wild ecosystems on the planet.

Tackling the problem of man-made, or anthropogenic, climate change, which is global in scope and whose impacts will stretch over centuries, will require the construction of institutions and analytic approaches well beyond those used for other environmental problems such as acid rain. The apparently successful international effort to control ozone depleting gases, codified in the Montreal Protocol,[1] gives us a model with which to work from, but the costs and uncertainties of climate change are several orders of magnitude larger than in this case. All this poses problems for the economist, used as we are to working inside well-defined legal systems where information is considered perfect, and actors are assumed to see into the future with certainty. While these assumptions may give insights into the workings of national-scale market activity, an economic approach to climate change will have to consider the full complex interactions of scientific uncertainty, international political relations and economic decision making into the very long run.

In this book we aim to address many of these issues, but focusing on the causes of conflict, dispute and trade-off that will occur in the next 10–30 years as the world (especially the industrialised countries) tries to take the first steps towards substantively dealing with the problem of climate change. As befits our intellectual background we focus on the economic forces which will drive international negotiations to limit greenhouse gas emissions, and concentrate less on the process of legal, institutional and political regime building. While political scientists may find this approach naive, focusing as they do on the actualities of decision making, we consider that stepping back and trying to quantify some of the major determinants of agreement and disagreement allows a different, more long term, perspective on the problem (for more process oriented studies see Young 1993, Vogler 1995, Hourcade 1993, Greene 1993 and FIELD 1994).

Below we fill in some of the scientific, institutional and methodological context of our work and describe the current state of negotiations on controlling climate change. We then outline the structure of the book and give a short description of the content of each chapter.

THE SCIENCE OF CLIMATE CHANGE

The Earth receives short wave radiation from the sun (including the visible part of the spectrum), one-third of which is reflected while the rest is absorbed by the atmosphere, ocean, ice, land and biota. The energy absorbed from solar radiation is balanced, in the long term, by outgoing radiation from the Earth and atmosphere. Terrestrial radiation is emitted in the form of long wave infra-red energy. The balance between energy absorbed and emitted as long-wave infra-red radiation can change due to a number of factors: changes in the sun's energy output, slow variations in the earth's orbit and the greenhouse effect. The greenhouse effect is one of the most important factors and is one which humankind has the capacity to change.

Short wave radiation can pass easily through the atmosphere, whereas long wave terrestrial radiation emitted by the warm surface of the Earth is partially absorbed by a number of trace gases in the atmosphere. These trace gases are called greenhouse gases (GHG), the main natural GHGs being water vapour, carbon dioxide (CO_2), methane (CH_4), nitrous oxide (N_2O) and ozone (O_3) in the troposphere and stratosphere. In the absence of these greenhouse gases the mean temperature of Earth's surface would be about 33°C lower than it is today.

However, human activities are increasing the concentration of these naturally existing GHGs, and adding new ones such as halocarbons (HFCs). If anthropogenic GHG emissions increase the atmospheric concentrations of these gases they will raise global average annual mean surface-air temperatures (referred to as global temperatures) through an enhanced greenhouse effect. Other potential indirect impacts of global warming are changes in precipitation quantity and pattern, changes in vegetation cover and soil moisture, increased intensity of tropical storms and a rise in sea level due to thermal expansion of water and the melting of Antarctic ice sheets. From the economic and social point of view these indirect impacts will be more important than direct changes in temperature.

Present measurements show that atmospheric concentrations of GHGs have been increasing since the industrial revolution due to human activities. Table 1.1 summarises present and pre-industrial concentrations and rates of change, as well as the atmospheric lifetime of the main greenhouse gases. It should be noted that not all of the emitted quantities will contribute to increased concentrations because a proportion are immediately absorbed in the oceans and biosphere, so-called greenhouse gas 'sinks'. The rate of absorbtion of some pollutants, especially CO_2, may even increase with temperature and concentration, and so there will always be an equilibrium level of emissions where atmospheric concentrations will not increase.

Carbon dioxide, methane and nitrous oxide all have significant natural and human sources whereas CFCs are only produced industrially. Water vapour and ozone are two greenhouse gases that have not been included in any study. This is because concentrations of water vapour are determined internally within the climate system, and it is difficult to quantify changes in

Table 1.1 Pre-industrial and present concentration of major greenhouse gases

	Greenhouse gases (atmospheric concentrations)[a]				
	CO_2 (ppmv)	*CH_4 (ppmv)*	*CFC-11 (pptv)*	*CFC-12 (pptv)*	*N_2O (ppbv)*
Pre-industrial (1750–1800)	280	0.8	0	0	288
Present day (1990)	353	1.72	280	484	310
Current rate of change per year	1.8 (0.5%)	0.015 (0.9%)	9.5 (4%)	17 (4%)	0.8 (0.25%)
Atmospheric lifetime (years)	50–200[b]	10	65	135	150

Source: IPCC 1990

Notes

a ppmv = parts per million by volume;
ppbv = parts per billion by volume;
pptv = parts per trillion by volume.

b The way in which CO_2 is absorbed by the oceans and the biosphere is not fully understood, so that a single value cannot be given.

Table 1.2 Anthropogenic sources of greenhouse gases

Activities	*Carbon dioxide*	*Methane*	*Nitrous oxide*	*CFCs*
Energy use				
Coal production	√	√		
Coal, oil and gas combustion	√	√	√	
Gas venting and leakages		√		
Industry				
Cement manufacture	√			
CFCs				√
Landfills		√		
Agriculture				
Animal husbandry		√		
Wet rice cultivation		√		
Fertiliser use			√	
Biomass combustion		√	√	
Deforestation and land use change	√	√	√	

the concentration of ozone as a result of human activity. Though the changes in atmospheric concentrations of greenhouse gases are reasonably well established, there are major gaps in understanding the flow of these gases between their sources and sinks. Table 1.2 gives the anthropogenic sources of the main greenhouse gases.

Higher GHG concentrations increase radiative forcing, the capacity of the atmosphere to absorb heat. For simplicity most researchers have expressed the total forcing in terms of CO_2 concentration which would give that forcing, termed the *equivalent* CO_2 concentration. By this measure GHGs have increased since pre-industrial times by an amount equivalent to about a 50 per cent increase in CO_2, although CO_2 itself has increased by only 26 per cent. The relative radiative effect of different gases depends on the atmospheric lifetime of the gas and the difference in their radiative forcing. The concept of Global Warming Potential (GWP) has been defined to account for these differences; the GWP index defines the time-integrated warming effect due to an instantaneous release of unit mass (1 kg) of a given greenhouse gas in today's atmosphere, relative to that of carbon dioxide.[2] Though there are some doubts emerging about its usefulness it remains a first-order yardstick with which to assess the importance of different gases.

In addition to direct radiative forcing there are indirect effects. Direct forcing occurs when the gas itself is a greenhouse gas, indirect forcing occurs through chemical reactions between various constituents of the atmosphere, which produce or support the formation of a greenhouse gas. Because of incomplete understanding of chemical processes most estimates of indirect effects are substantially erroneous, though the sign of the indirect contribution for some greenhouse gases has been identified reasonably robustly (IPCC 1992). Table 1.3 gives the IPCC's 1992 estimates of the direct GWP for a hundred year time horizon, and the sign of the

Table 1.3 Direct GWPs for 100 year time horizon

Gas	*Global warming potential (GWP)*	*Sign of the indirect component of the GWP*
Carbon dioxide	1	none
Methane	11	positive
Nitrous oxide	270	uncertain
CFC-11	3400	negative
CFC-12	7100	negative
HCFC-22	1600	negative
HFC-134a	1200	none

Source: IPCC 1992

indirect effect. Since this research was published it has been observed that the negative warming effect of CFCs, which results from the destruction of stratospheric ozone – a strong greenhouse gas, will cancel out all the direct warming effects from these chemicals (Daniel *et al.* 1995). As CFCs and HCFCs will be eliminated under the Montreal Protocol, future warming will depend on which chemicals are used to replace them in refrigeration and other applications. If HFCs are used global warming will increase, as obviously these chemicals do not destroy ozone; however, other possible substitutes, such as hydrocarbon coolants, will have no warming implications.

Although carbon dioxide is the least effective greenhouse gas per unit emitted, the overall radiative effect also depends on the level of emissions. Thus CO_2, with large emission levels and a long lifetime in the atmosphere is the largest single contributor to the greenhouse effect. Counteracting the effects of these warming gases, anthropogenic emissions of aerosols – especially sulphates – act to cool the atmosphere by absorbing and reflecting the sun's energy. However, these gases have a very short lifetime in the atmosphere, about six days, and are likely to diminish as countries tackle the local and acute problems of acid rain and urban air pollution.

MODELLING THE PHYSICAL PROCESSES OF CLIMATE CHANGE

The climate is a highly complex non-linear system and so predicting the impact of increased greenhouse gas concentrations is far from simple. The basic tools used for this are three-dimensional mathematical models of the climate system known as General Circulation Models (GCMs); GCMs, coupled with ocean models (CGCM), synthesise the physical and dynamic processes in the overall system and allow for complex, iterative reactions. However, the descriptions of many processes are crude, and so considerable uncertainty is attached to their forecasts. Results from different GCMs give an accepted range of a 1.5 to 4.5°C increase in surface air temperature following a doubling of CO_2 equivalent concentrations, with a value of

2.5°C chosen as the best-guess estimate (IPCC 1992). Under present best-guess future emissions scenarios CO_2 doubling is predicted to occur in 2030–2040, but the temperature increase at this date will be less than the predicted equilibrium temperature due to thermal inertia in the oceans.

Although different models generally agree in qualitative terms over the regional distribution of warming, there are gross disagreements over regional distribution of changes in precipitation because of the higher degree of complexity involved (Solow 1991). Uncertainties in climate predictions arise from imperfect knowledge about future anthropogenic emission rates, how these emissions will change atmospheric concentrations of GHGs (i.e. what proportion of emissions are absorbed by sinks) and the response of climatic processes to these changed concentrations; for example, the magnitude of the cloud feedback effect and the transfer of energy between atmosphere–ocean–land surfaces. A number of research programmes are ongoing or planned to address these key areas of scientific uncertainty.

To summarise we can say that pollution resulting from human activities has substantially increased atmospheric concentration of greenhouse gases from their pre-industrial levels, and that this process is accelerating. Modelling studies indicate that a doubling of CO_2 equivalent greenhouse gases will result in a global mean surface temperature increase in the range of 1.5 to 4.5°C; therefore, based on current concentration levels, even if we stopped emitting all GHGs immediately we are already committed to an equilibrium temperature rise of 0.4 to 1°C in the future.

The impact of socio-economic uncertainty

Of course policy makers are not that interested in the impacts at an arbitrary point such as CO_2 doubling. They wish to know the timing and rate of warming that is likely to occur in the future, as this will greatly determine the feasibility and cost of adapting to, or preventing, a changing climate. The timing of impacts will depend on the physical lags and inertia mentioned above, and the growth in GHG emissions over the future. In the absence of policy intervention emissions will depend on population growth, per capita productivity growth, changes in fossil resource use, land use changes and advancements in energy technology.

The Inter-governmental Panel on Climate Change (IPCC) has peformed a reasonably comprehensive analysis looking at the feasible ranges these parameters might take, and has combined them into a set of six alternative scenarios (IPCC 1992). Scenario outputs are not predictions about the future, rather they illustrate the effect of a wide range of economic, demographic and policy assumptions. The details of these scenarios are given in Appendix 1.1; Table 1.4 gives the summary of assumptions in the six IPCC updated scenarios, and Table 1.5 gives some key results.

The best estimate for CO_2 emissions from fossil fuels in 1989 and 1990 is 6.0 ± 0.5 GtC. Uncertainties in estimating carbon dioxide emissions from deforestation and land use change are large, and IPCC puts a tentative figure as 1.6 ± 1 GtC during the 1980s. For methane there are still many uncertainties; however, an amount of 500 Tg can be deduced from the magnitude of its sinks, and combined with its rate of accumulation in the atmosphere. Recent methane isotopic studies suggest that 20 per cent of methane is of fossil origin. Greater uncertainties exist with respect to nitrous oxide and other greenhouse gases.

The figures in Appendix 1.1 clearly show the importance of fossil CO_2, as opposed to deforestation and natural methane, as the main geenhouse gas; this is mainly because biologically sourced emissions are limited by the extent of extra land clearance which is possible, restrictions that are far tighter than those on the resource levels of fossil fuels. Cumulative CO_2 fossil emissions in the atmosphere increase from a range of 285–311 GtC over 1990–2025, to a range of 672–2050 GtC over 1990–2100. The main differences between scenarios are caused by assumptions surrounding population growth and GDP growth, and the actual out-turn is likely to be nearer the high end of the range where world population increases to 11 billon people and economic growth is between 2.3 and 3.0 per cent per annum.

The exponential nature of this growth in emissions, most of which will occur in the developing world, is the nub of the problem of controlling global warming. Fossil fuel use is ubiquitous in industrialised economies and pervades every part of modern life. Therefore, the legitimate aspirations of the world's poor mean the earth is on a path which would imply a warming commitment far exceeding CO_2 doubling (perhaps 5–10°C) by the end of the next century.

INTERNATIONAL INSTITUTIONAL RESPONSES TO CLIMATE CHANGE

In response to the perceived threat of climate change, international organisations have begun to act to co-ordinate co-operative multilateral action; most of this action has preceded public opinion and so has been driven by environmental groups, policy makers and scientists, not large scale political forces.

The first official action was the establishment, by the United Nations Environment Programme (UNEP) and World Meteorological Organisation (WMO), of the Inter-governmental Panel on Climate Change (IPCC) in 1988. The IPCC formulated different Working Groups to carry out a scientific assessment of the likelihood of anthropogenic climate change, study its potential impacts and identify response strategies to prevent and adapt to the impacts of climate change. From this activity, the United Nations International Negotiating Committee on Climate Change, containing 105 countries, was formed in 1991. A year of negotiation culminated in the first step towards a substantive international agreement to control the causes of greenhouse warming, when 155 countries signed the Framework Convention on Climate Change (FCCC) at Rio de Janeiro in June 1992.

Though this is only a framework for future actions, it does contain some detailed aims and objectives which are described below in Chapter 2; the basic form of the treaty also gives a guide to its potential evolution in the future. The essence of current commitments are that the developed countries agreed to stabilise GHG emissions at 1990 levels in the year 2000, but no later targets were specified, and developing countries only had to control emissions to the extent that these activities were funded by the developed nations through the Global Environment Facility (GEF; see Sjoberg 1993).

The large uncertainties involved in climate change have meant that the initial stages of international negotiations have been driven by the need to learn about, and gather information on, potential impacts and costs of control. However, now it seems that the negotiations are becoming a more traditional bargaining process where governments are making trade-offs and aiming to maximise utility in some way. Since the initial meeting in Rio the Council of Parties (CoP) of the FCCC has met once, in Berlin in 1995. The aim of this meeting was to set out procedural rules which would allow the targets of the FCCC to be revised and extended by agreeing an additional Protocol by 1998, an objective included in the original treaty text. However, the obstructiveness of the oil producing nations, led by OPEC, meant that agreement was not forthcoming and the future of the process is still in balance.

Despite this current setback the structures of the FCCC are part of an important progression from past international environmental co-operation, which has tended to be regional or bilateral, to a truly global regime; in fact, from 1973 to 1989 only 37 per cent of international environmental treaties contained more than sixteen parties (Haas and Sundgren, 1993). Discounting controls on nuclear and biological weapons, large (40+ parties) multilateral environmental treaties have included the Tropical Timber Agreement (1983), the UN Law of the Sea (1983) – though this was not ratified by national governments, the Montreal Protocol (1987), and the Convention on Biodiversity (1992). Apart from the Montreal Protocol none of these agreements have really enforced substantive controls for environmental preservation. The sucessful evolution of the FCCC into 'hard' international legislation will depend on many factors, notably: the political process of agreement, the perceived risks of climate change, the costs of controlling climate change and the distribution of emissions and impacts.

Of these we would argue that, given climate change is seen as a significant problem, it is the distribution of the costs and benefits of control between countries that will determine the success of any agreement. In international agreements there is no compulsive enforcement mechanism, so polluters cannot be forced to stop activity that harms other nations even if notionally they have such an obligation in international law. Agreement must proceed by consensus, hegemony or sanction (Young 1993). In the case of climate change there is no global power big enough to form a hegemonic control of the process in the long term, though among the developed countries the USA could probably perform this role if it wished.

Therefore, the only way forward is to have a consensus between countries, or for the countries who wish to control to impose some kind of sanction on those that want to pollute. International sanctions are perfectly legal, and have a precedent in the rules of the Montreal Protocol and the World Trade Organisation, which both allow trade sanctions to be levied as a compliance mechanism for their agreements. However, to date the parties to the FCCC have been very reluctant to adopt such measures, and the text of the agreement stresses co-operative action throughout.

In the simplest analysis, to achieve co-operative action all parties to an agreement must feel either that they are directly gaining from participating or that any adverse effects they suffer are legitimately imposed and equitably shared among all parties. The current format of the FCCC achieves some of these goals: all developed countries have to hit the same stabilisation target, which seems to approximate equity – though in reality these countries have very different levels of initial pollution per unit of economic output; developing countries face no restrictions because climate change damage is low on their list of priorities, and the majority of past emissions have come from the North; ex-communist countries have had their targets temporarily suspended, but, due to their economic contraction, are currently emitting quite small amounts of CO_2 compared to their 1990 levels, and so are in compliance anyway.

The general tone of the agreement is that the developed countries must demonstrate that they are serious about limiting emissions, and must make up for past pollution (possibly by funding reductions in the developing world), before the developing world will even consider adopting unilateral targets or restrictions on emissions. This is an equitable stance, but is slightly short sighted as developing countries, with their reliance on agricultural produce and location in the Tropics, are more vulnerable to climate change damage than developed countries. Therefore, all other things being equal – which of course they are not – they would wish to have larger abatement, more quickly than developed countries (Fankhauser 1995). This division of responsibilities means that in the short to medium term (10–20 years) the most important players in the agreement process will be the developed countries; they must agree emissions targets, and stick to them, for there to be any chance of limiting climate change to a level in the region of CO_2 doubling.

Despite their similar levels of industrialisation the developed countries are a far from homogeneous group when it comes to current emissions of GHGs, and perceptions of potential climate change damage in their own territories. Some countries, such as Canada, may even think that a small rise in temperatures will be beneficial, so achieving consensual *obligations* may be highly difficult. To make such countries join a binding agreement it may even be necessary to compensate them for their abatement efforts. This contradicts the Polluter-Pays-Principle (PPP) enshrined in the FCCC, but the PPP was originally designed for use inside countries where coercion is possible, and has only been applied internationally for regional pollutants where significant co-operation is already present between countries (e.g. sulphur dioxide in the European Union, or between Canada and the USA).

The above view of the FCCC process stems from an economic theory of negotiation where consensual, non-coercive, co-operative action is only considered likely to be sustained if it is in the interests of all parties, and none can gain by leaving, or threatening to leave, the agreement (so-called 'free-riding'). This type of analysis abstracts from political institutions and ignores historical, ethical and reputational reasons why parties may adhere to an action which makes them materially worse off. More concretely, economic analysis of international agreements has also tended to analyse each issue seperately, without considering win–win linkages which could be made between areas such as environment, aid and trade. The reason for this has been the desire to quantify effects, which precludes considerations of hard to measure – if potentially important – intangibles, and an attempt to avoid the complication of parallel optimisation of objectives. Perhaps there has also been an underlying cynisism among economists that, for all the fine and noble talk, in the end it is national material interests that guide international power politics and not ethical constructs.

THE APPROACH, METHODOLOGY AND OUTLINE OF THIS STUDY

This book focuses on using quantitative, econometrically-based techniques to model the cost to the developed countries of complying with future commitments under the FCCC, and how the distribution of costs will affect the potential for agreement in the short to medium term. However, we consider this economic approach to be complementary to political and legal analysis, not a replacement for it. Anybody who has delved into the complexities of multi-player, repeated game theory will know that the assumptions about information, commitment and opportunities for action – which essentially define the political process – are critical for determining long run economic equilibria.

Therefore, unlike other studies of global warming economics, we are not concerned with finding the 'optimal' level of climate change, or trying to predict how very long run changes in energy markets can reduce CO_2 emissions; many such studies have already been done and the latter topic seems to be better suited to technological assessment, rather than empirical economic modelling. Rather we aim to provide a *descriptive* analysis of the consequences of different ways of controlling climate change, and to embed these quantitative results inside the economic, political and legal framework of the FCCC process.

In the period in which we are interested (the next 20–30 years), the cost of controlling carbon dioxide emissions will depend not on the technical development of radical new technologies but on the ability of the market economy to react to changed prices, policies and regulations. This is because the majority of fossil energy is used inside machines or structures that are usually replaced only every 10–100 years; the shorter time for cars and some industrial equipment, the longer time scale for transport infrastructure and housing. In such a timeframe econometric analysis can give an accurate assessment of the costs of saving energy because it accounts for all the dynamics of sunk costs, investment, consumption and technical diffusion which are critically involved; assuming of course that the past is a reasonable guide to future behaviour.

The econometric approach means that many parameters which are the subject of speculation and sensitivity analysis in other models are here directly measured from observed data. This fact, and the relative computational unwieldiness of a large econometric model, means that we do not undertake exhaustive senstivity analysis. Instead we try to define the ranges of uncertainty, when they are considered critical, and assess whether these magnitudes produce significant changes in our results; for example, do they make the difference between disagreement or agreeement, or make one policy instrument more efficient or useful than another? The econometric approach also means that rather than focusing on parameter sensitivity we can look at the impact of different *structural* modelling assumptions on our results, and we have found this generally more illuminating than performing thousands of different modelling simulations.

Chapter outline

In Chapter 2 we give a qualitative analysis of some of the main policy issues we will be examining, and show the need for quantification of different effects. This discussion motivates Part II of the book, which describes the economic technicalities of constructing EGEM (Environmental Global Econometric Model), the model used in this study. An understanding of this material is not essential to interpreting the policy analysis in Part III, but where critical assumptions might be of interest the reader will be refered back to the relevant chapter in Part II.

Chapter 3 is a critical review of other models of the costs of controlling GHGs, describing the methodologies, assumptions and limitations that distinguish the main studies to date. Chapter 4 describes the formulation and estimation of the energy sector in EGEM, and explains how a model of price-driven endogenous technical progress was estimated and incorporated into the structure. Chapter 5 explains the larger macroeconomic structure of EGEM, and compares different approaches to modelling the effect of energy price rises on the productive sector. In particular the role of capital-energy substitution is examined, and past econometric estimates of this effect are reviewed. A supply-side model incorporating endogenous labour productivity, driven by capital accumulation, is then estimated and compared to a traditional production function approach. The structural assumptions of each are interpreted in light of their simulation properties and conclusions drawn as to the biases present in each modelling technique.

Finally, Chapter 6 shows how the cost of carbon abatement in developing countries can be modelled, taking into account their particular economic and structural context. A mixed macroeconometric and microeconomic model is constructed for India, and the costs of several future abatment commitments assessed.

This type of modelling, based as it is on observable data, allows investigation of many interesting effects governing the potential for climate change agreement. The policy analysis in Part III aims to outline the theoretical questions involved in each issue, and then to pinpoint the particular quantitative measurements that are vital in deciding which effect will dominate in practice. We then model, or illustrate, these effects using EGEM and draw conclusions as to the policy impact of each issue.

Chapter 7 considers the theoretical question of optimising carbon dioxide abatement, placing this simple economic approach inside the complex context of sustainability, uncertainty and irreversibility which surrounds the climate change problem. Concluding that the current state of knowledge and methodologies are insufficiently developed to perform a robust cost/benefit analysis, in Chapter 8 we instead model how uncertainty, learning and strategic behaviour interact to influence the potential for agreement and the benefits of global co-operation to control emissions.

Chapter 9 analyses the consequences of the FCCC's commitment to unilateral control of carbon emissions in the the OECD: will this harm national competitiveness, and will emissions increase in non-committed countries thus significantly offsetting developed country abatement? The influence of competitiveness effects, and of industrial relocation, on the stability of OECD agreements is also investigated and these strategic interactions modelled inside EGEM.

As a contrast to the explicit international focus of the first three chapters, Chapter 10 looks at the domestic political economy of limiting carbon emissions, as this will determine the ability of governments to garner democratic support for achieving international commitments. We analyse the potential for ecological tax reform, that is the shifting of the tax burden from labour to pollution, and measure the effects of this policy on growth, employment and the distribution of income over the range of abatement likely under the FCCC. The demand-side macroeconomics of carbon taxes, how they affect incomes, inflation, interest rates and investment, are considered also under different wage setting assumptions, and the sensitivity of results to changes in labour market conditions is determined. All of these factors have significant impacts on the likely timing of the policy instruments governments may use to contol CO_2, which in turn affect the cost of abatement and thus the type of international burden sharing that would be considered equitable.

Finally Chapter 11 looks at how different policy instruments – targets, international taxes and tradable permits – can contribute to co-ordinating an efficient and stable carbon abatement treaty between the major OECD countries. This involves looking at the welfare, income and distributional impacts of each mechanism and seeing if the agreement between countries is actually compatible with the type of efficient policy instruments that economists are always championing.

APPENDIX 1.1 SOCIO-ECONOMIC SCENARIOS CONSIDERED BY THE IPCC

Table 1.4 Summary of assumptions in the six IPCC 1992 alternative scenarios

Scenario	*Population*	*Economic growth (%)*	*Energy supplies*[a]	*Other*[b]	*CFCs*
IS92a	World Bank 1991: 11.3 B by 2100	1990–2025: 2.9 1990–2100: 2.3	12,000 EJ Oil. 13,000 EJ Natural gas. Solar costs fall to $0.075/kWh. 191 EJ of biofuels available at $70/ barrel.	Legally enacted and internationally agreed controls on SO_x, NO_x and NMVOC emissions.	Phase out of CFCs in non-signatory countries by 2075.
IS92b	World Bank 1991: 11.3 B by 2100	Same as IS92a	Same as IS92a.	Same as IS92a, plus many OECD countries stabilise/reduce CO_2 emissions.	Global compliance with scheduled phase out of Montreal Protocol.
IS92c	UN medium low case: 6.4 B by 2100	1990–2025: 2.0 1990–2100: 1.2	8000 EJ Conventional oil. 7300 EJ Natural gas. Nuclear costs decline by 0.4 per cent annually.	Same as IS92a.	Same as IS92a.
IS92d	UN medium low case: 6.4 B by 2100	1990–2025: 2.7 1990–2100: 2.0	Oil and gas same as IS92c. Solar costs fall to $0.065/kWh. 272 EJ of biofuels available at $50/ barrel.	Emission controls extended worldwide for CO, NO_x, SO_x and NMVOC. Halt deforestation. Capture emissions from coal mining and gas production.	CFC production phase out by 1997 for industrial countries. Phase out of HCFCs.

IS92e	World Bank 1991: 11.3 B by 2100	1990–2025: 3.5 1990–2100: 3.0	18400 EJ conventional oil. Gas same as IS92a. Phase out nuclear by 2075.	Emission controls (30% pollution surcharge on fossil energy).	Same as IS92d.
IS92f	UN medium high case: 17.6 B by 2100	Same as IS92a	Oil and gas same as IS92e. Solar costs fall to $0.083/kWh. Nuclear costs increase to $0.09/kWh.	Same as IS92a.	Same as IS92a.

Source: IPCC 1992

a All scenarios assume coal resources up to 197,000 EJ. Up to 15 per cent of this resource is assumed to be available at $1.3/gigajoule at the mine.

b Tropical deforestation rates (for closed and open forests) begin from an average rate of 17 million hectares/year (FAO 1991) for 1981–90, then increase with population until constrained by availability of land not legally protected. IS91d assumes an eventual halt of deforestation for reasons other than climate. Above ground carbon density per hectare varies with forest type from 16 to 117 tonnes C/hectare, with soil C ranging from 68 to 100 tC/ha. However, only a portion of carbon is released over time with land conversion, depending on type of land conversion.

Table 1.5 Selected results of six 1992 IPCC greenhouse gas scenarios

Scenario	Years	Decline in TPER/GNP[a] (av. % p.a.)	Decline in Carbon intensity (av. % p.a.)	Cumulative net Fossil Carbon emissions (GtC)	Tropical deforestation (Mha)	Cumulative net Carbon emissions (GtC)	Year	Emissions per year: CO_2 (GtC)	CH_4 (tg)	N_2O (TgN)	CFC[c] (Kt)	SO_x (Tg S)
IS92a	1990–2025	0.8	0.4	285	678	42	1990	7.4	506	12.9	827	98
	1990–2100	1.0	0.2	1386	1447	77	2025	12.2	659	15.8	217	141
							2100	20.3	917	17.0	3	169
IS92b	1990–2025	0.9	0.4	275	678	42	2025	11.8	659	15.7	36	140
	1990–2100	1.0	0.2	1316	1447	77	2100	19.0	917	16.9	0	164
IS92c	1990–2025	0.6	0.7	228	675	42	2025	8.8	589	15.0	217	115
	1990–2100	0.7	0.6	672	1343	70	2100	4.6	546	13.7	3	77
IS92d	1990–2025	0.8	0.9	249	420	25	2025	9.3	584	15.1	24	104
	1990–2100	0.8	0.7	908	651	30	2100	10.3	546	14.5	0	87
IS92e	1990–2025	1.0	0.2	330	678	42	2025	15.1	692	16.3	24	163
	1990–2100	1.1	0.2	2050	1447	77	2100	35.8	1072	19.1	0	254
IS92f	1990–2025	0.8	0.1	311	725	46	2025	14.4	697	16.2	217	151
	1990–2100	1.0	0.1	1690	1686	93	2100	26.6	1168	19.0	3	204

Source: IPCC 1992

a TPER = Total primary energy requirement.

b Carbon intensity is defined as units of carbon per unit of TPER.

c CFCs include CFC-11, CFC-12, CFC-113, CFC-114 and CFC-115.

2

INTERNATIONAL CO-ORDINATION OF CLIMATE CHANGE PREVENTION

INTRODUCTION

The problems associated with controlling emissions of greenhouse gases, principally carbon dioxide, are economic and political not technical. Technologies exist commercially, or are in an advanced development stage, which could greatly reduce the world's reliance on fossil fuels, but the questions remain: how much will this cost? how much abatement is economic? and who should reduce their consumption of fossil fuels, and by how much?

As the greenhouse effect is a global problem these issues of cost effectiveness and cost allocation must be dealt with at the inter-governmental level, as unilateral action is not a viable option. Unlike national environmental problems there is no existing, legitimate, international decision making body which can debate and evaluate such trade-offs, and no coercive legal mechanism at the global level to enforce communal decisions once taken. The only way to co-ordinate international action is through the negotiation of consensual treaties, which include all parties material to the success of abatement efforts, both now and into the future. It is obvious that this constitutes a far harder task than the setting of Pareto optimal taxation inside an existing compliance mechanism.

In this chapter we review the only existing international agreement to limit climate change, the Framework Convention on Climate Change; we analyse its provisions, extract the economic logic which underlies its legal statements and identify three areas crucial to the overall effectiveness of implementation: achievement of stated goals, efficiency of implementation mechanisms and stability of negotiated agreements. The remainder of the chapter then analyses the different policy instruments (taxes, tradable permits, targets, etc.) which have been proposed to co-ordinate international action, in each of these three areas of effectiveness, in order to give an insight into the trade-offs which exist inside the negotiation process.

THE FRAMEWORK CONVENTION ON CLIMATE CHANGE

The only multilateral agreement addressing carbon dioxide reduction that has been agreed and enforced[1] to date is the Framework Convention on Climate Change (FCCC), signed by 155 parties at the UNCED conference in June 1992. In Article 2 the stated aim of this convention is: 'to achieve, . . ., stabilization of greenhouse gas concentrations in the atmosphere at a level that would prevent dangerous anthropogenic interference with the climate system'.

Stabilisation is to be achieved using a combination of controls on sources of greenhouse gases (GHGs) and enhancement of carbon dioxide sinks such as forests and oceans. Befitting its status as a framework convention the FCCC is vague about defining a target for stabilisation and a timetable for reaching it. However, it does assign some obligations to the parties which differ between developed and developing countries.[2]

The thirty-five developed and former communist countries (referred to as 'economies in transition'), listed in Annex I of the convention,[3] account for approximately 66 per cent of global fossil fuel based carbon emissions (WRI 1990) and are committed under the FCCC to starting to control emissions, 'with the aim of returning individually or jointly to their 1990 levels these anthropogenic emissions of carbon dioxide and other greenhouse gases not controlled by the Montreal Protocol' (Article 4.2.b).

This weak target is further qualified by clauses which allow economies in transition some leeway in meeting their targets (Article 4.6), though in practice the fall in industrial production in these countries means they are all currently in full compliance. The remaining parties to the convention are committed to reducing emissions of GHGs and enhancing sinks but are not given a specific target level or date for compliance. Those parties which qualify as developing countries only have to reduce emissions under the following conditions:

> The extent to which developing country Parties will effectively implement their commitments under the Convention will depend on the effective implementation by developed country Parties of their commitments under the Convention related to financial resources and transfer of technology.
>
> (Article 4.7)

The commitment of the developed countries in Annex II of the convention is that they must pay the 'agreed incremental costs' of actions by developing countries to comply with the convention; this money to be transferred through a multilateral institution, the Global Environment Facility (GEF; Mintzer 1993). The voting rules in the GEF mean that a group of countries representing over 40 per cent of the donated funds can veto any proposal or commitment. So in the short to medium term the amount of global GHG abatement will essentially be determined solely by the group of developed countries in Annex II. They will decide how much money to transfer to the developing countries for CO_2 control, and on any further binding targets for themselves. This group of twenty-four countries produces approximately 43 per cent of global fossil fuel derived carbon emissions and of this 88 per cent is produced by the countries of the G7: USA, Japan, Germany, France, Italy, UK, and Canada (WRI 1990).

The future of the FCCC

The targets and conditions of the FCCC must be reviewed by 1998 at the latest (Article 4.2.d). Already the secretariat is considering draft amendments, or protocols, to the convention which will affect the conditions of enforcement and the strength of commitment of the parties (for example, the Draft Protocol of AOSIS 1994). However, the passage from a framework convention to a 'hard' treaty which has environmentally significant targets and objectives is likely to be a long one, and enforcement will be based more on consensus than on legal obligation (Greene 1993). Indeed the first meeting of the Conference of Parties in 1995 failed to even agree on the voting rules for future decisions, because the oil producing countries in OPEC vetoed any procedural amendments which moved away from consensus.

There is a symbolic importance in how the FCCC is applied to the developed countries. As the developing countries grow more populous, and with luck more wealthy, their share of global carbon dioxide emissions will rise and it will be impractical for the developed countries to subsidise all their abatement measures. These countries will eventually have to enter Annex I and be obliged to control emissions at their own expense. However, there is a great distrust of the North's motives in applying such environmental treaties, with many Southern countries seeing this as a way of slowing their economic development thus relieving competitive pressures on developed economies. Successful extension of the FCCC Annex I obligations to all countries will depend on whether the developed countries can persuade the developing countries they are sincere in their attempts to control emissions. The successful design of a system of international co-operation to control emissions between the Annex I countries will therefore be judged on both the direct level of GHG abatement and the way it is perceived by the other parties to the FCCC.

THE ECONOMIC THEORIES OF CO-OPERATION

Access to the global atmosphere is completely uncontrolled and so must be formally classified as an open access resource allocation problem (Pearce and Turner 1990). It is a widely known theoretical result that co-operative management of these types of resources is better than allowing non-cooperative exploitation by individual actors which would result in the 'Tragedy of the Commons' (Hardin 1968). The 'tragedy' referred to is the open access management regime, and the solution is to agree to manage the global atmosphere as a common property resource. Much of the empirical and modelling work on climate change to date has concentrated on calculating the direct economic gains from the polar cases of perfect co-operation and unilateral action. Though work is still very much ongoing, especially on the empirical side, most results, and the very existence of the FCCC, show that co-operation will give large global benefits (Clarke *et al.* 1993).

The concept of co-operation used by economists in these studies is a very limited one, and involves all countries acting as if they shared a common welfare function, implying a single government. In the real world there are many different degrees of co-ordinated action which are termed 'co-operation', and great care must be taken to specify exactly what is being enacted in an international agreement. In this chapter co-operation will be used in a broad sense to indicate any solution countries reach which involves a negotiated agreement codified in a treaty valid under international law. Under this definition non-cooperative agreements exist only when there is no formal negotiation procedure between parties, and no use of co-ordinating policy instruments.

In terms of formal game theory every interaction between countries, negotiated or not, is a gaming process, in which each party acts strategically to gain maximum advantage for itself, unless 'perfect' co-operation exists. Perfect co-operation implies that each country internalises the full external costs of climate change and so reduces emissions until its marginal cost of abatement equals the marginal *global* benefit of its emissions reductions. This will always maximise the sum of global welfare if no country has decreasing marginal utility of income; if there is decreasing marginal utility the efficiency gains must be redistributed through side payments in order for global welfare to be maximised. The contrasting non-cooperative solution exists when each country equates its marginal abatement costs with marginal *national* benefits.

Perfect co-operation can be a stable outcome (Nash equilibrium) of the open access management game, in that no party to the agreement has an incentive to renege unilaterally of their commitments (so called free-riding), if the 'Folk Theorem' is applicable. That is, the game is repeated often enough, or over a long enough period, that any non-zero penalty imposed for reneging on the co-operative agreement will eventually outweigh the rewards from free-riding. If discount rates are high and/or penalties low this will not be true and the non-cooperative outcome is the only Nash equilibrium for a simple game (Rasmusen 1989). Between the two polar cases of 'perfect' co-operation and non-cooperation countries could agree to internalise costs partially, to abate collectively and to punish free-riders or countless other permutations of institutional design which all have their justifications in terms of equity, enforceability and political convenience.

Certainly international co-operation to date has fallen well short of complete global internalisation of costs. The FCCC has only produced co-operation in the sense of discouraging free-riding by some developed nations, encouraging co-operation on abatement projects, transferring money to developing countries and laying the foundations for future commitments. This narrower form of co-operation, bringing correspondingly smaller gains than the full form, is still important and will make stricter targets in the future more likely; this is because if the costs of abatement are seen to be equitably shared the co-operative equilibrium will be more stable (Rabin 1993).

Effectiveness of international action

From the above description of the current institutional framework, and the perceived benefits of increased co-operation, it is obvious that the FCCC will have to develop into a stronger and more defined treaty in the future if its stated goals are to be achieved. To this end, much work has been carried out by economists and political scientists aiming to define a policy framework which will produce effective co-ordination of international action on climate change; where effectiveness can be considered as having three distinct parts:

Achievement Policies should lead to the achievement of stated goals, by providing the correct behavioural incentives and being compatible with appropriate and feasible monitoring, dispute resolution and compliance regimes.

Efficiency Actions or obligations assumed under the treaty should lead to economically efficient achievement of the stated goals.

Stability In the absence of coercive sanctions on non-complying parties to a treaty the policy framework should provide incentives for long term co-operation and reduce the rewards of non-compliance.

Different policy instruments will have different strengths in each category and finding the most effective policy will involve trade-offs between characteristics such as efficiency and stability.

There are four main classes of policy instrument that have been suggested to operationalise international co-operation on CO_2 emissions:

International emission targets This is the co-ordination mechanism currently implied by the FCCC, and commits countries to achieve specific CO_2 emissions targets by a specific date. The FCCC requires the same per centage reduction in each country relative to the baseline year (1990), but there is no reason why targets have to be the same in each country and they could be expressed in different ways such as per capita or per unit GDP.

Joint implementation A mechanism already included in the FCCC where countries may pay for carbon abatement projects in other countries and count the reduced emissions against their own limits.

Harmonised domestic or international carbon taxes National or international taxes are levied on the carbon content of each fossil fuel; these could levied on producers or consumers and revenues collected and redistributed nationally or internationally.

Tradable carbon permits/quotas Each country is given a quota of CO_2 it may emit each year distributed according to equity or other political considerations; permits may then be bought and sold between countries to match their actual emissions.

The next sections review the effectiveness of each of these policy instruments using the categories defined above and then summarises their advantages and disadvantages.

ACHIEVEMENT

The stated goal of the FCCC hinges on the definition of a 'dangerous' level of climate change. The focus of much debate has been whether it should be interpreted in a purely physical sense or if economic costs and benefits should be taken into account. The text of the convention is ambiguous as it calls for intervention to allow 'ecosystems to adapt naturally to climate change, to ensure that food production is not threatened and to enable economic development to proceed in a sustainable manner' (Article 2).

This emphasises the physical effects of climate change; however, all mitigating actions must also be 'cost-effective so as to ensure global benefits at the lowest possible cost' (Article 3.3).

Given the large scientific uncertainties surrounding the potential effects of current emissions, an additional precautionary principle is included in the convention which logically overrides strict cost/benefit calculations: 'Where there are threats of serious or irreversible damage, lack of full scientific certainty should not be used as a reason for postponing such measures' (Article 3.3).

Given these agreed principles it seems likely that the aim of climate change mitigation will focus on achieving emission and concentration targets of GHGs in the atmosphere, rather than the more conventional economic objective function of maximising global welfare. The efficiency aspects of this policy are discussed in the next section. Of the policy instruments described above, all are capable of co-ordinating action to achieve an emissions target in a verifiable manner except for joint implementation between partners with different abatement commitments, and internationally set, but domestically collected, emissions taxes.

Joint implementation

Joint implementation (JI) allows countries to gain 'credit' (sometimes 1:1 but usually lower) by investing in CO_2 abatement (source reduction or sink enhancement) outside their own country; this is encouraged under the

FCCC[4] and some schemes have already been initiated (Barrett 1994d). As joint implementation allows countries to 'shop around' for the lowest way to reduce emissions it offers potential for reducing the costs of GHG stabilisation.

Bohm (1993, 1994) argues that extensive use of JI on a project based approach is inherently open to monitoring and verification problems as the 'baseline' emissions (i.e. what would have happened without the project) are counter-factual and so cannot be measured. This is especially important for sink enhancements such as reforestation and land set-aside, as the different potential uses of the land (crops, ranching or fallow) greatly affect the net carbon emissions from the project. This problem is enhanced by the incentives in JI which encourage both the 'buyer' and the 'seller' to exaggerate the amount of CO_2 saved to the international monitoring authority.

The monitoring and verification problems are exacerbated in the more general case of JI between nations (Bohm 1994), when some countries have committed to emissions limits and others have not. Under these conditions JI between committed countries is efficient and monitorable, but JI with an uncommitted country is unmonitorable, because there is no agreed baseline for emissions. The overall conclusion is that JI is of some limited use at the moment, but cannot form a long term basis for achieving emission targets unless all countries involved have enforced national emissions limits.

International carbon taxes

The use of carbon taxes as a control instrument has been extensively studied by Hoel (1992, 1993) both theoretically and in the context of the proposed EU carbon/energy tax (CEC 1992). A uniform (in each year) tax level applied to all countries will achieve welfare efficient abatement if the tax is set to the marginal damage cost of CO_2 emissions, but uncertainty in energy elasticities and technological substitutes makes the actual magnitude of emissions reductions very difficult to calculate a priori. Domestically collected taxes are also open to verification problems, because any increases could be potentially offset by reducing existing domestic energy taxes leading to free-riding while in full compliance with the treaty (Hoel 1993)! Alternatively an international carbon tax could be levied, with the revenues paid to an international agency and reimbursed to states on some reciprocal basis, perhaps founded on emission reductions. Though this scheme has economic merit it has been considered politically difficult, because tax collection is a sovereign right, and such rights are explicitly protected in the FCCC. Experience with other environmental treaties points to a general reluctance to devolve sovereign powers to supra-national agencies, though co-ordinated restrictions on sovereign actions, which are controlled nationally, have been relatively common (Haas and Sundgren 1993).

Carbon leakage

The difficulty of accurately achieving emission targets using international carbon taxes is exacerbated by the reaction of energy markets to the decrease in fossil fuel use. As carbon emissions fall and consumption of fossil fuels drop, competition in energy markets should cause the price of traded fuels to drop, thus giving an incentive for higher fossil fuel consumption. If the agreement to limit emissions is incomplete non-participating countries in particular will benefit as prices drop; in addition differential fuels costs give an incentive for fossil energy intensive industries to migrate from committed to uncommitted countries. The net result is a shift of carbon emissions from controlled to uncontrolled countries, so called 'carbon leakage', a rise in total emissions and an increase in the cost of reaching emission targets in committed countries.

The magnitude of carbon leakage is much debated, but in the short to medium term both simple energy price and relocation effects are generally thought to be small (10–15 per cent of abatement offset by leakage) by most analysts (e.g. Oliveira-Martins *et al.* 1993, Smith 1994), though some have estimated the effect to be as large as 80 per cent (Pezzey 1992).

Fuel markets and strategic behaviour

More complex theoretical models of fuel markets, which include rent seeking and collusive behaviour, give further results that highlight the complications in setting an international tax to achieve specific emissions reductions. Wirl (1994) models a world of price setting producers and governments that impose energy taxation; rational producers will preempt carbon taxes by raising fuel prices to anticipated post-tax levels in order to capture some, or all, of the tax rent. The extent of this capture depends on the market power of the producers, and the predictability of the tax increases. Similar issues are looked at by Ingham *et al.* (1993), who model the difference between a tax imposed by producing countries (i.e. well-head tax) and that by consuming countries (i.e. a fuel consumption tax). The producer tax has essentially the same economic effect as the complete rent capture described in Wirl, and results in large amounts of tax revenue flowing to fuel producing countries; as energy consumption is reasonably income elastic this can lead to significant changes in global emissions. In practice the post-oil shock energy markets are probably too competitive to support large, long term price manipulation by producers, and it is very unlikely that a tax levied and collected by producers would be agreed internationally because of the distributional consequences.

Both Sinclair (1992) and Ingham *et al.* (1993) explicitly model the effect of including the exhaustibility of fossil fuel resources when calculating the optimum level for a carbon tax into the future. Both find that the taxes should fall over time (in a rational expectations, perfect information and infinitely malleable world) if the resources are near to depletion. An intuitive explanation of this result, given by Sinclair, is that if taxes rise over time producers will expect demand to fall in the future, and so will mine their resources more quickly (extraction costs are zero and the carbon stock does not decay) in order to earn more profit. This leads to CO_2 emissions and subsequent damage occurring earlier than if the tax decreased over time, and because the future is discounted this re-timing of impacts is sub-optimal. The situation is more complex when extraction is costly and the carbon stock decays over time. The optimum path is ambiguous and the tax will tend to rise initially when exhaustibility is not imminent, but will always end up falling as the stock of fossil fuels nears depletion. Ingham *et al.* interpret this last result as reflecting the fall in the marginal damage cost of emissions as emissions fall and fossil resources near depletion. A balancing force against a tax which is high in the short term and then falls, is the wish to avoid any extra costs incurred by scrapping of existing capital before the end of its useful life; such minimisation of transition costs requires that the tax be forecastable (to encourage up-front R & D investment) and that it starts low and rises slowly over time.

Golombek *et al.* (1993) analyse the effect of carbon taxes on different fuels in different types of agreement. They find that in a complete global agreement the tax per unit of carbon should be equal across fuels. However, if the agreement is incomplete taxes should be differentiated across fuels; this is because different fuels have different quantity/price elasticities and so as the price of traded fuels drops the increase in consumption in non-participating countries will be heterogeneous across fuels. Therefore, an optimum tax regime which accounts for participating and non-participating countries will levy different taxes per unit of carbon. Ingham *et al.* (1993) further argue that the change in prices across fuels will be complicated by substitution effects raising the demand for non-carbon intensive fuels such as gas; this could lead to an increase in the producer price of gas even as total fossil fuel use declined. Such effects make the a priori calculation of total abatement and carbon leakage very difficult, and the observed heterogeneity of responses to taxes can be seen in Chapter 4 where an empirically estimated model of inter-fuel substitution and conservation is examined.

The complex issues of imperfect fuel markets, rent seeking and carbon leakage complicate any simple procedure for setting carbon taxes internationally. It is unlikely that any of the above issues could be empirically measured with enough accuracy to ensure long term targets were met, given that the marginal costs of fuel production and market power of the oil producers are unobservable.

Summary

Given that the quantitative aims of future climate change agreements are likely to be expressed in terms of a time path for emissions and concentrations of GHGs, neither joint implementation nor international taxes seem to allow accurate achievement of the stated goals. JI is a useful short term, or pilot, measure and could contribute to abatement when used between Annex I countries; however, as long as two distinct groups of committed and uncommitted countries are differentiated by the treaty it will be open to monitoring and verification abuse. International taxes are easy to set but have unpredictable consequences and the desired emission reductions can be *legitimately* avoided by countries altering their internal taxation schemes. It should also be pointed out that any international co-ordination of enhancing carbon sink activity is fraught with difficulty, because of the counterfactual baselines involved. The easiest way to ensure achievement of a specific target is to use a quantity based scheme such as tradable emission permits, as these avoid the complexities of predicting the price reactions of countries; though of course they still do not solve the problem of carbon leakage if the treaty is incomplete.

As long as international obligations to reduce CO_2 emissions are limited to a few countries the problems of carbon leakage through energy market responses and industrial relocation will remain an obstacle to successful environmental protection. The evolution of the FCCC into a globally inclusive treaty is therefore imperative, and must be co-ordinated with future increases in fuel use in currently uncommitted countries.

EFFICIENCY

The FCCC currently contains references to achieving its aims in a 'cost effective' manner, but this falls short of stipulating strict economic efficiency where the costs and benefits of mitigating anthropogenic climate change would be weighed against each other. The simple conditions for an economically efficient solution would involve setting emissions reductions to a level where the marginal cost of emission reduction equals the discounted expected damage caused by the next unit of emissions.

The lack of emphasis in the final FCCC text on equating marginal costs with marginal benefits comes partly from the large uncertainty surrounding the cost and dynamics of damages from global warming, especially the possibility of catastrophic events (Fankhauser 1994a). This uncertainty is embodied in the 'precautionary principle' (Article 3.3) which implies that no meaningful probabilistic value can currently be placed on the likelihood of many future events which the treaty aims to prevent; therefore, standard cost/benefit techniques which rely on probability weighted future values cannot be used (see, for example, Vercelli 1994 for a review of these issues). Other reasons for the lack of explicit cost/benefit analysis are the conflict between sustainability and efficiency, and matters of inter- and intra-generational equity; these issues are discussed in more detail in Chapter 7.

Despite the absence of explicit cost accounting in the current treaty, it is obvious that countries will be taking these into account when negotiating targets on emissions and concentrations. Therefore, the use of a policy instrument which efficiently achieves stabilisation of atmospheric carbon dioxide concentrations will not only reduce global compliance costs but also decrease the long term concentration level, and thus damage, considered economically acceptable.

The most general condition for the efficiency of a policy instrument, in controlling emissions to absolute level, is that the marginal cost of emission reductions be equal in all complying countries. Therefore, there are no opportunities for reducing total costs by shifting a unit of emission reduction from one country to another. Under this criterion only tradable permit systems and JI between committed countries can be considered efficient policy instruments; while international flat taxes and targets are inefficient or, in the terminology of the FCCC, not the most cost effective policy instruments.

Uniform emission targets

As currently defined, the FCCC sets out a co-operative agreement with uniform emissions limits for each Annex I country; these are expressed as a per centage of 1990 emissions. This type of agreement will only be economically efficient if all countries have identical abatement costs; heterogeneous abatement costs mean that some countries' marginal abatement costs at equilibrium will be higher than others, which is Pareto inefficient (Hoel 1992, 1993).

As is detailed in Chapter 4, econometric estimates of long run energy elasticities in the main developed countries (the G8) have shown that large differences do exist in the price sensitivity of fossil fuel consumption, and thus the implied marginal cost of emission. These differences seem to be largely due to existing price differentials in fuel taxation and supply prices, and less dependent on institutional, geographic and technological differences. Therefore, emissions targets would only approach efficiency in these countries if energy taxes were harmonised beforehand. Given the different national policy objectives which existing energy taxes reflect, such as fuel security, revenue raising, transport policy, local pollution prevention and industrial policy, this is unlikely politically and would not be optimal for each country as it would elevate climate change mitigation to a privileged place above all other national energy related interests. If the FCCC expands to include commitments in less developed countries, technological and geographic differences would also rise in importance, increasing the inefficiencies caused by uniform limits, even if taxes were harmonised internationally. Of course different limits could be imposed in each country but these would require very detailed knowledge of national energy markets, at the international level, in order to be efficient.

International taxes

The theoretical argument for the welfare efficiency of environmental taxes is that they internalise an external cost (classic 'Pigouvian' taxation), and the optimal tax rate is the future expected damage cost caused by the last unit of pollution. As described above, the problems with measuring the cost of climate change, and the subsequent inclusion of the precautionary principle, means that taxes that have been proposed in practice have been aimed not at optimising welfare but at controlling emissions to a pre-specified target.

The optimum level, over time, of a stabilisation tax is determined by the target date for stabilisation and the marginal cost of emissions control at this point (usually represented by a clean 'backstop' technology such as nuclear fusion/solar energy). The optimal carbon tax is equal to the shadow price of emissions to this point, which equals the discounted cost of the backstop technology per unit of emission saved (STAP 1993). The rationale for this scheme is best explained as a version of the famous 'Hotelling' Rule for depletable resources. The remaining amount of atmospheric capacity for absorbing GHGs up to the concentration target is the depletable resource, and so the shadow price of that resource should rise at the rate of discount, until it reaches the price of the cheapest substitute, that is, non-fossil, energy (Anderson and Williams 1993); at this price no fossil fuels will be used because non-fossil energy will be cheaper.

As with uniform targets, uniform international taxes would be efficient if existing energy taxes were the same across countries. However, because taxes are different and the carbon taxes apply to input fuels, a uniform tax will be inefficient because the cost of energy, and so the incentive to abate emissions, will be larger in some countries than in others. This would not be the case if energy taxes were harmonised between countries, or levied on the pollution output (as it would be for other combustion products such as sulphur dioxide), and carbon scrubbing (post-combustion) or sink enhancement formed the majority of abatement efforts. Botteon and Carraro (1993) argue similarly that uniform carbon taxes are not cost effective because of cross country and sector differences. If there are significant rigidities in energy markets then the tax needed to stabilise emissions may be much higher than the marginal damage costs of climate change and so be welfare inefficient. Given such distortions they suggest an approach based on stimulating energy saving innovation and imposing multi-pollutant 'toxicity based' taxes would be better in the long term.

Tradable permits/quotas

In a situation where the costs of abatement are known with greater certainty than potential damages and there is a significant probability of catastrophic damage, it is generally acknowledged that a system of tradable quotas of emissions is a more effective policy instrument than emission taxation because environmental quality is guaranteed (Weitzman 1974, Pearce and Turner 1990).

In a tradable permits scheme each country would be allocated a quota of CO_2 emission permits which can be traded with other countries. If in any year they emit more CO_2 than they hold permits for, a penalty charge for each excess emission is levied by an international body (probably about ten times the market price of a permit). This framework will stimulate a market in permits which should automatically clear at the global marginal cost of CO_2 control, thus eliminating the centrally determined estimates of such costs needed to impose an international tax. The number of permits in circulation would be controlled over time in order to reach the stabilisation target. A futures market in permits would also develop, which would guide longer term infrastructure decisions, in the same way that committing to increase a carbon tax over time would (for a detailed discussion of the institutional issues surrounding tradable permits, especially the establishment of forward markets, see UNCTAD 1994). As an alternative to freely distributing the permits, global carbon emission allowances could be auctioned off to the highest bidder, with the proceeds being distributed to the poorest countries. However, given the large wealth disparities between developed and developing countries it is unlikely that such an auction would be seen as equitable, even though under some assumptions it is economically the same as distributing permits.[5]

A similar scheme to tradable permits is 'nationalisation' of the remaining carbon capacity of the atmosphere (Eckhaus 1993). In this scheme countries receive a stock of emissions capacity rather than a flow of permits, and therefore are free to decide for themselves on the most efficient intertemporal allocation on emissions; these stock rights could also be traded internationally in the same manner as emissions permits.

In a world of complete information, harmonised energy taxes, perfect markets and no transaction costs, permits and taxation are equivalent in terms of efficiency of *abatement*, but the permits scheme allows the maximum level of emissions to be set with certainty. The other great advantage of permits over taxes is that emissions reduction measures would be administered by states which can use a variety of instruments (taxes, regulation, direct investment, etc.) to meet the targets. This gives governments more flexibility to deal with market failures and to balance their conflicting policy goals.

MACROECONOMIC COSTS OF CARBON ABATEMENT

The superior efficiency of permits compared to non-harmonised taxes is a partial equilibrium result, which only takes into account the direct costs of emission abatement. In an efficient scheme, if the stabilisation target is set correctly, there should be a direct increase in global consumer welfare from introducing CO_2 abatement measures. However, because energy is both a direct consumption good and an input to production, there will also be broader macroeconomic impacts from introducing a tax. Macroeconomic impacts are defined as the effects of reducing fossil energy use on gross economic output, or the productivity of other inputs, such as labour and capital. These macroeconomic impacts will vary depending on the distribution of abatement costs between countries, which in turn are a function of which policy instruments are used to co-ordinate action.

An internationally set, but domestically collected and recycled tax (with no side payments between countries) results in no fiscal loss to an economy. Direct welfare losses come from a switch in consumption from direct energy use, or energy intense products, to less polluting goods. Indirect welfare losses arise from a decrease in total economic output. These output losses stem from reduced competitiveness in some industries and, perhaps, a move to a less productive economy as investment shifts from improving labour productivity to increasing energy efficiency. As fossil fuels will continue to be used in the foreseeable future, so carbon tax revenues will be substantial. If these revenues are recycled into reducing other economic distortions, such as employers' labour taxes (as opposed to being given back to households in a lump sum), then the net effect on macroeconomic output could be minimal or even positive (Barker 1994). The potential for positive output effects from recycling carbon taxes is theoretically contentious as some economic models deny such an effect is possible (Ligthart and Van der Ploeg 1994, Bovenberg and Goulder 1994). These negative results are usually driven by an initial assumption that the labour market clears at the given wage (i.e. there is no involuntary unemployment) and existing mixes of taxation are roughly optimal. Relaxation of these assumptions can generate models which allow recycling of tax revenue to offset completely the direct costs of abatement, even without taking into account environmental benefits (Carraro and Soubeyran 1994). Therefore, there are three components to calculating the cost of controlling CO_2 at the macroeconomic level: direct welfare costs, macroeconomic impacts and revenue recycling benefits. The modelling of these macroeconomic impacts inside EGEM is detailed in Chapter 5, and full assessment of the merits of different revenue recycling methods is given in Chapter 10.

In a tradable permits system buying significant numbers of permits on the world market results in a fiscal outflow, and less money to recycle through the economy. Depending on the initial distribution of permits, it is possible that the decrease in recycling benefits will outweigh the efficiency gains of using permits, rather than an international tax, to reach the same global emissions target; therefore, the permit scheme will be less cost effective than the flat international tax. If carbon taxes were collected internationally, then the revenue could theoretically be redistributed so as to produce the same fiscal flows as a given distribution of tradable permits; in this case a permit system would obviously remain the most cost effective instrument.[6]

For stabilisation of G8 emissions at 1990 levels the size of fiscal flows can be considerable. If a permit system only operated between the developed countries, and permits were distributed based on average CO_2 per unit GDP, the size of outflows to pay for permits would be comparable in some countries to current estimates of the macroeconomic costs caused by decreased energy use in the productive sector (≈ 1–3 per cent of GDP, Mabey 1995b). Tradable permit schemes between Annex I countries have been suggested with a variety of distribution rules: per capita, per unit of GDP, by energy use. Permits distributed per capita tend to benefit poorer countries; permits distributed per unit GDP benefit CO_2 efficient countries and permits distributed by energy use benefit countries with initially low energy costs and high energy use per unit GDP.

Conditions for macroeconomic efficiency

Given an understanding of the value of recycled revenue, it is likely that governments would take such costs into account when deciding how much CO_2 to control nationally, and how many permits to buy or sell on the international market. The inclusion of recycling benefits into the total cost calculations means that the initial distribution of permits may affect the total cost of reaching the emissions target, and thus the overall efficiency of the permit mechanism. This result is in contrast to that of the partial equilibrium analysis, where allocation of permits only affects the distribution of costs and benefits, and not the overall efficiency of the agreement.

Using public revenue to buy permits on the open market incurs extra costs, due to either the distortions that taxation places on the economy (lump sum taxes are not available) or the opportunity cost of not using the revenue raised from non-distortionary energy taxes to reduce other taxes in the economy. Therefore, the cost of buying a permit to a government is larger than its 'face value', and it is optimal for a country to decrease its emissions at a macroeconomic cost which is higher than this. Eventually, each country will control emissions until the marginal *total* cost of emission reduction is equal between countries (see Chapter 11 for a mathematical model of this situation).

A simple example of this effect is given by considering trade between two countries A and B, where the marginal cost of abatement is constant in both countries and always higher in A than in B. If A were given all the initial permits then A would not trade with B and all emission reduction would be carried out by B. If the permits were distributed equally between each country, and the revenue cost of buying a permit exceeded the difference between the two countries' macroeconomic abatement costs, then again no trade would occur as it would be cheaper for A to abate emissions than to buy permits. As a result of the redistribution of permits the optimum abatement levels in each country have changed, and the overall cost of reaching the target has been increased.

In the most extreme case these interactions could mean that the only way to ensure least cost compliance would be to distribute the permits optimally to begin with so that no country would wish to trade based on abatement costs. This reduces much of the attraction of a permits scheme as efficient distribution does not allow equity considerations and involves the same amount of centralised information as an international tax. However, the strength of the permit distribution effect will depend on many different factors including the size and heterogeneity of the benefits from recycling carbon tax revenues, and how this effect changes with the amount of revenue raised. Macroeconomic modelling of these effects in EGEM seems to show strong heterogeneity, with little correlation between costs of controlling emissions, macroeconomic costs and potential benefits from recycling. For example, while France and Japan have similar long run energy elasticities (0.645 and 0.507 respectively), the macroeconomic costs of reducing energy use are much higher in Japan, as are the benefits of recycling revenue into labour taxes; contrastingly Italy has similar macroeconomic costs as Japan but lower benefits from revenue recycling. The heterogeneity of effects mean it is impossible to make general predictions of the total macroeconomic costs of a permit system compared with a international flat tax, unless the size of these different effects are empirically modelled. EGEM is used to model these interactions in Chapter 11.

Summary

In a heterogeneous world, where existing energy taxes and energy elasticities differ significantly between countries, both emission limits and flat rate international taxes are less cost effective, in terms of the direct costs of emission reductions, than tradable permit systems or JI between countries committed to emission targets. However, when revenue recycling, fiscal redistribution between countries and macroeconomic costs are taken into account, this efficiency result becomes more ambiguous and only empirical modelling of the effects of specific proposals can show which instrument produces the most efficient result.

Tradable permits are unambiguously more efficient if permits are granted to, and traded between, private firms rather than governments because the distortionary costs of raising revenue are avoided (assuming companies all borrow on the same capital markets). The problem with this type of scheme is that it is limited as to the amount of emissions which can be controlled, and it gives perverse incentives for companies to relocate their factories to uncontrolled areas while retaining nationally held permit rights.

STABILITY

International agreements such as the FCCC are binding in international law, but in reality compliance with the terms of the treaty is only likely if countries think it is in their own best interest. There are no effective international mechanisms to force compliance, and the FCCC only binds parties to enter a dispute resolution procedure involving consensual discussions, not arbitration or sanctions.[7] The weakness of credible sanctions against countries reneging on their treaty obligations has led many commentators to argue that international agreements can never enforce significant global emissions reductions (Haas and Sundgren 1993). This result stems from the fact that most research in this area has assumed that agreements must be formed between large numbers of countries, or smaller numbers of homogeneous countries; see Barrett (1995) for a survey of this work. The possibility of agreement being dependent on strategic interactions between a small number of coalitions, which are of heterogeneous size and interests, has only begun to be investigated inside the economics literature (e.g. Botteon and Carraro 1995), but is explored below.

Forming stable agreements between large numbers of countries

Barrett (1994a, c) analyses this problem using a game theoretic model of agreement participation. Each country has the choice of either free-riding on the abatement efforts of committed parties or participating in the agreement. Countries only accede to the treaty if the net benefits (benefits minus costs) of co-operation outweigh the net benefits of free-riding. A 'self-enforcing' agreement is defined as the equilibrium number of countries where no party to the agreement wishes to leave, but no uncommitted country wishes to join. In this model the only credible reaction of the co-operating countries to a country leaving the agreement is to raise their combined emissions, and thus global damage, which reflects the smaller number of countries co-operating to reduce emissions. This limitation on the type of credible punishments by signatories on defectors removes any 'Folk Theoretic' solutions that could sustain a full co-operative outcome.

Barrett concludes that, given a large number of potential participants (30+), the maximum number of countries in a self-enforcing agreement (SEA) is small (2–3), with and without side payments, and this conclusion holds over several different assumptions about the form of country abatement cost and environmental benefit functions. The intuition of this result is that when there are a large number of potential participants the effect of one party's leaving produces a minimal decline in global emissions, and in the free-rider's own environmental benefits, while reducing its abatement costs substantially; therefore, there is a strong incentive for individual countries to free-ride. If the optimal non-cooperative and co-operative abatement levels are close, then an agreement could be sustained by a large number of parties; of course, in this case the gains from co-operation are small and the existence of a stable treaty is non-critical. The conclusions of this analysis are therefore that non-coercive international agreements are unlikely to produce significant environmental gains, over and above unilateral action, if the number of significant actors is greater than three.

Barrett (1994b) extends his analysis to include the case when pollution is associated with a good traded between oligopolistic (Cournot competitive), homogeneous countries. He shows that, for similar functional forms as the previous example, full participation in a SEA is possible if countries can enforce trade sanctions against free-riders in the goods and processes that are being controlled.[8] However, equilibrium only occurs at full participation or no participation. The intuition of this is that, while the number of participants is small, trade sanctions affect only a minority of world trade and the effect on non-signatories is small. As more participants join the treaty the decline in free-riders' trade increases, and, because all countries are the same, there is a point where they would all wish to accede to the treaty. At any point below this critical number of participants, however, the treaty would completely dissolve as there is a positive incentive to free-ride.

This scenario works because the participants' harsh reactions towards free-riders are credible as they raise *national* welfare by increasing the profits of domestic producers. If governments have no regard for domestic company profits when calculating net welfare, sanctions will not be used because they will decrease national consumer welfare, and therefore no stable agreement can be formed. In practice trade sanctions should never have to be enforced because the credible threat of sanctions will give an enforceable agreement containing all the parties. This is the most desirable outcome, as all forms of punitive sanction used to enforce collective agreements in the international arena, such as trade embargoes or import tariffs, impose costs on both parties. Therefore, no first best solution (i.e. optimal levels of co-ordination with zero transaction costs) to the co-ordination problem is possible, unless the mere threat of sanctions is enough to deter defection.

The use of import sanctions in this model makes it very relevant to the Montreal Protocol, but rather less so to the FCCC. This is because, unlike CFCs, the use and production of fossil fuels is ubiquitous throughout the world economy and cannot be categorised by a few major producers and technical applications. Therefore, it is very difficult to construct a simple and enforceable system of sanctions, or import duties, based on the carbon content of goods and processes. Such duties would probably also come into conflict with the provisions of the GATT, because the scope of levies would make them highly susceptible to protectionist manipulation for non-environmental reasons.

A more general model by Mabey (1995a) analyses the stability of existing carbon abatement treaties when a difference in abatement levels between two countries leads to a change in their terms of trade; a country imposing limits on energy use raises the costs of its exports and so becomes uncompetitive relative to an uncommitted, or free-riding, country. This effect could be either permanent or transitory, depending on the influence of transition costs, scale effects or other market rigidities (Ulph 1994). If exchange rates are flexible the terms of trade effect will eventually disappear, but the abating country still loses the equivalent amount of welfare due to the devaluation of its currency. In an incomplete treaty this effect increases the costs of abatement to countries which co-operate and also increases the benefits of free-riding. Using similar assumptions to Barrett it is shown that, if such trade effects are significant relative to net benefits from abatement, a self-enforcing agreement is possible which contains a large number of countries, and produces significantly more abatement than the non-cooperative case. The intuition of this result is that, as countries successively free-ride from an agreement, and trade related costs rise quickly, there may be a point where the costs of trade losses outweigh the environmental benefits from co-operation and all co-operating countries will leave the agreement. No country will rationally free-ride past this point because the breakdown of the agreement removes any incentive to do so; therefore the agreement is stable at this point. If trade costs are large, stability can occur at a high level of participation and abatement. This is generally consistent with Barrett's result because it depends on the sanction of treaty breakdown being in the co-operating countries' interest, and therefore credible. Along with carbon leakage we use EGEM to model this process in Chapter 9.

An example of a similar type of process is the GATT negotiations, where the large threat of an international trade war prevented any party unilaterally refusing to sign over a specific issue. If the threat of a trade war had not been credible, then the type of analysis outlined above would predict that every country would try to gain individual advantage from the treaty because it does not anticipate others doing the same. The sanction of a trade war is not institutionally levied by collective decision but is the inevitable outcome of economically rational decisions by individual countries. Therefore, it has a large degree of credibility as it is not dependent on co-ordination or enforcement measures that are counter-productive to those enforcing them.

These results argue for the importance of choosing a minimum level of ratification before the treaty enters into force, as this removes the disincentive of being the first country to accede to the treaty while all other countries free-ride on its efforts. Black *et al.* (1992) calculate optimum minimum ratification levels when countries can commit to abatement when acceding to a treaty, but have incomplete information about the benefits of co-operation. However, the analysis of such cases where countries commit to abatement on joining the treaty, and do not reoptimise when other countries free-ride, side-steps the question of self-enforcement by assuming that the legal commitment will hold whatever the actions of other countries outside the treaty (see also Carraro and Siniscalco 1993).

The limitations of the above work are both specific (i.e. the precise functional relationships between costs and benefits used) and methodological (i.e. the use of narrow direct cost comparisons). The results obtained are largely dependent on the use of very individualistic assumptions about country behaviour, where each party acts independently with no regard to reputation, status and other factors outside the current issue. In this environment, as Olsen (1971) showed in his classic analysis, meaningful co-operation is only possible between large numbers of actors if credible group sanctions exist. Recent work in game theory has looked at how past behaviour and reputation can lead to stabilising co-operative equilibria in a game context (Rabin 1993), and how related negotiations (in, for example, trade and aid) can produce equilibrium outcomes which are not stable in the isolated game context.

These extensions of stability analysis show the importance of institutional design and process in the formation of a productive agreement. Indeed, the only possibilities of a high level of co-operation from the models above occur when there is an existing binding agreement, and effective monitoring and reporting procedures for compliance. Even without large external costs from free-riding, or enforceable sanctions, the effects of reputation and associated negotiations may work to make agreement more likely, and more fruitful, than the above analysis would suggest.

Stability in strategic games

The nature of treaty stability changes when the number of active participants falls to a point where countries, or coalitions of countries, start to anticipate each other's actions. The classic analysis of games with many players assumes that free-riders believe their defection from the treaty will not have a substantial effect on other countries' co-operative behaviour. That is, other countries may alter their abatement levels slightly, but there will not be drastic reductions and nobody else will leave the agreement. This must be the case because every free-riding country, if it saw that every other country was a potential free-rider, would rationally predict the future breakdown in agreement (by 'backward induction' of breakdown) and either would never sign in the first place or would agree to coercive sanctions to preserve the gains from co-operation (Rasmusen 1989). However, this 'myopic' free-rider decision rule can only be justified if one or more of four conditions hold:

- The free-riding country is so small, relative to the rest of the players, that its actions are of no relevance to the co-operating parties.
- The free-rider is ignorant of the co-operating countries' reaction functions.
- Co-operating countries cannot observe the free-rider's behaviour.
- Other countries have no incentive(s) to free-ride which would be increased if another country left the agreement, e.g. the free-riders gain no competitive advantage by leaving.

None of these conditions holds when an international treaty is in place and the number of significant actors is small. Though 155 countries signed the FCCC only twenty-four have currently binding commitments under Annex II, and the majority of these are likely to follow the lead of the three main negotiating blocks of North America, the European Union and Japan. Therefore, in the short to medium term, the problem of stability revolves around the strategic reactions between these blocks of countries.

Given a small number of significant participants the dynamics of treaty stability become harder to generalise, and depend on the incentives for particular blocks of countries to co-operate. These incentives depend on the distribution of costs and benefits between countries which, as described above, are affected by the choice of policy instrument and the use of side payments between countries. Strategic interactions based on deception and failure to implement obligations should probably be minimal, as the FCCC process explicitly requires that all countries must submit regular reports on abatement measures, and there will also be third party monitoring of some extensive activities, for example fossil fuel imports, deforestation and land use (UNEP 1992). This information reduces the ability of countries to free-ride and if countries do renege on their commitments, doubtless complying countries will state their reactions to it publicly (Greene 1993).

Side payments and distributional issues

With full common information about each country's activities and degree of compliance strategic deception is impossible, and the stability of any treaty will depend on the real benefits each coalition gains from co-operating; which in turn will depend both on its abatement costs and benefits, and the value of its participation to the other parties.

It is simple to show theoretically that if all countries value emission abatement equally, and there are no competitiveness effects from differing abatement levels, then countries with the lowest direct abatement costs should abate most in an efficient agreement. In such an agreement, the countries which abate most will also gain least from participating, because their abatement costs are higher but their benefits are the same as in the other countries. If there are no side payments to compensate the low cost countries for their extra effort, it is to be expected that those gaining least from the treaty will consider free-riding first. Therefore, the parties with the largest incentive to defect will also produce the largest fall in abatement if they do so. This distribution of net benefits gives the lowest cost countries the negotiation power (which can be formally calculated as a Shapely Value) to argue for a redistribution of benefits in their favour. If countries have heterogeneous expectations of damage costs then negotiation power will also accrue to those who pollute a lot, but think their damages from climate change will be low.

The value of each country's contribution to the agreement, and thus its negotiating power, depends not only on its costs and damages, but on the order in which it and the other countries act. In co-ordination games it is often the last country that joins an agreement which can demand the largest share of communal benefits, especially if its non-cooperation leads to significant costs such as changes in terms of trade. If this is the case, an agreement is unlikely to form sequentially because each country will be waiting for the other to move. As with the many player games analysed above, a co-operatively agreed ratification level is needed to bring such an agreement into force. If there are advantages in beginning to abate emissions first, such as the ability to develop and then sell energy efficient technology or being able to avoid the costs of rapid changes in energy use, then a stable treaty should be able to form sequentially. To date the conventional view has been that acting first is not advantageous, but recent work which incorporates the effect of technical spillovers on economic growth has begun to challenge the theoretical foundations of this view (Steininger 1994, Naqvi and Schneider 1994).

Unlike the many party case, if there are only a few significant actors then predicting the outcome of any negotiation becomes a complicated empirical task in which all permutations of accession and defection, and the effects of different policy instruments, must be modelled. A preliminary assessment of these effects is given in Botteon and Carraro (1995), where the size of a stable coalition between five regions of the world is found to be dependent on the basis for bargaining over division of the surplus produced by co-operation (Nash Bargain or Shapely Value), and the distribution of damage costs; however, full participation was seen to be possible with side payments. Though side payments can equalise benefits between countries and so help to support stable agreements, they are administratively incompatible with the use of international flat tax which is domestically collected (e.g. the proposed EU carbon/energy tax) or the system of emissions limits currently outlined in the FCCC. With these instruments the cost of emission reductions is not readily measurable at the international level, and so the calculation of compensating payments is very complicated. Internationally collected taxes and permit systems allow easy redistribution of benefits, but this is not automatically in the low cost countries' favour. For example, an international permit system which allocated the permits based on GDP, regardless of actual energy use per unit of GDP, is equivalent to a system without side payments. This allocation would cause energy intensive and low cost of abatement countries, such as the USA, to abate a lot relative to other countries or be forced to buy extra permits on the international market. However, a permit system which distributed permits based on existing energy use would give a subsidy to low abatement cost countries, as they would have surplus permits to sell on the open market; this is equivalent to recycling revenues from an international tax based on emission abatement in each country.

From the above analysis it can be seen that different distributions of costs can serve three purposes: firstly, as a mechanism for insuring efficiency in a tradable permits system, secondly to address equity considerations between countries by equalising the macroeconomic costs from acceding to the treaty, and thirdly, as a way of ensuring the long term participation of all parties in the agreement by rewarding parties commensurate with their *net* contribution to the agreement. If tradable permits are used, the permit distributions needed to cause agreement are unlikely also to produce an efficient outcome; international flat taxes are never efficient but side payments can be used to ensure stability. In both cases there will be different marginal macroeconomic abatement costs in each country, because of the demands of the largest abating coalitions for a greater share of total benefits. This will result in a trade-off between efficiency, stability and final abatement levels which will fall short of the theoretical 'optimum' which could be imposed if the world were governed by a single authority.

Summary

The stability of international treaties to control carbon dioxide emissions depends on the number of negotiating blocks which are party to the process. Simple models which analyse each country as an individual actor, and assume a similar scale for each one, predict that maintaining a global treaty which produces significant increases in abatement levels will be difficult, and stability is independent of which policy instrument is used. Extensions to this type of model show that if there are large competitiveness effects between committed and free-riding countries, then inclusive and substantive co-operation may be possible due to the threat of the whole agreement collapsing if some countries defect to gain competitive advantage. If competitiveness effects exist, but are not large relative to the net benefits from emission abatement, then a stable treaty is again unlikely between a large number of countries.

More complex analysis, which allows for coalitions of countries reducing the number of actors to a point where strategic interactions are likely, changes the possibilities of stability. In this case the process of negotiation, and the monitoring procedures in the treaty, become more important for maintaining agreement, as do concurrent negotiations between the parties in other areas such as trade and development. Because the type of policy instrument used affects each country's contribution to abatement, and the practical potential for side payments, in a strategic game context they will also affect treaty stability. If efficient instruments, such as permits, are used then it is possible that the countries with the lowest abatement costs will gain least from the treaty but abate most. They therefore have the negotiating power to insist on side payments from the other countries to increase their payoff in return for co-operation. The policy instrument which allows side payments to be made most easily is an internationally collected carbon tax, but the initial distribution of carbon permits may also be used to produce a similar, if indirect, effect.

CONCLUSIONS: RELATIVE EFFECTIVENESS OF DIFFERENT POLICY INSTRUMENTS

By looking at the three different constituents of effectiveness – achievement, efficiency and stability – some conclusions may be reached over the suitability of the various policy instruments that have been proposed.

As a long term instrument for co-ordination, joint implementation is flawed because of problems with monitoring; however, it is probably a useful step to an international permit based scheme. Internationally set but domestically collected taxes are open to abuse as other energy taxes may be lowered to offset their effect; this problem is smaller if the tax is collected internationally. However, any use of taxes makes guaranteeing the achievement of an emission target difficult to accomplish owing to difficulties in modelling the global energy market. Tradable permits are therefore the best way of assuring accurate and verifiable compliance with the treaty aims.

Joint implementation and tradable permits are both efficient co-ordination instruments when direct costs are considered. Emissions limits and international taxes which are constant between countries are not efficient ways to meet a pre-set emissions target. Harmonised taxes will be, but their wider suitability will depend on their interaction with other national policy concerns. However, if all macroeconomic costs, including revenue recycling and raising public finance, are taken into account, the cost-effectiveness of tradable permits and emissions taxes is harder to compare as it will depend on the distribution of costs and side payments between countries.

For an agreement involving a large number of countries (30+) the choice of policy instrument has little effect on treaty stability because any agreement is prone to breakdown unless there are large competitiveness effects. If only a small (<10) group of countries, or coalitions of countries, is involved in the agreement then the distribution of costs between countries will affect the stability of any agreement. Tradable permits and internationally collected taxes both allow relatively transparent redistribution of costs, and so would be the best instruments to ensure the stability of any agreement. However, it is unlikely that the distribution of costs which assures agreement also provides the most efficient abatement when total macroeconomic and welfare costs are considered, so a trade-off must be made between efficiency and stability.

If abatement costs and energy prices are very heterogeneous between countries then tradable permits give large efficiency gains over an internationally set, but domestically collected, flat tax and will support an agreement if they are distributed to the countries which abate most. This will result in a flow of funds from the high cost countries, but if the cost of raising public funds is low this cost will be compensated by the increased efficiency of international abatement. If abatement costs and energy prices are similar across countries, and the benefits of recycling tax revenue are large, then an internationally set and domestically collected flat tax, or even emissions targets, will be more stable and probably nearly as efficient as a tradable permit system.

The interaction of macroeconomic and energy sector characteristics in determining the effectiveness of policy instruments complicates any decision on which should be used to co-ordinate climate change mitigation into the next century. Evaluation must be based on both theoretical insights and empirical measuring of variables such as the heterogeneity of energy elasticities, the size of competitiveness effects and the distribution of macroeconomic costs. These economic considerations are inescapably bound up in the political processes for ensuring agreement, as these will have a large impact on the distribution of cost and benefits and therefore the efficiency of the whole process.

One of the main insights to take from this analysis is the importance of countries working together to produce a measure of global benefits, while avoiding an overly self-interested negotiation stance which could lead to a complete breakdown in co-operation. The uncertainties in climate change impacts, in both magnitude and distribution, mean that there is a common cause for countries to aim for; this 'veil of uncertainty' over future outcomes should be a spur for meaningful co-operative action as there is no guarantee it will happen otherwise.

Empirical modelling of carbon dioxide abatement treaties

The theoretical analysis given above provides a strong motivation for the empirically based modelling of carbon abatement treaties, because the relative size of macroeconomic effects in the different countries will radically alter the potential for a meaningful and effective agreement.

Most previous macroeconomic modelling (as opposed to bottom-up engineering studies) of the costs of controlling carbon dioxide emissions has not disaggregated its results along political boundaries, but instead has aggregated the world into regions (see Chapter 3 for a review of previous modelling work). This is appropriate when considering the very long run (200+ years), when currently less developed countries will be major emitters, but gives little useful information about the short to medium term choices facing today's big polluters. The political and ecological dynamics of climate change control require that substantive action must take place soon in the developed countries if industrialising countries are to be persuaded to curb their emissions in the near future. The persistence of CO_2 in the atmosphere means that substantial climate change will occur if developing countries only agree to control emissions after they have reached current OECD wealth levels. Therefore, analysis of the current political economy of CO_2 control between the industrialised countries gives important insights as to the likelihood of expansion and strengthening of such agreements in the future, and so the likely levels of future climate change damage.

Using the macroeconometric model EGEM, which has detailed models of each of the industrialised countries, we can explore all the questions raised above in detail, and produce estimates of the magnitude of different influences on the political process. The details of EGEM's modelling structure are given in Chapters 4 and 5, and these are contrasted in Chapter 6 with the problems of modelling developing country economies. The various policy questions are then examined in Part III: the setting of economically optimal taxation levels, and different agreements are covered in Chapters 7 and 8; the problem of carbon leakage and its effects on agreement effectiveness are analysed in Chapter 9; the macroeconomics and domestic political economy of energy taxation are studied in Chapter 10 and, finally, the interaction of different policy instruments, treaty stability and strategic interactions between countries are analysed in Chapter 11.

Part II

ECONOMIC MODELLING OF CLIMATE CHANGE POLICY

3

A REVIEW OF MODELLING ISSUES AND PAST WORK

INTRODUCTION

The purpose of this chapter is to review some of the major global economic models which have been developed to study climate change related issues; concentrating on those that quantify the economic cost of control. This review aims to give a methodological and historical context in which to assess the strengths and weaknesses of the econometric approach that we have taken.

Of the many numerical models that have been developed to study the economic impact of mitigating greenhouse gas emissions most have focused on CO_2 emissions, because it is the most important gas in terms of its impact on climate change. Nordhaus (1991a) does look at other greenhouse gases by converting them to their CO_2 equivalent global warming potential. However, such elaboration is very prone to error, or irrelevant, because future production of CFCs will be controlled under the Montreal Protocol on Ozone Depleting Substances, and there are very large uncertainties regarding sources and sinks of other GHGs (for example, Methane and Nitrogen Oxides), which makes their economic modelling extremely difficult.

All models, economic or otherwise, are simplified representations of reality, but despite these simplifications most of the models considered here are still rather complex. Models are developed to capture important relationships in the economy, but incorporating all known features may make the model unwieldy, and difficult to understand.[1] Therefore, there is a trade-off involved between the extent of detail and the efficient modelling of relevant (from the modeller's point of view) relationships. Different types of models are most appropriate for assessing specific categories of economic impacts. As Boero *et al.* (1991) have stated, the right question about a model is not 'is it realistic?' but 'is it relevant?'.

Once a model's scope and structure have been defined a large number of parameters have to be estimated, and it is quite likely that these estimates may not be very robust. Parameters concerning the future are often guessed by the modellers, and so there can be no a priori validation of these numbers. In such cases, sensitivity studies have to be undertaken to determine how uncertainty in these areas affects the model results. These difficulties mean that studies of the economic impacts of climate change, and the cost of controlling it, are tending to diverge rather than converge as more work is done. Even the sign of potential economic impacts has come into dispute, as it is possible that some regions could benefit from global warming by, for instance, increased agricultural productivity in the colder areas of the world (Russia and Canada are often given as examples). Other studies emphasise the costs of global warming, such as greater incidence of drought in arid and semi-arid regions. A review of some of the more prominent economic impact studies is given in Chapter 8.

This high level of uncertainty and disagreement has prompted new work to develop 'integrated assessment models', which combine economic, climate and impact models of climate change (for example, PAGE model developed for the European Community and new modelling initiatives at Stanford and MIT). These models are highly aggregated, and designed to investigate the sensitivity of climate change, and the effectiveness of policy actions, under large ranges of uncertainty. This can be done in many ways, but the Stanford model, for example, uses Monte Carlo simulations to produce a probability distribution of outcomes given uncertainty in multiple parameters. As outlined in Chapter 1, this type of systematic sensitivity checking exercise is not our approach to modelling economic impacts, so we do not consider these structures here but instead concentrate on more comparable, and mature, modelling studies. However, it should be noted that these integrated modelling exercises have the potential to provide valuable information to policy makers as to the probability and magnitude of the risks they are facing in the very long term, even if their short to medium run relevance is rather low.

The first part of this chapter addresses the different methodologies used for modelling the economic cost of controlling CO_2, that is, microeconomic and macroeconomic models; the theoretical determinants of the costs of controlling CO_2 emissions are then discussed and the important areas of modelling highlighted. Six major macroeconomic models are then described, and their structural assumptions and parameterisation compared. The results of standard control scenarios run on each model are then given, and the determinants of inter-model differences outlined. At the end of this chapter we outline the modelling methodology used in this study, which is then described in detail in Chapters 4 and 5.

We review six major global models, namely, the IEA model (IEA), the Global 2100 model developed by Manne and Richels (MR; Manne and Richels 1992), the Edmonds and Reilly model (ERM; Edmonds and Reilly 1983), the Nordhaus model (Nordhaus 1991a), the GeneRal Equilibrium ENvironmental (GREEN) model (Burniaux *et al.* 1991b), and the Whalley-Wigle (WW) model (Whalley and Wigle 1991). These were selected because they have mature publication and documentation records in the public domain. A number of variants of some models exist, for example, the MR model has been modified by Rutherford to look at trade in carbon rights (Rutherford 1992). Besides these global models there are a large number of country specific models, but the focus of this review is limited to global models.

The global models differ in their structure, underlying assumptions, objective and sectoral disaggregation. We will categorise models by type, look at the objective or key theme each model addresses, report on the model results and assess the differences in results. There have been several surveys of the results from the models mentioned above: Boero *et al.* (1991); Cline (1991); Hoeller *et al.* (1991, 1992); Nordhaus (1991a); Beaver and Huntington (1991). Most of the surveys have compared the costs of reducing greenhouse gas emissions from various models, and have identified important parameters which explain the variance in costs and carbon dioxide emission estimates. In the OECD studies (Hoeller *et al.* 1991, 1992), attempts were made to standardise the basic underlying assumptions such as population growth rate and the baseline rate of growth of the economy; the same target reductions in emissions were then imposed in the counterfactual scenario to make the results from different models more comparable.

MACROECONOMIC MODELLING APPROACHES

The nature of CO_2 as a pollutant means that models with a global scope are needed for analysing different CO_2 emissions paths and their associated impacts (both physical and economic), assessing the effectiveness of different policy instruments for greenhouse gas control and studying the distributional impacts of mitigating greenhouse gas emissions. Global models must, by their very nature, be highly aggregated if they are to be manageable. An alternative approach to estimating the costs of control is to use microeconomic and engineering data to provide estimates for various technological options. The methodology underlying this approach is described later in this chapter, but these models, or studies, do not lend themselves to global simulations due to the detailed information required. To overcome this, UNEP (acting for the Global Environment Facility) has instigated the PRINCE programme, which collates micro-level studies from every party to the FCCC using a common methodology so that these may be used to perform global analysis in the future (UNEP–CCEE 1992).

The two main types of macroeconomic models are resource allocation models and econometric models. Resource allocation models can be further split into general equilibrium or partial equilibrium models.

Resource allocation models

Resource allocation models are theoretical constructs of the economy which are calibrated to observed data using a number of different techniques. The choice of model structure is determined by theoretical consistency, not statistical fit to measured behaviour, and they may use unobserved variables such as the social utility of consumers. The differences between resource allocation models are determined by their theoretical assumptions and the scope of the economy they cover. In terms of scope there are two main types: general equilibrium models which calculate prices and quantities in all relevant markets in an economy, and partial equilibrium models which concentrate on one specific sector and view the rest of the economy as exogenous and unchanging.

General equilibrium (GE) models have an explicit (aggregated) representation of all types of economically active agents in the economy, and the linkages between these agents. The price mechanism produces a market clearing equilibrium in all markets (which are a priori complete and perfect). In GE models the prices, the quantities and the growth of output are endogenous, whereas model parameters such as preferences, technology and policy are exogenous. Household product demand and factor supply functions are consistent with utility maximisation subject to a budget constraint; product supply and factor demand function of producers are consistent with profit maximisation subject to technology constraints. An important strength of GE models is that Hicksian welfare measures of equivalent and compensating variation can be calculated.

GE models usually use constant elasticity of substitution (CES) production functions to describe the underlying technology of the economy (or sectors). A typical two factor CES production function is defined by:

$$X = H(\alpha L^{\rho} + \beta K^{\rho})^{1/\rho} \qquad (1)$$

where: X = final output, L = labour, K = capital, H = total factor productivity.

ρ is the substitution parameter which determines the value of the elasticity of substitution between inputs (σ); taking logged first derivatives of (1) with respect to factor inputs and dividing to give $\delta \ln L/\delta \ln K$ it is easy to prove that $\sigma = 1/(1 - \rho)$. The elasticity of substitution is constant because σ depends only on the ρ parameter and this holds as long as technical change is Hicks neutral; that is, it affects all factors equally. H, α and β are the coefficients reflecting technological growth; H represents Hicks neutral growth, capital saving technical progress is given by changes in β, changes in parameter α would indicate labour saving technical progress. From the form of the equation it is obvious that one of these technical change parameters is redundant, because the term inside the brackets can always be divided through by α or β; when econometrically estimating factor demand equations, based around CES functions, this leads to identification problems, which are detailed in Chapter 5.

In most models constant returns to scale is imposed on the underlying technology of production. This is an important feature because it ascribes all trend changes in factor productivity to technical change, rather than scale effects; over the long term this will affect the simulation properties of the model and the ability of policy to influence the growth path. CES production functions do not model substitution among multiple factors (e.g. capital, labour, energy and material inputs) very well, as constant elasticity of substitution among more than two inputs implies that elasticity of substitution among all inputs must be the same (Jorgenson and Wilcoxen 1992). However, more flexible forms of production function, such as the trans-log, are harder to parameterise and so are not used in GE models which have not been econometrically estimated.

Parameterisation of GE models

The general specification of a GE model can be expressed as:

$$F_i\ (Y_1, \ldots\ldots, Y_n;\ X, B, e) = 0$$

where Y_i are the endogenous variables, X the exogenous variables, B unknown parameters and e the vector of stochastic disturbances.

The calibration method assumes that all components of e are zero and solves B on the basis of a single realisation of $Y_1 \ldots\ldots Y_n$ and X. Calibration forces the model to replicate the data of the base period (usually a year) by varying the parameters to reproduce base data as an exact solution to the model. This is only valid if the economy in the base period chosen can be considered to be at equilibrium, and not in the dynamic transition following a shock. However, to the extent that B has more than n components, extra information is needed to determine (m − n) of the unknown parameters. This implies an extremely strong assumption that observed values of the endogenous variables are determined only by the factors explicitly included in the model.

The use of CES production and utility functions means that it is possible to identify all parameters of the model on the basis of a set of extraneous elasticity estimates and one single observation of the economy. For CES functions, the factor demand functions have only three parameters: the scale factor H, the substitution parameter ρ and the distributional parameters α or β. With information on the value of one of these parameters, a single observation on equilibrium prices and quantities is sufficient to determine the remaining two parameters. This approach gives researchers the freedom to utilise econometric estimates of crucial elasticity parameters, and calibrate the rest of the parameters to the data. A brief review of econometric work on labour-energy and capital-energy elasticities in Chapter 5, shows the dependence of these values on a priori assumptions about the underlying economic structure. Therefore, there is great danger of 'spurious empiricism' if these numbers are naively used in a GE model which has a completely different structure and set of underlying assumptions.

Thus, even though a GE model has the potential of providing numerical estimates of various effects, the actual measurement may be too constrained. These simple functional forms fail most econometric tests, and in addition it is not possible to obtain a measure of the accuracy of the model and its predictions. A possible solution to this problem is to estimate the parameters through econometric models, but this requires highly sophisticated techniques. Jorgenson and Wilcoxen (1992) have used flexible functional forms in a model for the USA. However, Boero *et al.* (1991) feel that more complex forms are not worthwhile in the modelling context. A less ambitious alternative is to use stochastically specified sub-models of production and consumption and use these as building blocks of the GE model (Bergman 1990).

Once built, most GE models employ comparative static analysis, where two equilibria solutions are compared but no insights are provided into the dynamics of the adjustment path (UNEP 1992). A restriction of the GE approach is that it presumes full equilibrium is reached at each point of its solution run; there are no transitional dynamics in the economy. Also, a deviation from a 'no distortions', full equilibrium base run by assumption involves economic costs, because there can be no existing sub-optimalities, such as involuntary unemployment, in these models (Boero *et al.* 1991). GE models are being extended to include development over time and model equilibrium development paths; Jorgenson and Wilcoxen (1992) have produced a dynamic, econometrically estimated GE model for the USA.

In contrast to the comprehensive scope of GE models, partial equilibrium models only explicitly address relationships in a sub-section of the economy. Considerable detail is given to the selected sector, which in this case is usually the energy supply sector, whereas prices and other inputs from the rest of the economy are specified exogenously. Within the specific sector more importance is given to technical and accounting relationships than to behavioural relationships. There is usually a sector-wide optimisation objective for the sub-section, such as minimisation of costs; generally using linear programming or non-linear techniques for optimisation. These techniques have limitations in the form of extreme 'corner solutions' which ignore differences in expectations among different individuals, and in addition understate adjustment costs. These problems can be avoided by imposing additional arbitrary constraints which influence the final solution but this restricts the degrees of freedom in the model.

The results from these partial optimisation models can be fed into macroeconomic models to estimate the impact on the total economy. However, these models still remain limited as this approach does not account for feedbacks between the rest of the economy and the explicitly specified sector.

Econometric models

Econometric models consist of sets of equations defining relationships between economic variables (for example, consumption and income), the structure and parameters of which are estimated statistically from time-series data of the observed economy. Econometric models differ greatly as to the extent of theoretical structure they embody, but the choice of equation structure is primarily determined by its statistical ability to explain the data and not a priori theoretical assumptions.

Historically, econometric models have focused more on the overall level of economic activity and less on efficient or optimal resource allocation; this reflects their original use as short term forecasting tools to guide fiscal and monetary policy. As econometric models do not usually model the economic feedbacks which keep economies close to market equilibrium, their results become unrealistic (or chaotic!) when used over the long time periods which resource allocation models simulate.

Macroeconometric models are more relevant for short run and medium term analysis, as they are able to model market imperfections and disequilibria such as unemployment and capital shortages. Therefore, they are more appropriate to study the adjustment period and adjustment costs, as well as issues such as the general tax structure and investment profiles. Their basis in measured economic data allows fairly robust economic predictions over this time span. However, econometric models can identify welfare only with aggregate consumption, or with GDP, because they are based on national accounts data and not on the utility optimisation of representative agents. Boero *et al.* (1991) argue that, given the time horizon involved in the greenhouse problem, modelling the process of price adjustment and the associated disequilibrium is not of high priority. But, as political responses to the greenhouse problem will tend to focus on the immediate impacts in the short to medium term, we would argue that the results from macroeconometric models have much relevance for decision makers now.

Recent developments in macroeconometric modelling have made modern econometric models more suitable to address climate change issues over the medium to long term. Modelling of the supply side of the economy has greatly improved, and this has been facilitated by advances in estimation theory. These advances have expanded the types of theoretical models which can be estimated, and made transparent the long run specifications and validity of these models. For example, cointegration theory has enabled the empirical identification of underlying long term trended relationships, and error correction models allow the characterisation of short term measured fluctuations as an adjustment process towards this long run equilibrium.

MICRO-LEVEL MODELLING

So called 'bottom-up' models are technology based models founded on engineering relationships; a good overview of the bottom-up modelling approach is given in UNEP (1992). These models tend to look at narrowly defined sectors in isolation, and evaluate the future technologies available within each sector. On the other hand macroeconomic, or 'top-down', models use aggregate economic indices without considering specific end-use details. Engineering models stress detailed technological information, rather than behavioural functions, and study the impact of specific technological options. This classification between bottom-up and top-down models is not very discrete, and there can be considerable overlap. For example, a macroeconomic model may have a very disaggregate energy sector incorporating technological relationships.

Bottom-up and top-down models generate significantly different cost estimates for the same economy. This is due to both the difference in the definition of costs and the method of estimating costs. The specific representation of technologies means that engineering models can be used to assess non-price based policies, such as energy efficiency regulations on appliances, subsidies for energy saving and energy labelling schemes; all these policies have political attractions as they avoid the regressive nature of an energy tax. One of the great limitations of macro models is their reliance on purely price based policies and only *ad hoc* additions of specific technologies (notably as backstops) can alter this.

As in the case of macroeconomic models, engineering models can be of different types. These can be classified as:

Partial forecasting models The main content is data on technical characteristics, investment, operating and maintenance costs as well as fuel costs. These are used to forecast energy supply and/or demand based on the above information.

Integrated energy system simulation models These are more complex and have detailed energy supply and demand representation. This enables a detailed analysis of abatement options at the energy production, conversion and use levels. The main limitations of this type of bottom-up model are the complexities involved in checking the consistency of the system, and in achieving the optimum results. Trial and error methods usually have to be used to arrive at the optimum for the system.

Energy system optimisation models These are similar to integrated energy system models, but linear programming is used to arrive at the optimum solution. Limitations associated with this type are that: linear representation can only describe constant returns to scale; small variations in input parameters lead to large changes in results (bang-bang effects); and the implementation of the most attractive technology to its full extent, followed by the next most attractive, is not representative of reality where technologies of different vintages exist side by side.

A number of limitations are associated with engineering models. In these models the costs are based on an idealised evaluation of technology, where hidden costs are ignored. Engineering models can illustrate and evaluate the potential for emission reductions, but not the means of achieving this potential. The hidden costs that might be involved in its realisation, such as market imperfections and other economic barriers which prevent the full potential penetration of technologies, are not accounted for except by *ad hoc* methods. Macroeconomic indicators, impacts and relationships are also not included in these models; for example, any multiplier effects, price effects, structural effects and impacts on GDP and employment of different technological options.

Results of bottom-up models

The most important output from bottom-up or engineering models is the listing of technological options available, their associated costs and impact on carbon emissions. From this information it is possible to derive cost of abatement curves, which are an aggregation of many different technologies and structural changes in the energy system. In the context of global warming studies, these curves are formed from various carbon abating measures, weighted for proportional impact on the studied economy, ordered from least cost to highest cost.

Cost curves from engineering models address direct energy system costs, but do not incorporate transaction and other hidden costs. The value of each technology is assessed independently, as it is assumed that each technology results only in a marginal adjustment in the system; this causes problems when measures which interact in a non-linear fashion in the real world are combined linearly in deriving the cost curve (e.g. installing an efficient heating system and then adding extra insulation means that the original system will be over-sized, raising its associated cost of abatement). The assumption of partial equilibrium does not hold because a given level of investment can influence the total economy, and thus the greenhouse gas intensity in that economy.

Cost curves can be generated in four ways from a portfolio of specific technological options:

Partial solution Different technologies are evaluated separately for costs and greenhouse gas reduction with respect to a reference technology. The next step is to look at the incremental changes in greenhouse gas emissions and costs and rank these according to increasing costs. The limitation of this approach is that it ignores interdependence.

Retrospective system approach A simple or complex energy system framework is used to evaluate interdependencies in the system. The results of sequential incorporation of least cost technologies are compared. A limitation of this approach is that once an option is included it is a permanent part of the subsequent scenario.

Integrated system approach This requires the existence of a reference case, and fully defined energy system model, in which all system parameters can vary. Solution involves the choice of lowest costs for a given reduction in greenhouse gases, and this approach accounts for all interdependencies.

Its limitation is that several energy system solutions, which are economically equivalent, are feasible for the same level of greenhouse gas abatement. However, it is not possible to identify a unique technical energy solution at all points on the cost curve.

Multiple integrated system approach A few energy systems are selected using the integrated systems approach, and these are investigated using the retrospective method with respect to robustness and timing.

Given the respective strengths and weaknesses of macroeconomic and engineering models, there is often a need to integrate the two types of modelling approaches. Cost curves provide aggregated, highly detailed information, in a two-dimensional numerical form, and therefore are potentially an important means of communication between engineering and macro models.

It is generally assumed that bottom-up models tend to underestimate costs, whereas top-down models overestimate costs of abatement. Engineering models underestimate costs because they consider only direct costs and not indirect costs. Macro models on the other hand do not include negative costs of abatement, so called no-regrets technological options, which would yield carbon savings at negative or zero costs. Many of these negative cost options are not implemented to their full extent due to a number of factors, often assumed to be market failures but in some countries also government failures (Lovins and Lovins 1991). Obstacles to implementation of these negative, low cost options could be poor access to information; difference in discount rates used by the actual consumers (very high) and that used in economic analysis; preference for low first-cost rather than low life-cycle cost options (most individuals do not have information to do the necessary life cycle cost calculations); existence of energy subsidies and market distortions and other factors such as the perverse incentives to landlords and tenants in the case of buildings.

SUMMARY OF MODEL TYPES

Each type of model has its strengths and weaknesses, depending on the questions it seeks to answer. GE models have explicit structures which are internally consistent and accord with standard economic descriptions of a 'perfect' economy. Computationally they can be solved over very long time periods because they have an explicit (analytic) solution at each point in time; this gives them great advantages when considering the long run impacts of climate change. The weakness of this highly structured approach is that at no time will the model's results be in accordance with the real measured economy, with its transition effects, dynamics, hysterysis loops, and persistent disequilibrium in labour markets. Therefore, their parameters cannot be directly estimated using the model's structural equations but must be taken from econometric studies, which may have used very different structural equations in estimation. These exogenous figures are then used to calibrate the model to a single year's data, assuming that the economy was at equilibrium at that time.

In contrast modern econometric models have a strong empirical basis and model the full range of dynamic and disequilibria effects in the economy. If long run cointegration analysis is used they will have structurally determined long run solutions in many areas, which increases their consistency with economic theory in the medium term. However, because the model as a system is not necessarily consistent, it may or may not have a stable long run solution; chaotic divergence after a perturbation is always likely in such closely coupled systems of non-linear equations. This property means that econometric models are usually restricted to simulating the economy over the short to medium term (up to forty years).

The contribution of bottom-up engineering models is that they explicitly account for existing and potential technological options to save emissions. Of course they are limited to the future horizon of realistic technological options, which is probably about forty years, and contain no information about the overall economic impact of introducing technologies. However, by explicitly accounting for costs in different sectors, they can allow analysis of the type of incentives, both price and non-price, that would be needed to encourage adoption of carbon free technologies in a mixed economy characterised by widespread market failures in its energy sector. The difficulty in describing a single set of technological options, and their transition and diffusion path, in the short to medium term reduces the ability of bottom-up studies to be easily combined with macroeconometric models. However, because GE models are relatively unconcerned with transition effects, and generally define their conditions using *ad hoc* assumptions, it is easier to combine engineering data with this type of model. Therefore, the long run of many GE models is defined by a carbon free 'backstop' technology taken from engineering studies, the cost of which determines the level of carbon tax and thus macroeconomic impact of abatement policies.

CRITICAL DETERMINANTS OF CARBON EMISSIONS AND ECONOMIC COSTS OF ABATEMENT

Before we examine the specific structure of each model studied in this section we highlight the main parameters and assumptions which drive results and will be of interest. In a general sense future carbon emissions and the costs of curtailing them depend on the following key factors:

- Rate of growth of GDP and population.
- Energy use and the underlying fuel mix.
- Technological progress and its impact on energy supply and demand.

The simple relationship between these aggregate variables and carbon emissions is:

$$C = GDP * E/GDP * FF/E * C/FF$$

where: C = energy related carbon emissions in the economy (in mt/yr), GDP = gross domestic product (in $/yr), E = total energy used in the economy (in mtoe/yr), FF = fossil energy used in the economy (in mtoe/yr).

The product of GDP and the energy intensity (i.e. the energy consumed per unit of output) gives the level of energy use in the economy. A higher rate of growth of GDP implies more use of energy; a decline in energy intensity (E/GDP), given the sectoral mix, implies an increase in the productivity of energy use. Energy intensity thus reflects two factors: the sectoral composition of GDP and the efficiency of energy use in the economy. The share of carbon fuels or fossil fuels in energy supply is given by FF/E. The mix of these fossil fuels will determine the CO_2 per unit of fossil fuel energy, that is C/FF. The share of carbon based fuels and carbon intensity of fossil fuels together give the carbon intensity of energy use. And finally the product of energy use and carbon intensity of energy gives the level of CO_2 emissions.

Future levels of CO_2 emissions will depend on how technological progress affects energy supply and demand. Energy saving technological progress depends upon the composition of GDP (e.g. the mix of manufacturing, service and resource based industries), energy prices and non-price factors such as exogenous, as well as policy induced, technological change (e.g. efficiency regulations and government R & D in energy efficiency). Ideally, structural change and exogenous energy saving technological progress should be modelled separately. However, almost all studies combine the factors leading to a decline in energy intensity into a single parameter called the 'autonomous energy efficiency improvement', the AEEI. The value of this parameter is then specified exogenously, and values used range from 0 to 1 per cent per year. Few econometric estimates for the AEEI parameter are available, and modellers have had to use their judgement in selecting values. Unfortunately, apparently small differences in the AEEI selected become very large when compounded over long periods, and this is the cause of a significant part of the wide dispersion in future estimates of CO_2 emissions. Chapter 4 describes how the aggregate AEEI parameter was estimated in EGEM from economic data, and then disaggregated into price induced, structural and exogenous changes in energy intensity.

The other important class of parameters, which can explain the differences in results across models, are the substitution possibilities within fossil fuels, the extent of substitution between fossil and non-fossil fuels and between energy and other production factors as well as the substitution of energy intensive products by non-energy intensive products in the consumption mix. Cline (1992) lists the pathways for reducing carbon emissions as:

- Substitution into cleaner fossil fuels, or intra-fossil fuel substitution (IFFS).
- Substitution of energy by other factors of production (OFES).

- Consumption substitution into lower energy intensive goods, or product substitution (PS).
- Substitution of fossil fuels by non-fossil fuels (NFFS).

Intra-fossil fuel substitution has considerable scope for reducing carbon emissions in the short term. On an average, coal emits 1.04 tonnes of carbon per tonne of oil equivalent, oil 0.87 tonnes of carbon per tonne of oil equivalent, and natural gas 0.65 tonnes of carbon per tonne of oil equivalent. Thus, at current levels of emissions, with coal accounting for nearly 40 per cent of CO_2 from fossil fuels, a complete substitution of coal by natural gas could achieve about 20 per cent decline in total fossil fuel related carbon dioxide emissions. However, IFFS presents only a short term option for reduction of carbon emissions, because natural gas and oil reserves are much more limited than coal reserves, which would form the majority of future energy supply if there were no constraints on CO_2 emissions (see Table 3.1).

Substitution between other factors and energy (OFES) represents energy conservation (usually substituting capital goods, such as double glazing, for energy) and the reoptimisation of productive processes over time to use different input mixes. In aggregate models this effect is captured by the elasticity of substitution, σ. The lower the value of σ the more costly it is to substitute energy with other factor inputs; when the energy share is small, the price elasticity of demand for energy approximates the elasticity of substitution. Most models take the absolute value of the elasticity of substitution to be less than unity. Chapter 5 gives a detailed discussion and review of what OFES entails economically, and how it might be estimated from available data. This effect is often conflated with product substitution (PS), and no global model explicitly accounts for PS as a means for reduction of energy or carbon emissions, though it is implicitly present in econometrically estimated factor demand equations.

Along with OFES, substitution of fossil fuels by non-fossil fuels is the most important long term solution, as reduction of emissions by IFFS

Table 3.1 Ultimate recoverable fossil reserves (Gtoe)

Fuel	*Gtoe*	*Per cent*
Hard coal and lignite	3400	76
Conventional oil	200	5
Unconventional oil:		
Heavy crude oil	75	2
Natural bitumen	70	2
Oil shale	450[a]	10[a]
Natural gas	220	5
Total	4415	100

Source: World Energy Council Commission 1992

a These reserves are sub-economic under foreseeable market conditions and technology.

presents limited opportunities. Biomass, hydroelectricity, nuclear power, solar energy, wind energy, OTEC, tidal power, and so on are some possible non-fossil sources of energy. However, again there are limitations on the supply of these resources (for example, biomass supply is limited by availability of land), the speed of market penetration and technology diffusion or very large production/supply costs (for example, solar photovoltaics) when compared to coal. From an aggregate economic modelling point of view NFFS is identical with OFES, because a switch to the above technologies would increase the share of capital (and to a lesser extent labour) in the economy, at the expense of energy.

In the extreme OFES is often described as use of backstop technologies, non-carbon based fuels which can supply an infinite amount of energy, but at much higher costs than those of current energy sources. The date and cost at which non-carbon backstop technologies would be available are obviously critically important determinants of economic costs of greenhouse gas abatement. Most studies assume the availability of non-carbon backstop technologies at some point, and some models go into details about electric and non-electric backstops in order to define the end point of their scenarios.

The economic costs of reducing energy use

Significant carbon reductions imply availability of energy (or energy services) at a much higher cost, defined in the limit by the price of non-carbon energy. Assuming that the original mix was approximately optimal, this will result in a decline in output, because the changed factor mix will be less productive. In addition, there is a reduction in welfare associated with a shift to a less desired composition of products, though of course this will be balanced by the benefits of preventing climate change. Macroeconometric models give the losses in GNP due to a constraint on carbon emissions, but this only measures the decline in production. General equilibrium models often calculate estimates of the Hicksian equivalent variation (the increase in income that would be required to leave consumer welfare unchanged), which is a better measure because it includes the loss in welfare to the consumer from reduced use of energy.

Carbon reductions through intra-fossil fuel substitution (IFFS) and non-fossil fuel substitution (NFFS) do not reduce energy availability directly, but reduce the carbon intensity of the energy used. Indirectly, a shift to these more expensive sources of low carbon or non-carbon energy will reduce *total* energy use, the magnitude of this depending on the price elasticity of energy demand. Higher price of non-fossil energy technologies implies a drop in production as resources have been diverted into the energy sector and away from producing direct consumption goods.

The total impact on output, or the economic cost, associated with a carbon constraint in optimisation models is measured as the difference in output/welfare in the constrained scenario (with a limit on carbon emissions) and the scenario with no restrictions on carbon emissions. In production function models, the loss is measured in terms of reduced output; in models without an explicit production sector, the output impact can be inferred (Cline 1992). In this case, changes in carbon emission levels and energy consumption, in response to a carbon tax, reflect the opportunity cost of energy. It is therefore possible to integrate across marginal taxes to infer the production cost of the carbon constraint as this cannot exceed the amount of money saved by switching to low carbon technologies (i.e. price of carbon × the amount saved), if consumers and producers are acting rationally, and there are no other production externalities from energy use.

The economic costs of imposing a carbon tax can be overstated if the efficiency gains resulting from the recycling of carbon tax revenue (by replacing the most inefficient taxes on other factors of production) are not considered. The optimal mix of public revenue raising occurs when the welfare gain from increasing carbon taxes, and recycling revenues, equals that from decreasing taxation on any other factor of production/consumption. If the optimal level of energy taxation, *not taking into account climate change externalities*, is less than the Pigouvian environmental tax (i.e. direct marginal damage cost of CO_2), then the optimal tax, including the externality costs, will be below the Pigouvian level. This is because taxing energy involves a loss in income as well as a gain in environmental quality. The tax would only be set at the Pigouvian level if reducing energy use imposed a pure welfare cost, rather than a production externality. These issues are addressed in a few models such as the OECD–GREEN model and the Whalley and Wigle (1991) study.

The lower the substitutability between energy and other factors of production, the more steeply taxes will have to rise in order to stabilise emissions. Even with an elasticity of substitution equal to unity, the tax curve is non-linear, and taxes increase more than proportionately with the target cutback in carbon. If a model uses a discrete backstop technology then, in the long run, all energy will be produced by this technology (heterogeneity of end use, i.e. transport or heating, is usually not modelled for backstop technologies). As a carbon tax increases, the differential between non-carbon and carbon backstop technologies is neutralised and the tax rate settles at the difference in cost between the carbon free backstop and the carbon backstop. Thus availability of backstop technologies breaks down the monotonic relationship between required tax and the target reduction. Essentially the technical divorce between carbon reductions and energy usage produces much smaller economic costs than if there were no backstop technology.

In long run studies (100–200 years) the trend parameters of CO_2 and GDP growth, combined with the price and timing of the backstop technology, will completely dominate the results; making the sophistication of the remaining model structure rather redundant in its influence on policy prescriptions.

OVERVIEW OF GLOBAL MODELS OF CONTROLLING CO_2 EMISSIONS

A comparative overview of models is given here. The first section describes the overall structure of the different global macro models; the second section discusses differences in structural assumptions (e.g. time scale, technologies, scope and disaggregation) and in critical parameter values (especially the value of AEEI and ease of factor substitution). Model results are then compared, using tests where the baseline inputs and scenario design have been standardised as far as possible. The final part addresses the limitations of the models, and how far their assumptions pre-determine the results they produce. This work draws on the surveys by Boero *et al.* (1991) and Cline (1991, 1992) and the OECD surveys by Dean and Hoeller (1992) and Hoeller *et al.* (1991, 1992).

Description of different models

IEA

The IEA model is an econometric energy sector model constructed till the year 2005. The macroeconomic indicators are exogenously determined, and most of the parameters are estimated econometrically over the period 1965–89. Adjustment factors for technology, saturation of markets and resources complement these economic forecasts. The strengths of the model include the detailed modelling of end-use consumer energy prices and their link to primary prices, and the incorporation of the rigidities of the current energy systems of OECD regions. The latter, however, limits its use for very long term policy analyses (100+ years). The IEA model does not account for feedbacks from the energy sector into the rest of the economy, therefore cost results can only be given in terms of the size of carbon tax needed to reach a specific target, and not in terms of GDP loss.

MR model (Global 2100)

Global 2100 is a dynamic GE optimisation model developed by Manne and Richels (MR) as an analytical framework for estimating the costs of carbon emission limits. The MR model combines a detailed energy sector process model with a macroeconomic production function model; this maximises the discounted value of consumption utility over time, subject to specified carbon constraints. The focus of the model is GNP loss under different assumptions of emission constraints, costs and availabilities of energy supply technologies, inter-factor substitution, exogenous energy efficiency improvements and price induced substitution. The model applies a nested production function approach, which combines two composite intermediate goods consisting of Cobb–Douglas functions of labour/capital and electric/non-electric energy, inside a CES function with a non-unitary elasticity.

The model is solved as an inter-temporal system over eleven ten-year intervals (benchmarked against the base year 1990), assuming that producers and consumers have rational expectations about all future scarcities of energy and environmental restrictions. Factor supply and demand is equilibrated within each time period, but there are features that allow for interactions between periods, particularly for the depletion of exhaustible resources and for the accumulation of capital over time. Each of the regions faces an exogenously determined carbon emissions quota and an international crude oil price; a limitation of the model is that it neglects the possibility of trade in carbon emission rights. Later versions of the model (for example, Rutherford 1992) include such trade.

ERM

The Edmonds and Reilly model (ERM) has been described by Cline (1992) as an energy-carbon accounting framework, which calculates carbon emissions, by major world regions, at twenty-five-year intervals, from 1975 to 2100. The focus of this model is on estimating energy-related carbon emissions, and it has limited strengths in explaining the energy system's impact on the economy. ERM has nine energy types: conventional oil, conventional gas, unconventional (synthetic) oil, unconventional gas, coal, biomass, solar electricity, nuclear electricity and hydroelectricity. The model applies iterative price adjustments to achieve equilibrium between supply and demand for each fuel in each region.

Nordhaus: DICE

Nordhaus (1991a) presents a very simple general equilibrium model that links the economy, emissions and climate change; the model is aggregated at the global level. He also summarises empirical evidence on the costs of GHG abatement, and the damages from greenhouse warming for the year 2050. The model accounts for all GHGs by transforming each of the greenhouse gases considered into its CO_2 equivalent using its total warming potential. A simplified model is used for the change in atmospheric concentration of CO_2 equivalent GHGs and the associated change in temperature.

It is assumed that the economy has reached a resource steady state, and balanced resource augmenting technological change enables the economy to grow at a constant rate. The objective is to maximise the social welfare function; that is, the discounted sum of per capita consumption utility. Consumption is defined as the product of output with no emission reductions and no climate damage, and the difference in the steady state cost function and steady state damages from climate change. This framework is very aggregate, but its strengths are the more complete coverage by including all greenhouse gases, and the inclusion of the costs as well as the benefits side of the greenhouse effect and abatement measures.

GREEN

This applied general equilibrium model was developed by the OECD to study the economic impacts of CO_2 abatement policies (Burniaux *et al.* 1991a). The *G*ene*R*al *E*quilibrium *EN*vironmental (GREEN) model focuses on the energy sector, and uses government excise taxes or a carbon tax as a policy instrument to reduce CO_2 emissions. The model highlights the relationship between depletion of fossil fuels, energy production, energy use and CO_2 emissions.

The time horizon for the model is 1985–2020 and it is simulated for five-year intervals. The recursive structure of the model describes the economy as a sequence of single period static temporary equilibria. In each sector output is produced using the five energy inputs (coal, natural gas, crude oil, refined petroleum and electricity) which can be domestically supplied or imported, fixed factors (which are predetermined), capital, labour and intermediate goods and services (domestic or imported). The individual energy sectors produce a composite energy good through a CES production function. The model version reviewed here had no potential for backstop technologies, though these are included in a later version. Capital and sector specific factors combine to produce a capital–energy composite. This merges with labour (CES) to form a composite input which produces output subject to a (Leontief) input–output structure. The model assumes constant returns to scale, and a common production structure based on cost minimisation given the sectoral demand and relative after-tax prices.

WW

Whalley and Wigle (WW; 1991) use a global static general equilibrium model, which incorporates trade, production and consumption of energy (carbon based and non-carbon based) and non-energy products (energy intensive manufacture and other goods) in a number of countries/regions.

WW addresses the issue of how different countries might fare under a carbon tax adopted to limit the build-up of CO_2 and other greenhouse gases. Effects of the tax would depend upon a number of factors. Firstly, the burden would depend on whether the tax were imposed on consumers or on producers. Elasticity of supply would determine the burden sharing between producers and consumers – a low elasticity of supply favouring consumers. Other factors determining the effects of taxation would be the disbursement of tax revenues and trade in energy intensive manufactures.

The model is capable of looking at alternative forms of taxation as well as international trade effects and generates results for the period 1990–2030. As the model is static it does not have a time path.

Comparative summary of modelling structures

The structure of a model depends on the issues it was designed to study. The GREEN, MR and WW models examine the macroeconomics and trade impacts of CO_2 control. ERM and IEA were developed to give detailed predictions of energy sector behaviour, and have more limited interest in the economic impacts of abatement of CO_2 emissions. In contrast, the Nordhaus model has been developed to balance costs and damages of greenhouse warming, and to estimate optimal use of resources.

The GREEN model and the WW model measure economic costs in terms of loss in welfare measured by Hicksian compensating variation. Other models measure welfare by looking at GDP loss; the critical parameters driving these results are discussed later. The rest of this section compares the structural detail of the models in time horizon of simulation, regional disaggregation and sectoral disaggregation.

Time horizon

The time horizon studied by each model is shown in Table 3.2, and is linked to the objective of the respective model. Given the timescales of warming impacts from current emissions to occur (≈50 years), models studying the next 20 years can be described as short term models (though more conventional classification would refer to them as medium term models), which focus on transition costs and dynamic adjustment to price shocks. Models dealing with 21–50 years are medium term models, which focus on economic costs inside a period where discounted costs are still meaningful. Models running over more than 50 years are long term models. These look at long run equilibria and results are driven by the effects of discount rates and long run trends, with the economy being very flexible in its response to factor prices. Thus the IEA model (up to 2005) falls in the short term model class. The WW model and GREEN (a more recent version does go up to 2050, however) are medium term models, whereas the MR model, the Nordhaus model and ERM are long term models. Simulating over very long periods using parameters derived from empirical studies (e.g. elasticities) is probably misleading as the economy will be more malleable in the long term than it was in the estimation period. Contrastingly, small differences in assumed exogenous trends in productivity, demographics and discount rates have a large effect over the equilibrium solution of the very long run models.

Table 3.2 Main characteristics of the global models

Model	*Type of model*	*Time horizon*	*Regional disaggregation*	*No. of energy sectors*	*No. of industries*	*Comments*
IEA	Econometric	2005	10	5	9	Detailed econometric model for the energy sector.
MR	Dynamic optimisation	2100	5	9	-	Forward looking inter-temporal model; disaggregated energy supply sector with backstop technologies; international trade only in oil.
ERM	Partial equilibrium	2050	9	6		Energy–economy links simple; energy trade modelled.
Nordhaus	Optimal growth	2050	world	2	-	Looks at all GHGs and the interaction between climate change and economic growth; the model aims to maximise the costs and benefits of control.
GREEN	General equilibrium	2020	7	5	3	Dynamic structure through resource depletion sub-model and savings; full trade links; endogenous oil prices.
WW	Static general equilibrium	1990–2030	6	2	5	Trade links focus on international incidence of carbon taxes.

Sources: Hoeller *et al.* 1992, Dean and Hoeller 1992

Table 3.3 Regional disaggregation in models

Model	*Regions represented*
IEA	3 OECD regions (North America, Europe, and Pacific); Africa; Asia; Latin America; Middle East; East Europe; former Soviet Union; China
MR	USA; other OECD (OOECD) nations; the former USSR (SU) and Eastern Europe (EE); China; the rest of the world (RoW)
ERM	USA; OOECD West; OOECD Asia; centrally planned Europe; centrally planned Asia; Middle East; Africa; Latin America; South and East Asia
Nordhaus	Globally aggregated
GREEN	3 OECD regions (North America, Europe and Pacific); ex-USSR; China; energy exporting LDCs (EELDCs); an aggregate RoW sector
WW	EU; North America (Canada and USA); Japan; OOECD; oil exporters (OPEC and non-OPEC); rest of the world (RoW)

Regional disaggregation

Regional disaggregation again differs in the models from a single global sector in the Nordhaus study to seven regions in the GREEN model and ten regions in the IEA model (see Table 3.3). This makes comparisons of results for different regions across models more difficult and has not been attempted here (a detailed study has been done by OECD, Hoeller *et al.* 1992, and Dean and Hoeller 1992).

Of the IEA model's ten regions only the three OECD regions have been modelled in great detail, the other regions are modelled in lesser detail and China's energy system is imposed exogenously on the model due to data limitations. Each of the MR model's five major geopolitical regions has two sub-models with a two-way linkage between them, and a dynamic non-linear optimisation is employed to simulate either a market or a planned economy. In GREEN and WW all regions have the same structural characteristics and Nordhaus's model is globally aggregated.

Even when regions of the world are disaggregated the parameters in each region are often based on or refer to data from the OECD region (e.g. the GREEN model), which reduces the apparent heterogeneity available to the analyst. As many of the important features of international decision making are connected with division of abatement burdens, and relative abatement costs in different regions, this greatly reduces the ability of some of these models to model the effectiveness of different policy instruments.

Sectoral disaggregation

Since in all models the energy sector is obviously disaggregated from the rest of the economy, differences lie in the amount of disaggregation inside each energy sector (number of fuels and technologies available), and the number of other sectors, or productive factors, which can be substituted for energy.

The MR model has a detailed energy sector split into electric and non-electric energy, each of which has a menu of current and future conversion technologies or fuels to choose from; these energy supplies include exhaustible hydrocarbon resources and also 'backstop' technologies. Associated with each technology are the cost and carbon emissions per unit activity level, upper bounds on the speed of introduction of each new technology and lower bounds on rates of decline. Therefore, the MR model concentrates on adjustments through substitution among energy forms, both fossil and non-fossil, and accounts for substitution among energy and other factors through its production function. However, it does not account for changes in final demand composition.

Contrastingly, the GREEN model includes changes in the structural composition of the economy by using three separate sectors, namely, agriculture, energy intensive industries and other industries and services. The energy sector itself is split into five sub-sectors: coal mining, crude oil, natural gas, refined oil products and electricity gas and water supply (non-fossil energy). Therefore, this model while less detailed on the energy side can investigate the impacts of higher energy prices in different sectors of the economy, and thus trade.

In the WW model each region has four non-traded primary factors: primary factors excluding energy, carbon-based energy sources (deposits of oil, coal and gas), other energy sources, and sector specific skills and equipment in the energy intensive manufacturing sector. There are three internationally traded commodities: carbon-based energy products, energy intensive manufacturing and other goods (all other GNP) and two non-traded goods (non-carbon energy products and a composite energy product). There is domestic market clearing within each economy, with nested functional structures representing production and demand in each region.

The model uses CES production functions at each of the three stages of production: production of carbon or non-carbon energy from primary factors and respective energy resources; production of composite energy from carbon and non-carbon energy; and production of energy intensive and other goods from primary factors, energy, and sector specific factors. A carbon tax increases the price of carbon-based energy sources and composite energy products. The price change and the extent of substitution (between carbon-based energy sources and non-carbon energy products and between composite energy and other inputs) will depend upon elasticity values used at the respective nodes in the nesting of substitution possibilities.

Sectoral disaggregation depends upon the focus of each model. For example, those models (MR, IEA and ERM) developed as detailed energy models obviously incorporate a large number of energy sources. Results are sensitive to the number of energy sources included, the intra-fossil fuel substitution possibilities and the extent of substitutability between fossil and non-fossil fuel sources as well as the availability of backstop technologies. The WW model and Nordhaus assume only two fuels – a composite fossil fuel and one non-fossil fuel. Thus intra-fossil fuel substitution is not considered at all in these models; in the short run this biases costs for carbon abatement upwards in these models, as initially significant reductions in carbon emissions are possible by switching between high carbon intensity (such as coal) and lower carbon intensity (natural gas) fuels.

CRITICAL MODEL PARAMETER ASSUMPTIONS

In energy-economic models the link between energy use and macroeconomic variables is typically represented by the *energy intensity* coefficient, defined as energy consumption per unit of GDP. In industrialised countries energy intensity has been in steady decline since the early 1970s or before, and the rate at which it is expected to continue to fall in the base-case is critical to the predictions of models concerning greenhouse gas emissions. In calibrating the models to yield paths for energy intensity in the base-case, Whalley and Wigle (1991), Nordhaus and Yohe (1983) and Nordhaus (1990) assume neutrality in technological change, that is, it affects all factors of production equally; other studies allow for biases towards energy-saving technical progress. GDP growth is treated exogenously in the base-case in almost all studies and various assumptions are then made about the base trend of energy intensity, that is, the trend it would follow in the absence of policy changes.

Declining energy intensity is considered to be due to a number of factors, and here it is important to separate the factors that are not connected with movements in energy prices and those which are. Non-price factors which affect energy use are summarised by Boero *et al.* (1991) as:

- 'Exogenous' energy-saving technological progress occurring regardless of price changes.
- Changes in the composition of GDP.
- Policy-induced technological change.
- Elimination of existing inefficient technologies, that is, ongoing technical diffusion.

Almost all studies amalgamate the non-price factors into a single 'exogenous energy-efficiency' parameter, which applies equally in the base-case and in the simulations. The problem with this approach is that non-price factors are by no means 'exogenous' in the sense of applying equally in the base and the constrained cases, because policies to constrain emissions will affect the last three factors given above. It would be unwise to assume that policy-induced changes were the same between base-case and simulations. Indeed, a strong case has been made out for policy measures, such as tightened energy standards and government spending on energy conservation, to complement demand-side measures such as a carbon tax to stabilise greenhouse gas emissions (Lazarus *et al.* 1992).

If 'exogenous' changes in energy intensity are supposed to apply to the same extent in base-cases and simulations, then this may understate the potential for energy-saving and overstate the costs of abatement. For example, Williams (1990) argues that the 1.0 per cent annual decline in energy intensity postulated by Manne and Richels (1989) is too low because it underestimates the potential for policy-induced conservation measures. Therefore, a single parameter which has the same value in base and constrained cases is likely to imply either underestimates of emissions or overestimates of cost controls.

Hogan and Jorgenson (1991) criticise what they call the 'conventional wisdom' behind the assumption that, in the absence of relative price changes, energy intensity should decline. This is reflected in Table 3.4 by the assumptions for autonomous decline in energy intensity, for instance the 0.5 per cent p. a. for the USA of Manne and Richels (1990). Hogan and Jorgenson report estimates of sectoral productivity trends for the USA and

Table 3.4 Elasticities of substitution and the technical progress parameter

Model	*Inter-fuel or intra-fuel substitution*	*Between energy and other inputs*	*Energy own price elasticity*	*Energy intensity rate of decline (% p. a.)*	
IEA	−0.5 inter-fuel				
MR	−1.0 between electric and non-electric			*1990–2050*	*2050–2100*
USA		−0.4		0.5	0.5
OOECD		−0.4		0.5	0.5
USSR & EE		−0.3		0.25	0.5
China		−0.3		1.0	0.5
RoW		−0.3		0	0.5
ERM	varies			*1975*	*2050*
OOECD:					
residential			−0.9	1.0	1.75
industrial			−0.8	1.0	1.75
transport			−0.7	1.0	1.75
Other regions			−0.8	1.0	1.3
GREEN	−1.2 intra-energy	−0.3 K-E −0.6 L-E		All regions	1.0
WW	−1.0 between carbon and non-carbon	−0.7		Oil exporting	2.5[a]
				OECD	0.3[a]
				RoW	2.7[a]

Sources: Boero *et al.* 1991; for IEA only, Dean and Hoeller 1992

a Hicks neutral technological progress in all sectors, so growth rates in their models must represent technological progress.

find that, contrary to conventional wisdom, most sectors are 'energy-using' rather than 'energy-saving'. This means that there is an autonomous tendency for the share of energy to output to *rise* rather than decline. Such a trend, they say, has been swamped in the last twenty years by the effects of the large changes in relative price of energy, but in the long term it would prove significant. The implication of their findings, if true for countries such as the USA, is that the studies are substantially underestimating the costs of greenhouse gas abatement (they suggest by as much as half the total cost). This highlights the importance of isolating the relative price effects on energy demand from those of autonomous technological change.

As well as giving the technological (AEEI) assumptions in each model, Table 3.4 also summarises the assumptions made about the long run own-price elasticity of demand for energy, or the elasticity of substitution between energy and other inputs (this is an approximation to the former when energy share of GDP is small). Most studies make assumptions that the elasticity is less than unity, but nevertheless quite high with most greater than 0.5.

The MR study assumes relatively low price elasticities of 0.3–0.4 and an AEEI of 0.5 per cent annually in the USA and other OECD nations, only 0.25 per cent in the former Soviet Union and Eastern Europe (because of further industrialisation before moving towards a service based economy) and 1.0 per cent in China (because of the enormous potential for efficiency improvements). This model assumes that in the long term trade in technology means that the AEEI in these regions will converge to a single figure, and the only heterogeneity will be in the price response.

In the ERM, supply of resource constrained energy sources is determined by a logistic curve and does not respond to prices, but proceeds at a given extrapolated rate over time. In the Middle East, OPEC production is specified exogenously. Other than conventional oil and gas, all other energy sources are treated as backstop, with a family of horizontal long run supply curves at successively higher price levels and an upward sloping short-term supply curve that assesses a cost penalty if output is forced to rise faster than at the normal rate. Technological change is incorporated as a shift along the long run cost curve over time (Cline 1992).

Nordhaus has an aggregate model for the whole world with two factors of production – labour and energy. Carbon emissions are based on weighted averages from the present and the future expected composition of fossil fuels. The output effect of reducing carbon emissions is traced by a production function with an elasticity of substitution of -0.7 for energy and labour (equal to the price elasticity of demand for energy) and -1.2 for carbon and non-carbon energy (Nordhaus and Yohe 1983). Baseline prices of non-carbon energy depend on technical change, and are founded on existing studies. The price of carbon-based energy depends on technical change, resource depletion and taxation.

In GREEN, the AEEI is the same for all regions and energy is imperfectly substituted by other factors of production. Coal reserves are assumed to be infinite, while crude oil and natural gas supplies are linked to a resource depletion sub-model, which allows for some price sensitivity. Non-carbon energy (electricity) has a low supply elasticity. Energy exporting LDCs set world oil prices and other regions are price takers. Optimal combinations of inputs are determined simultaneously assuming competitive supply conditions in all markets conditional on the oil price, which is exogenous in the model.

Over time the economy is characterised by a sequence of period related but inter-temporally uncoordinated flow equilibria. Agents base their decisions on static expectations about prices and quantities. Dynamics in GREEN are associated with depletion of exhaustible resources and capital accumulation. In the resource depletion sub-model, potential supply is a function of proven resources, unproven resources or 'yet to find' resources, the rate of reserve discovery and the rate of extraction. The ultimate reserves, which are equal to the proven plus the unproven reserves, are predetermined in each period, but the rate of reserve discovery may be sensitive to the world oil prices. The model is dynamic, as saving decisions affect future economic outcomes through the accumulation of productive capital. Investment is computed residually and the model includes factor market rigidities making the capital sector specific, and drawing a distinction between old and new capital vintages.

In the WW model, demand and supply elasticities for carbon-based energy products are important parameters. These values are not directly specified. On the demand side they reflect preferences and intermediate production technology; on the supply side, the relative importance of fixed and variable factors in carbon-based energy production (oil in the ground versus extraction costs) and the marginal productivity of variable factors. Whalley and Wigle use a supply elasticity of 0.5 with sensitivity analyses over a range of 0.1 to 1.5 for carbon-based energy products.

The elasticity of input substitution in composite energy production is set equal to 1.0 in the absence of any estimates (justified by stating that it is relatively easy to substitute between fossil energy and electricity in the domestic and industrial sector; also recognising that this substitution is not so easy in transportation). The elasticity of input substitution in composite energy production and the ease of substitution between composite energy and the two non-energy products in consumption affect the elasticity of demand for energy.

The omission of backstop technologies in WW and ERM means that there is no ceiling to the carbon tax, which must continually rise in order to keep emissions stable. The IEA model has a detailed energy sector; however, the importance of backstops is limited before the end of its simulation period in 2005. GREEN and the MR models do look at backstop technologies, and the time of their availability and the cost at which these are available determine the carbon tax needed to control emissions.

RESULTS OF THE DIFFERENT MODELS

The OECD model comparison project (Hoeller *et al.* 1991, 1992) has attempted to standardise key inputs and reduction targets and to examine the difference in baseline CO_2 emissions, GDP, carbon taxes and economic costs for each of the main models. Table 3.5 shows that the business-as-usual projections for the different studies are within a narrow range till 2020, but by 2100 there is a difference of a factor of two. This is due to the fact that small differences in growth rates get compounded, and become very large in absolute values over long time horizons.

Baseline estimates for GDP and GDP loss associated with CO_2 abatement across different studies are shown in Table 3.6.

IEA

The IEA model was used to assess the carbon taxes needed to produce 1 per cent, 2 per cent, or 3 per cent annual reductions in emissions, as well as for stabilisation at 1990 levels. The major conclusion was that emission restrictions would require very high taxes (for example, $1,222 per tonne for Other OECD and $700 per tonne for NA OECD for a 3 per cent annual reduction in carbon emissions), given the rigidity in the energy sector in the short and medium term. Lower starting fuel prices, and higher carbon intensity in the North American power generation system, result in a carbon tax nearly double that for the same level of reduction in other OECD regions.

MR

In the MR model, baseline annual carbon emissions increase rapidly from 6.0 GtC to 39.64 GtC in 2100. In the restricted scenario, carbon emissions

Table 3.5 World-wide business-as-usual energy-related CO_2 emissions (GtC)

Study	*1990*	*2000*	*2005*	*2020*	*2050*	*2100*
IEA[b]	5.92	7.93	7.93			
MR	6.00	6.97		9.52	14.99	39.64
ERM	5.77		6.71	8.18	11.84	22.58[a]
N&Y[c]		5.50		10.30	13.30	20.00
GREEN	5.82	7.07	7.70	10.81	19.00	
WW[d]	Annual average for 1990–2100 is 25.2					

Sources: Dean and Hoeller 1992; for N&Y only, Boero *et al.* 1991

a 2095

b Excluding non-fossil solid fuels, bunkers and non-energy use of fossil fuels and petrochemical feedstock.

c Nordhaus and Yohe 1983.

d WW have a point estimate of 65.5 billion tonnes for 2100 giving an average annual growth of 2.3 per cent and an average annual emission reported above.

Table 3.6 Reduction in CO_2 and associated loss in GDP

Study	*Projection period*	*Region*	*CO_2 emissions % of baseline*	*CO_2 emissions % of ref. year*	*GDP changes % of baseline*
MR	1990–2100	USA		−20 (1990)	−3.0 (2030+)
		OOECD		−20 (1990)	−2.0 (2010)
		SU–EE		−20 (1990)	−4.0 (2030+)
		China		100 (1990)	−10.0 (2050)
		RoW		100 (1990)	−5.0 (2100)
		World	−75 (2100)	16 (1990)	−5.0 (2100)
ERM	1975–2050	USA	−60 (2050)	70 (1990)	−0.4 (2050)
		World	−40 (2050)	162 (1990)	−1.0 (2050)
Nordhaus	1990–2100	World	−50 (2100)		−1.0
GREEN	1990–2020	N America		−20 (1990)	−0.8 (2020)
		Europe		−20 (1990)	−7.0 (2020)
		Pacific		−20 (1990)	−3.7 (2020)
		EELDC[c]		+50 (1990)	−3.6 (2020)
		China		+50 (1990)	−1.5 (2020)
		USSR		−20 (1990)	−2.2 (2020)
		World	−37 (2020)	+17 (1990)	−1.8 (2020)
WW	1990–2030	World			
		NP tax[a]	−50 (2030)		−4.4
		NC tax[b]	−50 (2030)		−4.4
		Global tax	−50 (2030)		−4.2

Source: Boero *et al.* 1991
a National production tax.
b National consumption tax.
c Energy exporting LDC.

increase by 15 per cent over 1990 levels by 2030 and then stabilise at 6.6 GtC. These limitations require taxes which are initially steep, but despite the large amount of abatement needed they settle at a long term average of $250 per tonne of carbon, because this tax is equal to the difference between the carbon and non-carbon backstop technology (see Table 3.7).

For the USA, the tax peaks at $400 in 2020 before settling at $250 per tonne of carbon. It then declines as new technology comes on stream. Carbon taxes peak at $650 per tonne in SU and EE during 2020–2060. The reasons for this discrepancy is the lower substitutability in consumption and production and a lower availability of alternative energy sources (Table 3.7). Economic costs for the USA level at 3 per cent of GDP, at 2 per cent of GDP for OOECD (due to larger resources of oil and gas), 3 per cent of GDP for SU and EE. Losses are low for RoW till 2030 (due to generous carbon emissions ceilings and abundant availability of oil), but reach 5 per cent of GDP by 2100. The largest losses, 8–10 per cent of GDP, are

Table 3.7 Carbon taxes and CO_2 emission reductions

Study	*Projection period*	*Region*	*CO_2 emissions % of baseline*	*CO_2 emissions % of ref. year*	*Taxes ($/tC)*	
MR	1990–2100				*Peak*	*End*
		USA		−20 (1990)	400	250
		OOECD		−20 (1990)	>250	250
		SU–EE		−20 (1990)	700	250
		China		100 (1990)	>250	250
		RoW		100 (1990)	>250	250
		World	−75 (2100)	16 (1990)		250
ERM	1975–2050	USA	−60 (2050)	70 (1990)	100% coal;	
		World	−40 (2050)	162 (1990)	78% oil; 56% gas; 115% shaleoil[a]	
Nordhaus	1990–2100	World	*CO_2 reduction*		($/t$CO_2$)	
			−10 (2100)		20	
			−40 (2100)		100	
			−75 (2100)		250	
			GHG reduction			
			−10		3	
			−17		13	
			−25		38	
			−50		119	
GREEN	1990–2025	N America		−20 (1990)	209	
		Europe		−20 (1990)	213	
		Pacific		−20 (1990)	955	
		EELDC[d]		+50 (1990)	209	
		China		+50 (1990)	65	
		USSR		−20 (1990)	101	
		World	−37 (2030)	+17 (1990)	215	
WW	1990–2030	World				
		NP tax[b]	−50 (2030)		448	
		NC tax[c]	−50 (2030)		448	
		Global tax	−50 (2030)		439	

Source: Boero *et al.* 1991
a Per cent of fuel price.
b National production tax.
c National consumption tax.
d Energy exporting LDC.

projected for China due to the fact that coal is the dominant energy resource (see Table 3.6).

ERM

In the ERM, baseline annual carbon emissions increase to 22.6 GtC in 2100. ERM generates a menu of percentage cutbacks in emissions for specified carbon taxes. Several versions of the ERM are in use by different researchers, and the results vary depending on the assumptions made by different users. One set of results shows that there is a decline in global emissions by 40 per cent for a 1 per cent decline in GDP (see Table 3.6).

The economic costs of emission reductions given by the ERM should be interpreted with caution. The economic costs are the change of world GDP due to a rise in energy prices, calculated by the application of GDP feedback elasticities to the rise in energy prices (the feedback elasticity is equal to −0.1 for industrialised countries and −0.2 for developing countries). This method describes the situation which prevailed in the 1970s and the 1980s, where energy prices had a short-term Keynsian effect due to fiscal outflows and monetary tightening to prevent inflation. Using the elasticity approach for long term analysis would cause the estimates of GDP losses to be greater than the GDP losses estimated under a regime where there is no fiscal loss (energy taxes recycled), and agents anticipate higher energy prices. In other words, the elasticity approach accounts for limited substitution possibilities, and assumes a 'putty–clay' framework rather than a 'putty–putty' one in which the economy has more flexibility in terms of technological choice, which is more representative of the long run.

Nordhaus

Nordhaus and Yohe (1983) estimate baseline carbon emissions to reach 20 GtC by 2100, which is similar to the ERM studies but half that of the MR model. Nordhaus (1991a) synthesises carbon tax estimates from several existing models, and shows that these are relatively close to a central curve, which he takes as the representative marginal carbon tax curve. For 2050, estimates for marginal tax per tonne of carbon per percentage point cutback from baseline are $5.9 in the MR model, $4.9 in the ERM and $2.8 in the Nordhaus–Yohe model (Cline 1992). In another study, Nordhaus (1991b) concludes that, owing to foreseeable cuts in CFCs, little damage will take place due to other greenhouse gases, therefore preventive measures would cost very little. Given the paucity of data on damages of greenhouse warming and uncertainty over the difference between real interest rate on goods and growth rate, Nordhaus (1991a) takes costs of 2 per cent, 1 per cent and 0.25 per cent (0.25 per cent is the central estimate for damage costs for the US GNP in 1989), and discount rates of 0 per cent, 1 per cent and 4 per cent for a doubling of CO_2 equivalent GHGs. Nordhaus's results are broadly in the range of 1 per cent loss in GDP for 50 per cent abatement over baseline emissions. Table 3.7 gives some details of the carbon tax schedule derived by Nordhaus.

GREEN

In GREEN baseline carbon emissions are 10.8 GtC in 2020, which is 17 per cent higher than the emission level in 1990, and they rise to 19 GtC in

2050, the highest rate of increase in any of the models examined here. Carbon taxes are modelled as a fixed excise tax in absolute dollars per tonne of carbon emitted, thus its level does not vary with shocks to energy prices. It is fuel specific, and levied on consumers of primary fuel only. The tax is computed for each region, as the equilibrium shadow price for an additional tonne of carbon dioxide emissions when a given constraint on total emissions is imposed. GREEN highlights the extent of reliance on coal in different regions and the inter-regional divergence in existing prices of fossil fuels. These price distortions are important in determining the level of carbon taxes across regions.[2]

In 2020, GREEN freezes emissions at 1990 levels by imposing the following cuts – OECD and SU cut emissions to 80 per cent of the respective 1990 levels by 2010 and freeze emissions thereafter; China and energy exporting LDCs increase emissions by 50 per cent by 2010 and freeze emission levels thereafter (Table 3.7). The carbon tax is computed as the equilibrium price of carbon that achieves the required emissions constraint, and carbon taxes rise over time as the gap between restrained and unrestrained scenarios increases. There is regional variation in tax rates – $209 in North America, Europe, EELDC, $955 in OECD Pacific, $65 in China and $101 in the Soviet Union (Table 3.7). Carbon tax revenues in GREEN are less than 5 per cent of world GNP (WW calculate 10 per cent); global welfare loss (Hicksian equivalent variation) is 2.2 per cent of household real income. Again there is regional variation in welfare loss: 0.8 per cent in North America, 0.9 per cent in Europe, 2.4 per cent in OECD Pacific adding up to an aggregate loss of 1.2 per cent for OECD, 2.3 per cent in China, 0.6 per cent in SU and 7.5 per cent in EELDCs – where terms of trade losses double the underlying GDP loss. The economic costs are quite low considering the cutback in emissions is 43 per cent over baseline by 2020.

WW

The WW average for baseline emissions over 1990–2100 is 25.2 GtC per year. The model considers a 50 per cent cut in carbon emissions relative to their baseline over the period 1990–2030. The model determines endogenously the *ad valorem* carbon tax rate which applies to all fuels. Then the implied carbon tax in $/tonne of carbon is calculated. Economic costs estimates by WW are twice the average of 2 per cent in other studies – this is due to limited non-fossil fuel substitution and the absence of intra-fossil fuel substitution. Therefore, a 50 per cent cut in emissions requires a 47 per cent reduction in energy. Thus WW calculate that 94 per cent of carbon reductions must come from energy reductions, whereas Edmonds and Barnes (1990) place this at 21 per cent only. Note must also be taken of the fact that the WW model takes Hicksian equivalent variation concept of economic cost. This measure takes account of changing relative scarcities and relative prices, and gives larger proportionate reductions than does

GDP, which is an index of the volume of production at constant relative prices.

WW estimate the discounted value for a lower carbon energy base as 10 per cent of GNP. This is high considering that the share in GNP for all sources of energy is only 6 per cent, but could be plausible as the marginal productivity of energy will rise as use declines. WW results also give high carbon tax levels. One reason is the model's rigidity in providing low or non-carbon sources of energy. Cline (1992) explains that an additional reason may be the model's treatment of the shifting of the tax burden. The incidence depends on the elasticity of supply of the carbon resource; an inelastic supply implies that there is a reduction in the producers' rent rather than a price rise for consumers, and a consequently low cutback in use of carbon fuels. If supply elasticity is high, consumers bear the burden of taxation through higher prices and a significant reduction of carbon fuels use. The use of 0.5 supply elasticity implies that relatively high taxes must be imposed, because a substantial portion of the effect is neutralised by rent redistribution away from producers. WW emphasise these distributional implications of their findings, which are a consequence of the tax mode implemented.

Summary

Results from the global macro models are clustered around 1 to 2 per cent loss in GDP, for 40–50 per cent reduction relative to baseline emissions; there is an increasing dispersion of results as higher targets are set (Boero *et al.* 1991). As would be expected there is a tendency for costs to escalate at an increasing rate as reductions become more severe. The model results depend on a number of critical model parameters such as: the initial prices of carbon-based fuels in each region, the availability of backstop fuels or technology, ease of substitution between fuels and factors of production, growth of GDP, factor productivity growth, GDP feedback elasticities, the AEEI and the form of taxation or objective function of the model.

Business-as-usual emissions are different at the regional level across different models. The exception is for the OECD region, where the underlying assumptions are relatively close. In the non-OECD regions, baseline projections of emissions are very different across different models due to wide variations in the underlying assumptions. Besides differences in AEEI, growth rates and fuels use, emission paths are divergent because of differences in the technical specification of the link between emissions and economic activity. The wide difference in baseline emissions will produce differences in equilibrium carbon taxes, in the absence of backstop technologies, but otherwise will only affect the dynamics of carbon taxes on the way to equilibrium.

MR and GREEN incorporate non-carbon backstop technologies, whose price puts a ceiling on the required carbon tax, the value of which will equal the difference between the costs of clean and dirty backstop technologies.

The lower the cost of the dirty backstop technology, the higher the equilibrium abatement cost and tax, because it entails foregoing the use of the dirty backstop technology and using the more expensive clean backstop technology; in MR this translates into a uniform global tax of $250, which is the cost difference between dirty fuels and non-electric renewables. Non-availability, or limited supply, of clean fuels will increase the price of energy when carbon targets are imposed. What is important in the case of these technologies is not only the date that the technology is theoretically available but also the 'effective introduction' dates which depend on speed of diffusion (capital costs, scrapping of old plant, learning curves, etc.); this accounts for much of the dynamics of impacts and taxes in these models.

The lower the ease of substitution between factors the higher the tax needed for abatement, and the higher the GDP impact of reaching a specific target. The WW model has the highest rate of taxation, because of the aggregation of energy sources into clean and dirty fuels only, and an elasticity of supply of 0.5 for carbon and non-carbon based fuels. This requires high taxes to switch from dirty to clean sources. In general, substitution elasticities do not differ much across regions, but they differ considerably across models, showing the large amount of disagreement over empirical estimates of long run elasticities, and the paucity of such estimates for large regions of the world.

Initial energy prices forecast in the baseline case will determine the leverage for a given level of tax. Countries with high prices require higher taxes to reduce consumption, as they are already efficient in their energy use. Also, given the same elasticity to achieve a 1 per cent reduction in energy use, an *x* per cent change in energy prices would be required for all regions; in the real world one would expect elasticities to change with prices, but this is not compatible with a CES functional form, though it is with a trans-log. As elasticities do not differ much across regions, this would imply a higher level of absolute taxes in regions with higher energy prices. Expectations about future carbon taxes will affect the energy efficiency of capital installed, but models with dynamics mostly assume myopic behaviour. MR however, assume perfect foresight. With perfect foresight the aggregate costs (especially in the transition period) are likely to be lower, as decision making is based not only on current prices but also on future prices.

All of the studies impose percentage reductions in CO_2 on their models except Nordhaus (1991a), who determines the optimal rate of carbon taxation endogenously by equating the marginal cost of abatement to the marginal damage cost. As carbon is a stock pollutant, the marginal cost of abatement has to be equated with the marginal damage cost of current as well as future increases in concentration. This implies that an optimal tax will rise over time, in line with the rate of interest. Variations in carbon taxes over time in other models are caused by resource depletion, availability of backstop technologies and the rate of increase of baseline emissions and AEEI.[3]

SPECIFIC LIMITATIONS OF MODELS

This section discusses the specific limitations of the models, beyond the limitations of the general modelling methodologies which have been described above. In all models, econometric estimation of AEEI values has not been attempted to date, and non-energy productivity growth is also represented as an exogenous trend. With the exception of the IEA study and ERM, in the models listed above energy demand depends upon the aggregate constant elasticity of substitution (CES) production function.

The econometric nature of the IEA model precludes its use for analyses beyond the medium term, without significant adjustments in the model. The short time horizon of the model limits the availability of technology with low carbon intensity. Also, the semi-exogeneity of the macroeconomic inputs prevents changes in the composition of GDP towards less energy intensive industries with increasing carbon taxes. The model can determine GDP losses from increases of world oil prices through a simple feedback equation, but cannot assess the impact of a redistribution of carbon tax revenues. The non-fossil energy sources are fixed, which prevents marginal and backstop technologies from entering into the model as energy prices increase in the medium term.

Cline reports that the global average of AEEI in the MR study (weighted by base-year carbon shares) is 0.4 per cent which is much lower than the 1 per cent assumed in ERM. Thus MR has higher baseline carbon emissions, requiring larger reductions in emissions to achieve the target level and therefore higher costs. Manne and Richels set up an arbitrary figure of $250 per tonne of carbon as the differential in the carbon and non-carbon backstop technology. There are divergent views among other researchers on the figure of $250 – some find it too low, and some researchers find it too high. Introducing this non-carbon backstop sets a ceiling to the tax rate which would have been increasing otherwise. Thus the model relies heavily on non-fossil fuel substitution to limit carbon emissions. As mentioned before, the WW study treats all carbon energy as uniform, thereby ruling out inter-fossil fuel substitution. Non-carbon energy is limited thereby understating non-fossil fuel substitution (Cline 1992).

In the ERM the supply of resource constrained energy technologies does not respond to price, but proceeds at a given extrapolative rate over time defined by a declining logistic production function. This model was originally designed to examine the issue of carbon emissions, whereas in most other models the focus was on impact of energy availability and policy on the economy. This explains the limited strength of ERM in explaining the impact of reduced energy availability on the economy.

The Nordhaus model provides a structure with which to attempt to determine the optimal tax rate by maximising benefits; however, there are many major problems with attempting such a simple approach to a complex problem such as climate change, and these are detailed in Chapter 7. The model is useful for illustrating the influence of different parameters on the level of emission reductions which could be considered economically efficient. This type of structure can be elaborated and refined as and when better data is available, and in some ways the new Integrated Assessment Models mentioned earlier are attempts to do this, though they do not always explicitly attempt to optimise outcomes.

Though its structure seems to give much disaggregation and detail, data limitations exist for many of the parameters specified in the GREEN model. For example, GREEN requires the elasticity of substitution between labour and the capital–energy bundle, but these are not available; estimates are available for inter-factor elasticity of substitution between labour and capital. However, reported labour–energy elasticities indicate that these inputs are often substitutable to the same extent as capital and labour. Therefore, identical capital–labour and energy–labour elasticities are implicitly assumed in GREEN. Econometric estimates of inter-energy elasticities of substitution are scarce, not reliable, and they are sensitive to model specification. High substitutability is indicated in the literature between electric and non-electric energy and between natural gas and electricity, and very low substitutability between coal and natural gas (Burniaux *et al.* 1991a). As none of its equations is directly estimated, GREEN depends on such conflicting research to calibrate its complex structure.

A number of additional simplifying assumptions are made in GREEN: identical values for the CES functions were imposed in all regions, in production, in international trade and in the government sector; producers, consumers and the government sector have the same inter-energy elasticities of substitution; disinvestment elasticities and depreciation rates were assumed to be identical across sectors and regions. The use of identical assumptions reduces the advantages of the disaggregated approach, as it is mathematically identical to using a single agent or region for the world (everything else being equal).[4]

CONCLUSIONS OF MODEL REVIEW

The climate change problem is one with global dimensions. Many global models have been developed to address economic problems associated with preventing enhanced global warming, but these suffer from being highly aggregated. Specific country models can be more detailed and disaggregated, but these lack the global context which determines so many of the economic interactions; for example, international sharing of the abatement burden, international effects on fuel prices and competitiveness effects.

There is no model type which can address in detail all issues concerning regional as well as sectoral disaggregation, technologies, markets, trade, time dimensions and interlinkages. Thus the choice of model used has depended on the specific issues of interest to the researcher.

Between the different macro modelling approaches, macroeconometric models have some advantages over general equilibrium ones when the short to medium term is being studied, and adjustment factors and costs are important. Recent developments in econometric modelling such as error correction models and cointegration analysis have made enhanced theoretical consistency possible, without sacrificing empirical accuracy; this has made these models more suitable to address longer time horizons.

On the other hand, GE models are more suitable for very long term analysis because their interactions are more transparent, and their parameters are not limited to those derivable from short periods of historically observed data. Some of the limitations of GE models, such as comparative static analysis, and the lack of insight into the dynamics of the adjustment path, can now be resolved. Advances such as the development of dynamic general equilibrium models, and econometric estimation of their parameters (instead of using the calibration method) have made this class of models more powerful and credible in the short to medium term.

Macroeconomic modelling is necessary to study the impact of GHG abatement on the economy, as it provides a framework for studying structural effects, macroeconomic impacts, market constraints and responses in the short and long term. Engineering models are important from the point of view of technological possibilities, and the potential for efficient and cost effective emission abatement. The representation of technology in engineering models is very explicit, whereas in macro models it is almost absent. The specific representation of technologies means that engineering models can be used to assess non-price based policies, such as energy efficiency regulations on appliances, subsidies for energy saving and energy labelling schemes; all these policies have political attractions as they avoid the regressive nature of an energy tax.

For evaluating greenhouse gas abatement strategies, engineering studies are not sufficient by themselves as they suffer from a number of limitations; underestimation of economic costs due to non-inclusion of hidden and indirect costs, as well as the lack of a mechanism to estimate macro impacts, structural effects and price effects. Macro models on the other hand overestimate costs as they do not adequately model the response of technology to constrained scenarios, and ignore 'no-regrets' or zero/low cost abatement options. Integration of the macro and engineering approaches can bring about a representation of technology in the macro models, and improve the estimation of costs associated with greenhouse gas abatement policy. Some attempts have been made in this direction (for example, Global 2100).

One possible way to integrate the macro and bottom-up approaches in a macroeconometric model is through the cost curves generated from engineering models. These can be transformed into a simple functional form, and incorporated in the macro structure to account for technological options and potential; this approach is especially relevant in developing countries where the economic structure is changing too rapidly for structural econometric estimates to be useful. Chapter 6 details the construction of a combined macro/micro model for India, where cost curves are constructed from engineering data in order to calculate the demand for investment funds needed to save carbon dioxide. This investment/CO_2 abatement relationship is then embedded in a macroeconometric model in order to estimate the effect of investment diversion and price changes on the wider economy.

Importance of 'trend' parameters

More detailed empirical work on all critical parameters will improve the accuracy of the results further, but given the long timescales of forecasts the most important are the exogenously imposed trends in factor productivity, in both the energy and non-energy sectors.

The exogenously imposed trend rate of non-price decline in energy intensity is generally interpreted as being a function of general technical progress, and so is termed the Autonomous Energy Efficiency Improvement (AEEI) parameter. The AEEI parameter is an important determinant of CO_2 emissions in the baseline scenario. Higher values of AEEI give lower baseline CO_2 emissions, and therefore lower the cost of stabilising carbon emissions and concentrations into the future.

The AEEI in most studies is the same in the baseline scenario as in the constrained scenario. There is no reason why it should not be different in the two cases. For example, if in the base-case, energy intensity declines by 1 per cent per annum (or AEEI is 1 per cent per year), then with a constraint on carbon emissions the decline could be greater, say 1.5 per cent per annum, due to a greater emphasis on energy saving technological development.

Exogenously determined, arbitrary decline in AEEI in the baseline scenario is questionable by itself, but the process becomes more dubious if a new set of rates for AEEI are specified for the constrained case, again exogenously. The basis for the link between the change in AEEI and the emission abatement strategy or policy is remote. It is therefore necessary to endogenise the factors responsible for a change in non-price induced energy intensity in a consistent manner, and see how they change in response to different policy options. In practice empirically measured estimates of the AEEI capture the changes in energy intensity due to irreversible technological progress, reversible technical diffusion due to price rises, structural change in the economy and other non-technological factors such as consumption trends. It is important to separate the purely technical factors from the other structural effects to see if they are price sensitive, which would be an intuitive assumption, and would greatly affect the impact of taxation policies.

However, if technical progress increases in the energy sector it may decrease in the wider economy, due to diversion of R & D funds, resources and/or capital investment. Endogenising the interactions between the 'trend' rates of productivity growth in each sector of the economy is, therefore, a vital and understudied part of estimating the cost of CO_2 control for different policies (e.g. carbon taxes vs. R & D credits in energy saving technologies). One form of endogenisation is possible through the integration of results from micro level studies (bottom-up models) in macro models, but in the very long run detailed information of technological options may not be available to support this approach.

MODELLING APPROACH IN EGEM

As detailed in Chapter 1 the aim of this research is to examine the political interactions of developed countries which are faced with an immediate decision as to how much carbon dioxide to abate, and how to share the costs of reaching any global abatement level. We assume that governments act as rational actors, and make decisions on whether to join or leave international abatement agreements based on considerations of costs and benefits alone; there is no scope for altruistic action in this study. Therefore, to analyse these interactions we need to model the economic costs, to specific countries, of different abatement commitments over the short to medium term (5–35 years).

As well as the issues of stability of international agreements, Chapter 2 also showed how the choice of abatement co-ordination instrument (permits, taxes or targets) is ambiguous and depends on the direct and indirect cost of abatement in each country. To capture these effects the modelling framework must give details of the effects of different types of tax recycling, allow optimal trading of carbon permits and optimal control of emissions levels to reach prespecified targets.

The time span of interest, and the importance of accurate measurements of economic effects, means that econometric modelling is most appropriate for this study. As a basis for our modelling we have used an existing macroeconomic model GEM (Global Econometric Model), which has been used for short range forecasting of the global economy for several years (LBS 1993). This model is global in scope, but has most detail in the developed countries; it can be solved under many different objective function regimes including rational expectations, learning, optimal control and simple gaming between countries. To model carbon tax effects the forecasting span of GEM was extended to 2030, and the developed countries were augmented with detailed energy sector models (described in Chapter 4), and a consistent supply side incorporating energy effects (described in Chapter 5). Policy objectives and instruments were included by the addition of a greenhouse damage function (see Chapter 8) and various types of tradable permit scheme (see Chapter 11).

Microeconomic data was not used in the model, and so there is no explicit backstop technology cost for the tax rate to tend towards. In a conventional macroeconomic model this would mean that taxes have to rise continually to stabilise CO_2 emissions. The rise in taxes is not as great in this model as in other elasticity based models however, because the energy sector allows for price induced increases in the trend rate of energy productivity growth, which at a certain price level would outweigh growth induced increases in energy use thus giving an implicit backstop price. The one weakness of not explicitly including energy technologies is that abatement can only occur if taxes are imposed on energy, which leads to macroeconomic costs and regressive redistribution of the tax burden. Non-price policies such as efficiency standards and targeted investment subsidies cannot be examined, even though these would probably have smaller macroeconomic and welfare impacts (disregarding transaction costs) than a general energy tax, for similar savings in carbon dioxide.

Using an econometric approach does have the disadvantage of building in past economic behaviour and potentially biasing the cost of control upwards. Because the estimation period used (usually 1960–87 for macroeconomic equations, and 1978–90 for energy sector equations) is shorter than the simulation period (1995–2030), and relative factor price movements in the estimation period are likely to be smaller than proposed carbon taxes, the measured flexibility of the economy in adjusting to price shifts is likely to be lower than will actually occur in the future. With luck, the discrepancy will be small and the use of cointegration techniques to extract long run parameters should improve the accuracy of the estimation. The main advantage of the econometric approach is that it allows us to investigate the relative magnitude of the various influences on macroeconomic costs (AEEI, elasticity of substitution, etc.) inside a consistent, empirically based framework and without imposing a priori assumptions of the size of any one effect.

4

EMPIRICAL MODELLING OF ENERGY DEMAND RESPONSES

INTRODUCTION

EGEM is a macroeconomic/energy model designed to simulate the economic impacts of different international policies to reduce carbon dioxide emissions. It consists of a detailed econometric model of energy demand integrated into a more general international macroeconomic modelling structure, the background and development of which is described in Chapter 5. This chapter describes the estimation of the energy demand model for eight OECD countries, and discusses the forecast and simulation properties of the resulting equations in terms of the ability of carbon and energy taxes to control CO_2 emissions.

In each country equations are estimated for total fossil fuel energy consumption, and then disaggregate demand is modelled in terms of the changing shares of coal, gas and oil. Cointegration techniques are used to establish the long term equilibrium relationships between the variables. To capture the non-equilibrium, dynamic nature of instantaneous energy demand, error correction models (ECMs) are then estimated using the residual from the long run relationship; in this form the coefficient on the long run residuals represents the rate of adjustment towards the cointegrating equilibrium. This specification has the advantage of long term stability together with a rich dynamic structure.

The cointegration analysis of a conventional energy demand model showed that changes in energy intensity over time were mainly influenced by the path of exogenous technical progress, modelled as a time trend in the estimation. This implies that raising energy taxes will have little effect on demand, which is unlikely and an undesirable simulation property for the model. In order to investigate the components of the 'exogenous' trend an alternative formulation of the fossil fuel consumption equation was then estimated with the Kalman Filter, using an endogenous technological change term in place of the deterministic trend. This model allowed higher energy prices to induce greater technical innovation and explicitly included the influence of structural economic change; thus allowing assessment of the long term impacts of policy changes. Estimation with this model structure showed the long run factors to be the major determinants of past changes in energy intensity. In simulations of future carbon taxes innovation in energy efficiency also proved to be significantly more important than demand reductions from conventionally measured elasticities.

AGGREGATE ENERGY MODEL

For each country studied the analysis is divided into two parts: aggregate fossil fuel consumption, and a system of fuel shares. Two possible models are developed for the aggregate energy consumption: the first includes an exogenous trend in energy intensity, while the second attempts to endogenise this trend. In order to make the energy demand equations consistent with the production modelling used for the aggregate economy constant returns-to-scale were assumed. This is achieved by carrying out all estimation on energy intensity (energy use/GDP) rather than estimating the influence of output on gross energy consumption. This restriction makes explicit all changes in production and consumption patterns, whether these are price reactions, changes in technology or structural shifts due to manufacturing's declining share in advanced economies. This separation of influences allows the proportion of energy use which can be influenced by policy to be identified by estimation; not applying this restriction allows energy intensity to be *determined* by output levels, which 'black boxes' important influences which need to be simulated in a meaningful policy analysis.

Aggregate energy analysis using an exogenous trend

A two-stage procedure was followed, involving a long run cointegrating equation and a dynamic relationship in the form of an error correction model (ECM). The Granger Representation Theorem (Engle and Granger 1987) states that if a set of I(1) variables are cointegrated (so that the resulting residual is stationary) then there is a valid error correction representation. A range of diagnostics were used to test the order of integration of the variables and it was concluded that logs of price and energy intensity could be considered I(1) over the estimation period, which was 1978–90 (quarterly data). Full details of the econometric analysis including results and significance statistics are available in Boone *et al.* (1995) and Hall *et al.* (1993).

The Johansen procedure (Johansen 1991) was used to identify long run cointegration vectors between logs of fossil fuel intensity (demand per unit GDP), weighted average real fuel price and a time trend included as a proxy for technical progress and structural change (1). Generally four lags were included in the VAR, corresponding to the quarterly data, although some countries required up to eight to give a good result.

$$\ln(C_t/Y_t) = a + b.\ln(RP_t) + c.T_t + \varepsilon_t \qquad (1)$$

where: C = fossil fuel consumption, Y = GDP, RP = weighted average real fuel price, T = time, and ε = residual.

Instantaneous demand for fossil fuels, accounting for the dynamics of transition between long run equilibria after exogenous shocks, was then estimated in an ECM. Change in fossil fuel consumption is regressed against the residuals from (1), lagged once (the 'error correction term'), together with lags of differences of the other variables as shown in (2). Generally up to four lags in each were included at the outset and retained according to significance.

$$\Delta \ln(C_t) = \alpha + \beta.\varepsilon_{t-1} + \sum_{i=1} \gamma_i \Delta \ln(C)_{t-i} + \sum_{i=0} \psi_i \, \Delta \ln(RP_{t-i}) + \sum_{i=0} \phi_i \Delta \ln(Y_{t-i}) \qquad (2)$$

Including the residual ε_t allows past forecasting errors to be taken into account in forecasting the next change in fossil fuel consumption. The coefficient on this error correction term indicates the rate at which consumption adjusts towards the long run equilibrium.

In some countries two or more significant cointegrating vectors were found, in which case residuals from both were included in the ECM, and the most significant retained. In three cases (Japan, Canada and the Netherlands) one of these embodied a positive relationship between price and demand, that is, increasing demand increases prices. The lagged residual from these relationships, as expected, turned out to be non-significant in the ECM, indicating that causality from prices reduces energy consumption through the negative relationship.

The results from the long run analysis, summarised in Table 4.1, show a negative price elasticity was found in each case although the values are fairly low. All countries show strong declining time trends in energy intensity.

Estimating an endogenous technical progress model

The time trend in the above model is taken to represent technical progress, analogous to the 'autonomous energy efficiency improvement' (AEEI) used in other models, a term coined by Manne and Richels (1992). Technological improvement and increasing energy productivity provide the

Table 4.1 Fossil fuel intensity elasticities with respect to price and time trends

Country	*Price*	*Time (% p. a.)*
Canada	−0.101	−2.17
France	−0.147	−5.67
Germany	−0.063	−3.12
Italy	−0.130	−2.19
Japan	−0.133	−4.06
Netherlands	−0.085	−1.61
UK	−0.045	−2.58
USA	−0.159	−3.09

possibility of reducing energy use in the long term, without increasing costs. Expressing this rate as an exogenous factor severely limits the model's ability to represent how policy or macroeconomic changes may affect technological development. Studies to measure the AEEI empirically have been relatively few and the results inconclusive (Hogan and Jorgenson 1991).

Decline in energy intensity may be due to several factors: structural change in the economy, the development of non-fossil fuel sources, price-induced innovation and non-price policy interventions, in addition to autonomous technical progress. As the time trend is substantial, it is important to refine the model to endogenise some of the determinants of this trend.

To estimate this endogenised trend, a model analogous to the error correction representation above was supplemented by an endogenous trend term, including determinants such as non-fossil fuel supply, manufacturing component of GDP, trade, and investment. Price is also included as high prices may stimulate innovation. In addition there is a stochastic trend, which is taken to represent the autonomous improvement in technology. The model was estimated using the Kalman Filter, as described by Cuthbertson *et al.* (1992), expressed as equations 3a–c:

Measurement equation:

$$\Delta \ln(C)_t = \alpha + T_t + \beta_1 \ln(C/Y)_{t-1} + \beta_2 \ln(RP)_{t-1} + \sum_{i,k} \gamma_{ik} \Delta \ln(Z_{i,t-k}) \quad (3a)$$

where all variables are as before, $\Delta \ln(Z)$ are lags of differences in consumption, GDP, and price and T is the endogenous trend from the transition equations below.

$$T_t = T_{t-1} + \pi_{t-1} + \gamma \ln(RP)_{t-1} + \sum_i \eta_i X_i + \varepsilon_t \quad (3b)$$

$$\pi_t = \pi_{t-1} + \nu_t \quad (3c)$$

where π is an exogenous trend including a stochastic component ν_t and X_i are the structural determinants of the endogenous trend, as discussed above. In most cases the dependent variable used was total energy intensity (as opposed to fossil fuel intensity used in the initial model), including nuclear and hydro expressed as its fossil fuel replacement value. Although (by definition) increases in non-fossil fuel must be accompanied by decreases in fossil fuel for a given energy demand, this relationship was not always adequately represented by including the change in non-fossil fuel consumption in the endogenous trend. Using total energy consumption allows fossil fuel consumption to be calculated by subtracting exogenous forecasts of non-fossil fuel production.

It can be seen that this model subsumes the conventional 'elasticity plus trend' model above, with the additional price term in the transition equations implying long term irreversible improvements in energy efficiency as a result of higher prices. In a simple demand elasticity model, if the price were to be raised and later reduced, consumption would return to its original level; with endogenous technical change this is not the case.

When simulating policies to *stabilise* levels of carbon emissions into the future there is a more important difference between the two models. Taking the differential of (1) with respect to time, and omitting the log signs for clarity, gives equation (4a). It is obvious that, given continuing economic growth, stabilisation of energy use ($\partial C/\partial t = 0$) can only occur in this model if fuel prices constantly rise ($b<0$, $\partial RP/\partial t>0$), or if the exogenous increase in energy efficiency exactly matches aggregate growth in the economy ($-c = \partial Y/\partial t$). Discounting this unlikely coincidence this model has only two equilibrium solutions in the long run: either fossil energy use increases to infinity along with economic output, or it decreases to zero. The exogeneity of the efficiency trend means that government policy has little long run influence, because stabilisation is only achievable if energy prices tend towards infinity at $1/b$ times the underlying rate of growth of energy use ($\partial Y/\partial t - c$)! This unrealistic simulation property is why many of the GE models surveyed in Chapter 3 include a 'backstop technology' which caps the size of energy tax needed to achieve stabilisation (and elimination?) of carbon emissions. The short and medium term results of these models are therefore driven by the a priori assumptions of the date of introduction and rate of diffusion of the backstop technology. While the long run results are determined solely by the projected costs of these new technologies, and have nothing to do with consumer preferences and measured substitution behaviour.

$$\partial C/\partial t = b.\partial RP/\partial t + \partial Y/\partial t + c \qquad (4a)$$

$$\beta_1.\partial C/\partial t = -\beta_2.\partial RP/\partial t + \beta_1.\partial Y/\partial t - \pi^* - \gamma.RP_t - \sum_i \eta_i X_i \qquad (4b)$$

Equation (4b) shows the differential for the endogenous change model (3) assuming constant growth in the X_i variables and π^* being the average future rate of exogenous technical progress. This model has a long run stabilisation equilibrium for a non-growing real energy price because the *level* of energy prices, as well as the differential, enters the expression (4b). The equilibrium is defined by: $\gamma.RP_t = \beta_1.\partial Y/\partial t - \pi^* - \Sigma\eta_i V_i$. Therefore, for π^* and Z_i not changing over time (which is true for the simulations), there is a single energy price which will stabilise emissions for a constant *rate* of economic growth. This simulation property represents an implicit backstop technology which has the advantage of being estimated and consistent with the observed transition dynamics of the economy.

This model was estimated for the eight countries and the results are summarised in Table 4.2, together with 'implied elasticities' calculated by simulation as the percentage change in energy consumption caused by a 1 per cent tax. As shown, this value depends upon the length of time the tax has been in place, since with endogenous technical change the 'elasticity' continues to increase as long as the price remains high. This could be

Table 4.2 Price coefficients in endogenous technical change model

	Price coefficient		*'Implied elasticity'*	
Country	*Direct*	*Trend*	*5 yrs*	*35 yrs*
Canada	−0.171	−0.00253	−0.268	−0.393
France	−0.085	−0.00199	−0.115	−0.507
Germany	−0.248	−0.00231	−0.303	−0.570
Italy	−0.044	−0.00039	−0.050	−0.107
Japan	−0.016	−0.00238	−0.095	−0.645
Netherlands	−0.081	−0.00249	−0.127	−0.431
UK	−0.045	−0.00515	−0.126	−0.713
USA	−0.181	−0.00246	−0.224	−0.460

likened to conventional short and long term elasticities, but their long run behaviour is in fact quite different.

The main X variable in the model was industrial production as a proportion of GDP, which was found to be significant and to have the expected sign (energy consumption increasing with industrial production) for all countries except France and Germany; in the UK manufacturing was used instead. For the USA business investment was found to have a small negative correlation.

Figure 4.1 shows the projected fossil fuel intensities in each country over the simulation period which will be used for policy analysis (1998–2030), using the endogenous trend model and fuel prices calculated from the equations described below.

As would be expected from the estimated coefficients, fossil energy use per unit of GNP falls in all countries over time; this derives from exogen-

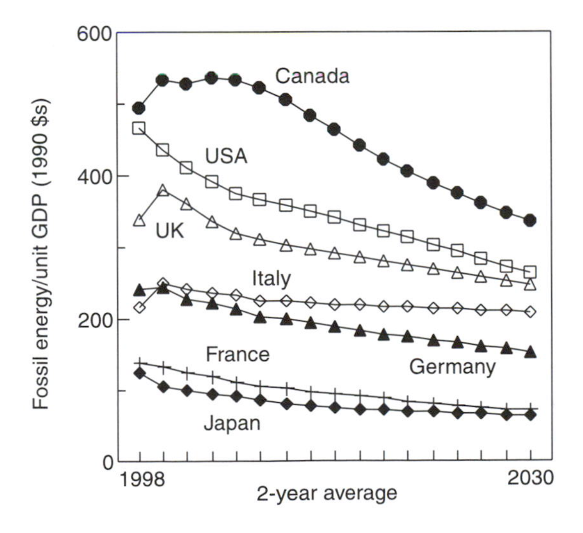

Figure 4.1 Fossil fuel use (toe) per unit GDP (millions 1990 US$), 1998–2030

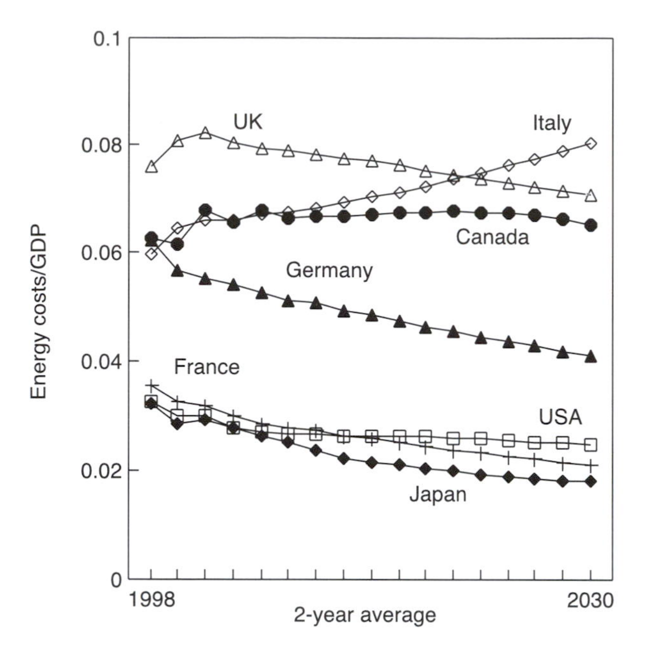

Figure 4.2 Projected energy costs as a proportion of GDP, 1998–2030

ous technical progress and a gradual increase in the price of fuels as global demand grows. The rate of decline is highest in those countries with initially the highest energy use, leading to a slight convergence in energy efficiency in the long run. This pattern is not mirrored in the cost share of energy in GDP however, because of the different taxes levied on fuels in each country. Figure 4.2 shows the share of fossil energy costs in GDP over the simulation period.

Fossil energy's cost share declines in all countries except Canada and Italy, but at a lower rate than energy volume measured in toe. This is because the international price of energy rises over time and even when combined with exogenous technical progress none of the long run price elasticities of energy is greater than one. Energy costs rise in Italy because its currency devalues relative to the US dollar due to its lower economic growth rate (see Chapter 5). This makes all fuels more expensive as they are related to the dollar price of oil. Energy costs rise in Canada because over time it substitutes into gas which has a higher cost per toe; this substitution is a feature of the share equations estimated below.

ESTIMATING FUEL SHARE EQUATIONS

In order to calculate carbon dioxide emissions the fuel mix of an economy, as well as total fossil fuel demand, must be known. Consumption of coal, gas and oil are defined as shares of fossil fuel demand by their primary energy values. These are expected to respond to changes in their relative prices, and to changes in total energy demand. Additional determinants include changes in GDP and non-fossil fuel supplies. Fuel markets are complex: the three fuels are not wholly substitutable, and the fuel mix will be influenced by factors such as industrial mix, technological change, regulatory changes corresponding to policy objectives such as security of supply and environmental control, and also discrete events such as price shocks and strikes. Each of these will be of differing importance in each country, and for each fuel.

As for aggregate analysis, a two-part procedure was used to estimate the equations, comprising a long run cointegrating relationship and a dynamic error correction model. This model is a modified Almost Ideal Demand System (AIDS), originally due to Deaton and Muellbauer (1980).

Long-run analysis proceeded by estimating functions for each fuel share, whereby shares s1, s2, s3 (for coal, gas and oil respectively) at time t are given by:

$$s_{i,t} = a_i + \sum_{j=2}^{3} b_{ij} \ln(p_{jt}/p_{1t}) + c_i \ln(C_t) + \sum_{k=1}^{n} d_{ik} Z_{kt} \quad (5)$$

where p_{jt}, j=1,2,3 are the prices of each fuel, C_t is total fossil fuel demand at time t, and the Z_{kt} are a set of n exogenous variables of interest. The Z variables are shown in Table 4.3 and include GDP, consumption of nuclear energy and hydroelectricity, and in some cases dummy variables representing various one-off shocks such as strikes and government policy, for example investment programmes in power stations.

The price variables are expressed relative to the price of coal as a restriction to ensure homogeneity in the system, that is, a rise in the price of oil has the same effect on shares as an equivalent fall in the prices of coal and gas, apart from the effect on the aggregate fuel consumption.

The sum of the fuel shares sums to unity ($\Sigma s_{it}=1$), and so for consistency there must be *adding up* across the share equations, that is:

$$\sum_i a_i = 1, \sum_i b_{ij} = 0, \sum_i c_i = 0, \sum_i d_{ik} = 0$$

This system is therefore singular, that is, only two of the equations can be estimated independently. Equations (5) for gas and oil were estimated using OLS, while coal's share was then obtained as the residual. As the three shares by definition add up to one, this procedure has no effect on the coefficients obtained or their significance, as mathematically the calculation is identical whichever fuel is left out of estimation.

Estimation of (5) is carried out with the aim of finding stable long run relationships where both sides cointegrate together; therefore the residuals were tested for stationarity using a number of diagnostics. It is clearly highly desirable a priori to obtain negative own price coefficients, and the best relationships were sought which embodied these. In some cases this presented difficulty because of the distortions in the energy markets under consideration, which was countered by the inclusion of dummy

Table 4.3 Summary of fuel shares equations, long run relationships

Country	Fuel share	Price and consumption coefficients[a]			Other explanatory variables
		$ln(p_2/p_1)$	$ln(p_3/p_1)$	$ln(C)$	
Canada	Coal		0.049	−0.056	Time
	Gas		0.008[b]	−0.274	Dummies in 1978, 1989
	Oil		−0.057	0.330	
France	Coal	0.036[b]	0.029[b]	0.067[b]	Dummy in 1986–88
	Gas	−0.271	0.169	−0.366	
	Oil	0.235	−0.198	0.299	
Germany	Coal	0.058	0.071	−0.278	Nuclear, hydro
	Gas	−0.062	0.093	−0.057	Dummy in 1981
	Oil	0.005[b]	−0.164	0.334	
Italy	Coal	−0.048	0.063	−0.187	Y
	Gas	−0.072	0.036[b]	−0.105	Dummies in 1980–1,
	Oil	0.121	−0.099	0.292	1982, 1984–6
Japan	Coal	−0.013	0.046	0.069	Ogive from 1978–90
	Gas	−0.122[b]	0.152	−0.063	
	Oil	0.135	−0.198	−0.006[b]	
Netherlands	Coal	0.102	0.009[b]	−0.011[b]	Dummies in 1983–5,
	Gas	−0.153	0.060	0.062	1988
	Oil	0.051	−0.069	−0.051	
UK	Coal	−0.016[b]	0.222	0.251	Y, time
	Gas	−0.062	−0.061	−0.025[b]	Dummies in 1984,
	Oil	0.078[b]	−0.161	−0.226	1983–5
USA	Coal	0.126[b]	0.032	−0.328	Y, hydro
	Gas	−0.014	0.039	−0.039	Dummy in 1986
	Oil	0.002[b]	−0.071	0.367	

a Owing to adding up, each group of three must sum to zero.
b Denotes coefficient not statistically significant.
p_i = price of fuel i (1 = coal, 2 = gas, 3 = oil). C = total fossil fuel expenditure.

variables as discussed above; for instance, in the UK, dummies for the miner's strike of 1983–85 explained the large drop in coal's share and the increase in oil.

In the long run relationships estimated, the oil share is found to have a significant negative own-price coefficient everywhere, usually with a value of 0.1–0.2. For gas the coefficients are lower, but all are significant with the exception of Canada, where the gas price was not significant for any fuel share, and Japan, where the own-price coefficient is not significant for gas. Cross-price coefficients generally are positive and smaller than own-price. Coal is the least price-responsive, generally with coefficients under 0.1, insignificant in the case of France, UK and USA, and negative signs in some cases. Total consumption was significant in all countries but with no consistent pattern, although in five out of the eight countries the share of oil increases with increasing consumption and has the largest coefficients;

gas is usually negative and has lower coefficients. GDP was significant in Italy, UK and USA. A time variable was used in Canada and the UK, and an S-shaped ogive in Japan, in each case correlated with the increase in gas use.

Residuals from the long run relationships are fitted into an ECM framework to produce a dynamic model, analogous to the ECM in the aggregate case, but consisting of three equations, one for the change in each share.

However, this is not a single equation but a system, which obeys adding up constraints and exhibits singularity. Hence it is not possible to estimate error correction equations for each share independently, nor is it possible to identify the dynamics or the adjustment coefficients of the system (Anderson and Blundell 1984, Barr and Cuthbertson 1991). Adding up implies that

$$\Delta s1_t + \Delta s2_t + \Delta s3_t = 0 \qquad (6a)$$

and

$$res1_t + res2_t + res3_t = 0 \qquad (6b)$$

for all t, where res1 is the residual from the long term cointegrating equation in s1, etc.

The resulting estimated system must obey (6a), which implies that all equations contain the same set of exogenous variables on the right hand side, with coefficients that sum to zero. Furthermore (6b) implies that the sum of the adjustment coefficients γ_{ij} (the coefficient on res_j in equation Δs_i) down each column must be the same for each column:

$$\sum_{i=1}^{3} \gamma_{i1} = \sum_{i=1}^{3} \gamma_{i2} = \sum_{i=1}^{3} \gamma_{i3} \qquad (7)$$

The most general specification for the residuals is therefore to choose *two*, for example res1 and res2, to appear in each of the equations for estimation. (This is equivalent to choosing three, as res3 = −(res1 + res2), but avoids having a linearly dependent set of independent variables). The matrix (γ_{ij}) of adjustment coefficients obtained is therefore (3×2).

It would also be possible to choose just one residual in each equation, for example res1 in the equation for $\Delta s1$ and so on. This is a more restrictive approach, yielding a matrix of coefficients that is diagonal with equal elements $\gamma_{11} = \gamma_{22} = \gamma_{33}$, all negative, that is, requiring that the shares adjust towards equilibrium each at the same rate. However, we have chosen the more general method, with res1 and res2 in each equation, allowing adjustment at different rates.

For stability in the estimated dynamic system, we require that the eigenvalues of the 2×2 matrix (γ_{ij}) $(i,j = 1,2)$ be negative. This is obtained if and only if:

$$(1)\ \gamma_{11} + \gamma_{22} < 0 \qquad (8a)$$

and

$$(2)\ \gamma_{11}\gamma_{22} - \gamma_{21}\gamma_{12} > 0 \qquad (8b)$$

A similar argument applies in the case of the lagged dependent variables $\Delta s(i)_{t-i}$. In this case we have chosen to restrict the dynamics of each share to lags of itself; for example, $\Delta s1_{t-i}$ only are included in the equation for $\Delta s1_t$ and so on, which implies each will have the same coefficient to ensure adding up.

The resulting system of three equations is:

$$\Delta s_{j,t} = \alpha_1 + \gamma_{j,1} res1_{t-1} + \gamma_{j,2} res2_{t-1} + \sum_{i=1}^{m} \mu_{\ i} \Delta s_{j,t-i} + \sum_{i=0}^{m} \beta 2_{j,i} \Delta rp2_{t-i} + \sum_{i=0}^{m} \beta 3_{j,i} \Delta rp3_{t-i} + \sum_{k=1}^{n} \sum_{i=1}^{m} \delta_{\ j,k,i} \Delta Z_{k,t-i} \qquad (9)$$

where each equation has the same dynamic structure and the same set of ΔZk. All coefficients on the residuals, the relative prices and the independent variables add to zero down the columns. The coefficients on the lags of the respective dependent variables are the same for each share. The system was estimated jointly using non-linear 3-Stage Least Squares.

ESTIMATING FUEL PRICE EQUATIONS

Any drop in the demand for fossil fuels will have a depressing effect on world energy prices, this in turn will decrease domestic prices thus raising emissions. In order to model this important effect the linkages between world prices and domestic prices in each country must be modelled; this link is not straightforward because the fuels are imperfect substitutes and oil products are often sold through restricted markets (especially in Japan). Therefore, prices for the three fuels, net of tax, were compared with movements in the world oil price, expressed in domestic currency. A relationship for each fuel price was estimated (10):

$$PL_{f,t} = \alpha + \beta.WOL_t + \Sigma\gamma_i Z_{i,t} \qquad (10)$$

where: $PL_{f,t}$ = log of net price of fuel f, WOL = log of world price of oil, and other variables included in Z might be the exchange rate, domestic price level (GDP deflator) and GDP, all in logs. Dummy variables may also be included. Stationarity of residuals was tested as an indicator of long term stability.

For the price of oil in the majority of countries, the estimated equations show the required characteristics. The price of oil in each country depends strongly on the world price, with other factors accounting for short term fluctuations and the 'smoothing' behaviour of oil companies, contracts, domestic sources of oil, and so on. As expected the coefficients for the world oil price, b, are positive and fairly close to one, and this is the most important explanatory variable. The use of domestic prices means that the domestic price level is also likely to be significant and positive. Exchange rates and GDP are also significant in some cases.

For gas and coal, other country specific factors, particularly the GDP deflator, tend to be more important than for the oil price equation although world oil price is still a major factor. The coefficient β would be expected to be less significant and lower than for oil but must be positive. However in some cases no relationship was found with the oil price: in Canada for both gas and coal, and in the USA for gas. This is thought to be because North American gas (and to a lesser extent coal) produced domestically is not closely linked to world markets, but prices are determined internally. By contrast, in Europe contracts linking gas prices to that of oil, and competition in electricity generation markets, provide stronger links with the oil price. Also in some cases the relationships include world price lagged by one year, indicating inertia in the response to world oil price fluctuations. In many cases, the relationships are very similar for gas and coal, as their prices are generally determined by similar factors.

Dummy variables were used to account for particular discrepancies. In particular, in 1986 the collapse of the world oil price was not fully passed on to the consumer, and the average price paid after 1986 continued to show a significantly increased differential over the world price. Gas and coal, especially where domestically produced, are susceptible to a range of policy factors affecting their prices.

Consumption levels and depletion are not taken into account, as they were not found to be significant in the estimation, nor considered likely to become so within thirty-five years although gas in particular may become supply-constrained in the very long term. Future prices use a forecast of world oil price in EGEM based on OPEC's cartel pricing behaviour which is fully described in Chapter 9. The model leads to prices generally increasing at slightly more than the domestic inflation rate, although in some cases the coal price falls in real terms.

SIMULATION PROPERTIES OF THE ENERGY MODEL

The energy model developed can be used to look at the impacts of different taxation strategies on fuel mix and carbon dioxide emissions. Changes in aggregate energy consumption due to price rises reflect non-fossil fuel substitution, product substitution, price induced technical innovation and pure substitution into other factors of production. All these different process are conflated in the aggregate model with an exogenous trend as supply and demand side changes are not modelled separately. The fuel shares model provides a means of looking at inter-fuel substitution, and its importance compared with overall energy conservation. Carbon and energy taxes can be compared, along with other tax regimes such as applying the same per centage rates of tax in each country.

A carbon tax would be expected to be more effective in reducing emissions than an energy tax, as it more accurately internalises the externality. To compare carbon and energy taxes, two scenarios were run, corresponding to the EC proposal of $3/bbl in 1995 rising to $10/bbl by 2002 (in real 1990 prices), but in the first scenario wholly raised on energy and the second wholly on carbon content of the fuel.

Many factors determine the overall elasticity of energy. Income levels are likely to play a part, together with the structure of the economy. Actual price levels vary widely between countries due to differences in existing taxes, which in the case of a carbon or energy tax implies that the percentage tax rate is higher in low-cost countries (USA, Canada) than those with high prices (Germany, Japan). The level of government intervention in such areas as efficiency regulation also varies widely. In a country where energy use is currently high and inefficient, there may be many low-cost opportunities for reductions, while countries that currently have lower energy intensity may find it harder to make further improvements.

Part of the difference between countries' price responses may be explained by the differences in prices and existing taxation in different countries (Hoeller and Coppel 1992). In order to assess the magnitude of this effect, an *ad valorem* tax scenario was constructed in which the same percentage rate of tax is levied in each country, at a level giving the same total revenue overall, in the proportions on each fuel as for a 50/50 carbon/energy tax in the UK. This translates to 20.8 per cent on oil, 28.47 per cent on gas and 83.95 per cent on coal.

Table 4.4 compares tax rates in the UK (comparable with most of Europe) and the USA, to illustrate the differences between scenarios and countries. The results from these three scenarios are compared in Figure 4.3, using the error correction model. These values represent the percentage reduction in CO_2 emissions in the year 2030 as compared to the base-case value.

The levels of abatement are divided into conservation – the amount caused by the absolute reduction in energy consumption – and substitution, caused by the shift in fuel mix from coal to gas. It can be seen that there are large differences between countries, both in the amount reduced and the role of inter-fuel substitution. As expected, for carbon or energy taxes, low prices in the USA and Canada lead to these countries having the most abatement, with Japan and Germany having the least.

Table 4.4 Prices and taxes on fuels in the three scenarios

	Price	*Tax*		
	£($)/toe	*Energy (%)*	*Carbon (%)*	*Ad val. (%)*
UK				
Coal	116	56.9	61.6	84.48
Gas	269	24.5	15.2	28.62
Oil	428	15.4	13.8	21.03
USA				
Coal	57	157.9	186.0	84.48
Gas	267	33.7	22.8	28.62
Oil	274	32.8	32.1	21.03

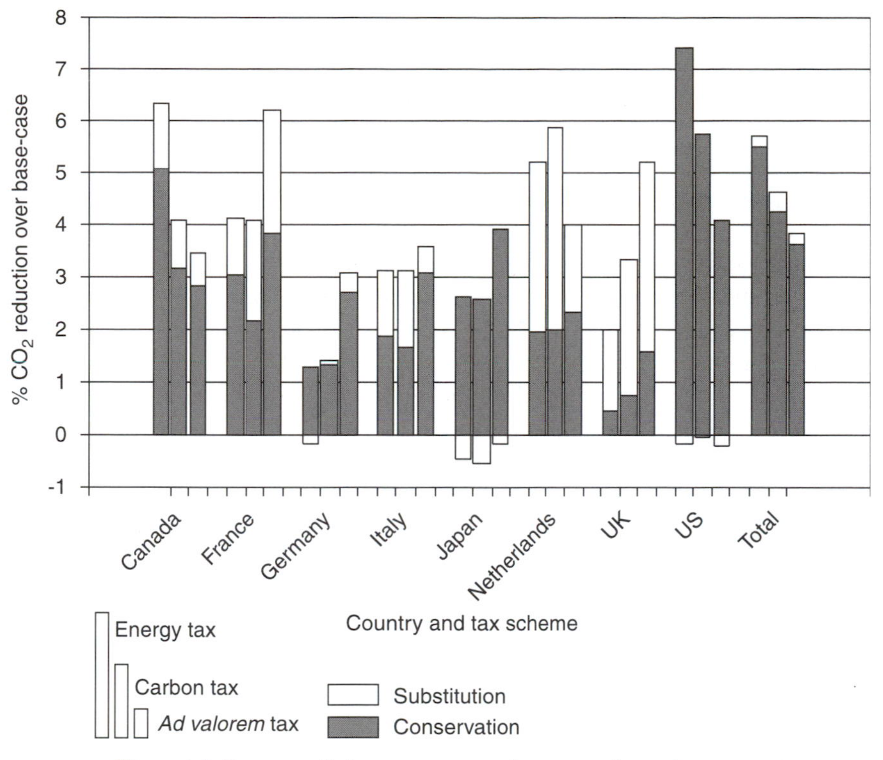

Figure 4.3 Impact of three tax strategies on carbon abatement

Canada, the Netherlands and the UK are the three major producers of natural gas amongst the eight countries, and they all show a significant reduction from substitution of gas for coal or oil, although this is subject to the perfectly elastic supply assumed in the price equation. France and Italy also show a high level of substitution, in this case away from coal. Germany, Japan and the USA however actually show negative substitution, indicating a shift away from gas and into oil, without a great reduction in coal. In the case of Germany this is probably due to the lack of price response in the coal industry due to government subsidy. In Japan, coal is relatively minor at present and gas is expensively imported as LNG: the system is highly constrained by their lack of resources and government intervention, again reducing the response to price.

Substitution between fuels is not always in the direction expected, due to differences in their price levels, and in responses to price. For instance, for the USA forecast in 2000, the prices of oil, gas and coal are \$205, \$186 and \$54/toe respectively (in 1990 \$). The combined carbon/energy tax level in 2000, once the carbon contents of the fuel have been applied, works out at \$33, \$23 and \$40/toe for the three fuels respectively, resulting in additional tax rates of 32 per cent on oil, 30 per cent on gas and a massive 135 per cent on coal. One would expect coal to lose some of its market share to both oil and gas. However the analysis demonstrates that the price elasticity of gas is less than that of oil, with respect to the price of either relative to coal. Hence oil takes over most of the share lost by coal, while the consumption of gas, due to the overall decrease in fossil fuel consumption, actually falls.

As compared to an energy tax, a carbon tax does not necessarily imply a greater reduction in carbon emissions although it will be more efficient in terms of cost per tonne carbon removed. In the case of an agreement consisting of an international tax, in addition to each country's agreeing to some stabilisation target, it can be seen that some countries would favour a carbon tax while others would benefit from an energy tax on the basis of carbon abatement. However this omits any consideration of the revenues raised from the tax, which will be different for each scenario.

The third *ad valorem* tax scenario was included in order to explain the extent to which differences in prices between countries determines their response. It can be seen that the response is rather more uniform, as would be expected, with Canada and the USA reduced to a similar level to Europe, and Germany and Japan increasing. This implies that the large differences between these countries seen in the carbon/energy tax scenarios is due more to differences in price than to behaviour or technical limitations.

Due to differences in fuel mix, consumption level and price, the tax burden imposed by each scenario will vary according to country. Table 4.5 shows the tax revenue per unit of GDP for each country in 1995; these rates will increase by a factor of 3.3 between 1995 and 2002, as the tax rates increase, and decline as energy intensity declines thereafter. The total tax revenue for 1995 and 2030 is also shown, in constant 1990 $ prices.

For the energy tax scenario, the rates reflect the fossil fuel intensity of each country. Those countries with a lower fossil fuel carbon intensity (i.e. those using more gas and less coal) will show a decreased revenue with a

Table 4.5 Energy/carbon tax revenue as a percentage of GDP and total for the eight countries, 1995

	Tax		
Country	*Energy (%)*	*Carbon (%)*	*Ad val. (%)*
Canada	0.74	0.61	0.58
France	0.18	0.16	0.27
Germany	0.30	0.30	0.61
Italy	0.30	0.28	0.41
Japan	0.14	0.13	0.22
Netherlands	0.66	0.58	0.67
UK	0.53	0.42	0.66
USA	0.50	0.57	0.42
Total tax revenue for the eight countries (million US$ 1990)			
1995	64,120	65,104	70,058

carbon tax, such as Canada, the Netherlands and the UK. The *ad valorem* tax will be lowest where current fuel prices are low, such as the USA.

Total revenue, while broadly similar for the three scenarios, changes considerably over time as each tax has a different effect on fuel mix and consumption. Thus it is difficult to compare directly the carbon abatement per $ of tax from the three scenarios. The magnitude of the revenue reinforces the importance of the '64 billion dollar question' of how these are recycled.

The above results however do not take into account the endogenised trend in energy intensity. The second model estimated including a term allowing price to increase the rate of technical progress, would be expected to forecast larger reductions in emissions in the long run. Results from the two models are compared in Figures 4.4 and 4.5, for abatement in 2005 and 2030. A base-case forecast is compared with a scenario corresponding to the tax of $3/bbl in 1995 rising to $10/bbl in 2002, split 50/50 between energy and carbon.

Comparing the two models, it can be seen that over this simulation period endogenising the trend does in most cases imply a greater potential level of carbon abatement; of course in the long run this will always be true as the endogenous trend will continue to grow. The models are broadly similar in 2005 but the difference becomes greater in the longer term. As the tax level reaches its maximum in 2002, for the error correction model

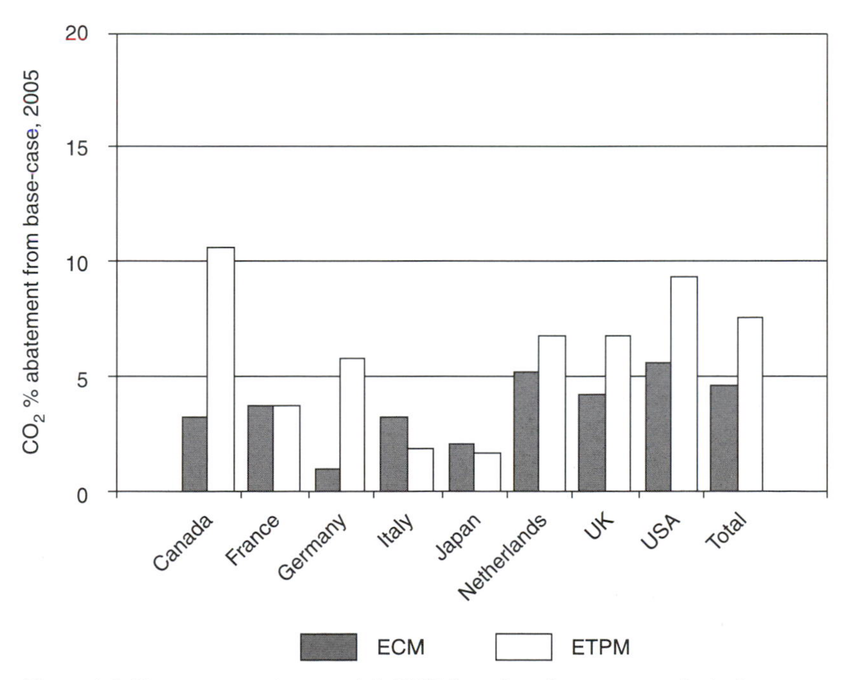

Figure 4.4 Error correction model (ECM) and endogenous technical progress model (ETPM): CO_2 abatement in 2005

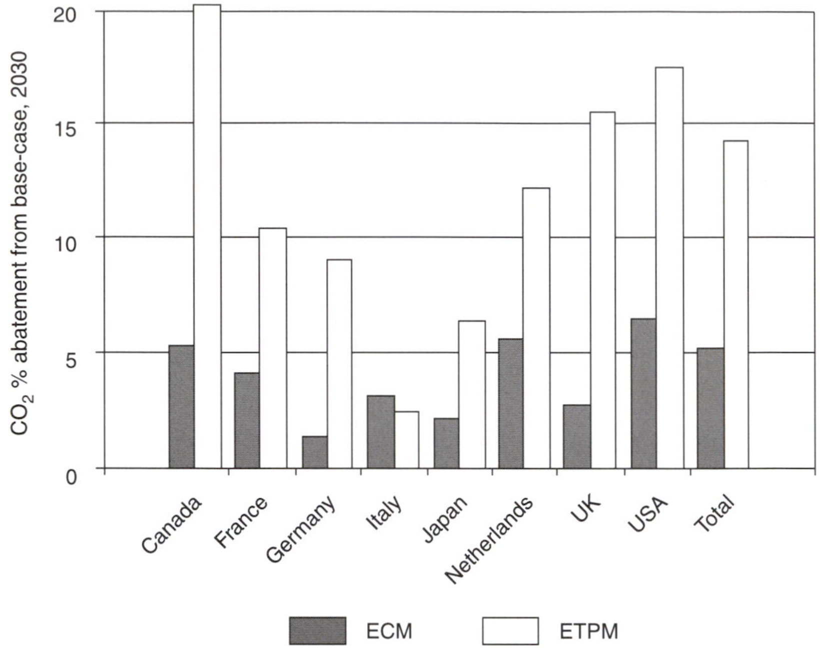

Figure 4.5 Error correction model (ECM) and endogenous technical progress model (ETPM): CO_2 abatement in 2030

there is little difference between the abatement level in 2005 and 2030, apart from some longer term adjustment to the higher price level. For the endogenous model however, the intervening twenty-five years at the high price level brings about continued reductions in energy intensity.

Figures 4.6, 4.7 and 4.8 show the effect of the tax in more detail on aggregated emissions in Europe, North America and Japan, using the endogenous technical change model. In each case emissions continue to rise – a tax of this size alone cannot stabilise or reduce emissions in the long run. The abatement relative to the base-case is greatest in North America, followed by Europe, and quite small in Japan. The relationship between energy consumption and CO_2 emissions is illustrated by the second line on each graph, representing the emission per toe of fossil fuel consumed, which depends on the fuel mix. In Europe, there is a significant decline in this ratio, brought about by the increased use of natural gas. In North America and Japan the figure increases, implying substitution away from gas.

These changes can be clarified by looking at the changes in consumption of each fuel, as shown in Figures 4.9, 4.10 and 4.11. In Europe, as one would expect, coal and oil consumption decrease from their base-case levels, with coal showing the greatest reduction, while gas initially increases, returning to its base-case consumption by 2030. This shift in fuel mix explains the decreased CO_2 intensity per unit of energy. However, in

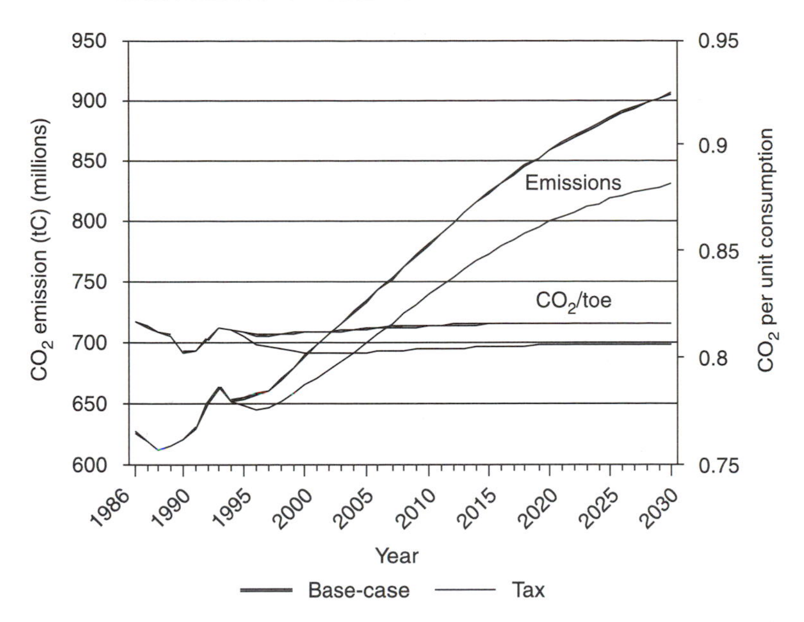

Figure 4.6 CO_2 emissions and intensity in EC5 (France, Germany, Italy, Netherlands, UK)

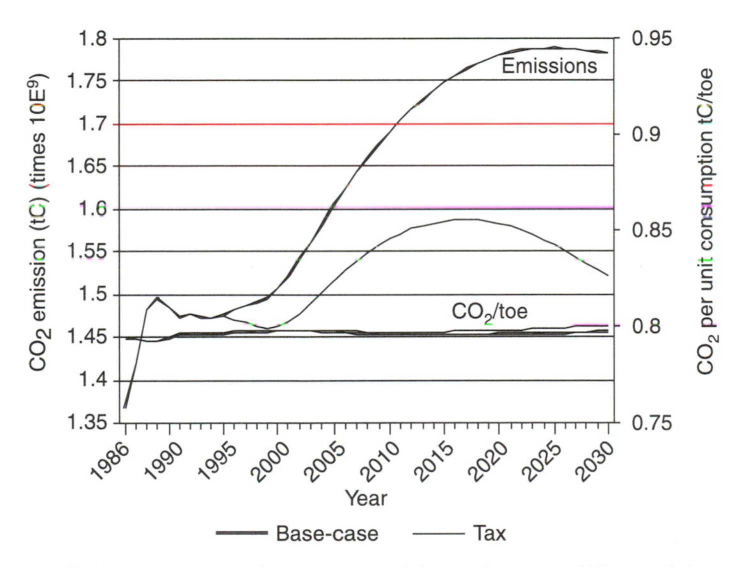

Figure 4.7 CO_2 emissions and intensity in North America (USA and Canada)

America, the absolute changes are large, and not in the direction expected. Although consumption of all three fuels declines and coal initially decreases most, gas also shows a large reduction while oil initially increases followed by a smaller decrease, leading to an overall increase in CO_2 intensity per toe. In Japan the changes are smaller, and coal and gas are reduced while oil

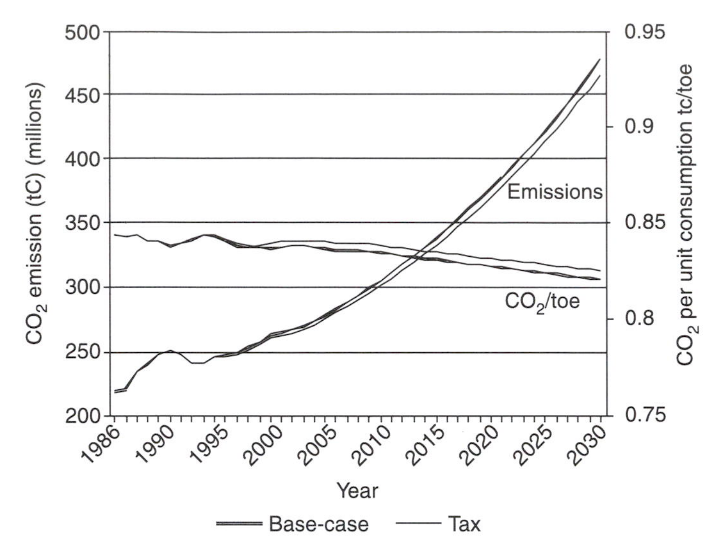

Figure 4.8 CO_2 emissions and intensity in Japan

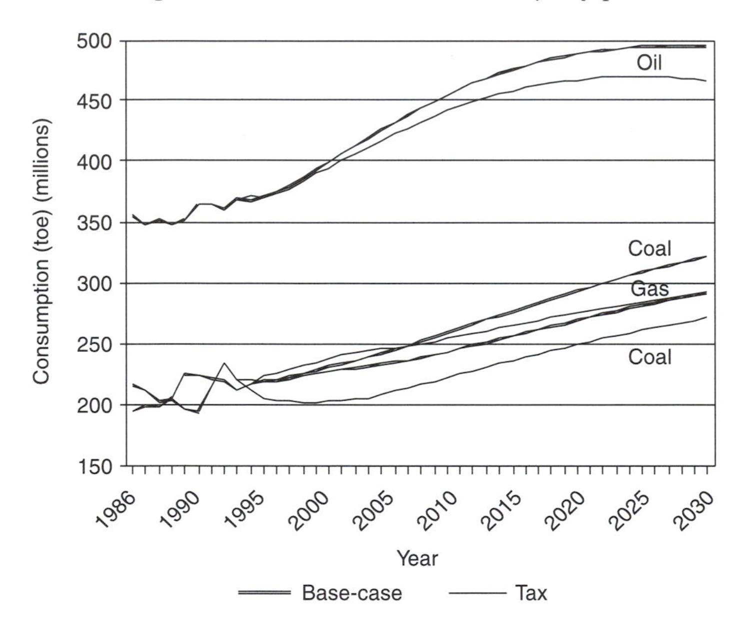

Figure 4.9 Fossil fuel consumption for EC5 (France, Germany, Italy, Netherlands, UK)

increases. As discussed above, this is due to differences in the prices and cross-price elasticities of the fuels.

In a uniform world a flat-rate tax theoretically should be an efficient means of reducing emissions at least cost (Hoel 1993) as compared to setting an equivalent uniform target. Consumers in the long term im-

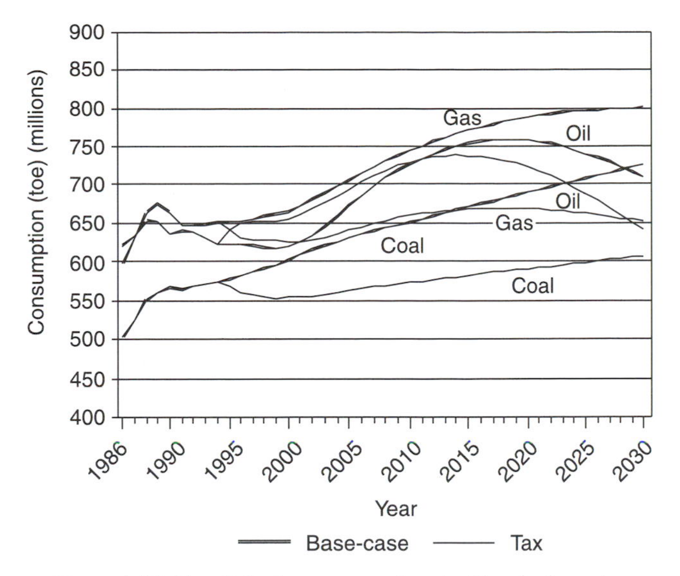

Figure 4.10 Fossil fuel consumption in North America

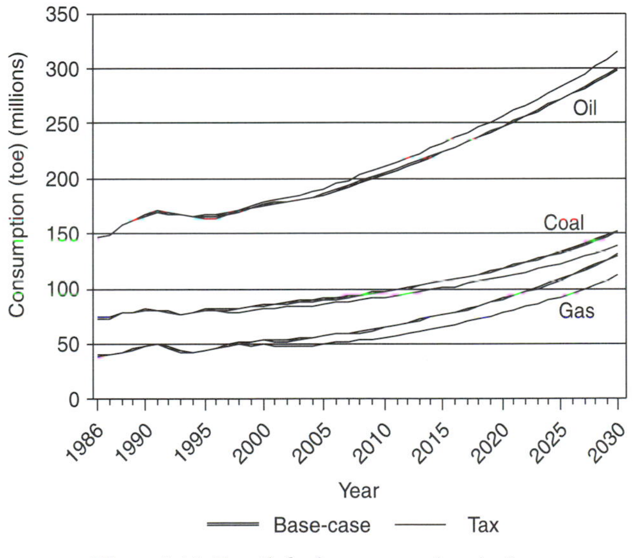

Figure 4.11 Fossil fuel consumption in Japan

plement those emission reduction measures costing less per tonne of carbon than the tax rate, so internationally the cost will be minimised. However the heterogeneity of the results above does not appear to tally with this theoretical framework, as the tax is only optimal in the case of perfect international markets (Golombek *et al.* 1993). The differences between countries are in part due to physical differences (resource base, infrastructure, industrial and fuel mix) and in part to other factors such as government policy (the structure of energy utilities, taxes and subsidies, efficiency regulation etc.), consumer behaviour and institutional structure. An efficient international agreement would compensate for the physical differences and the variations due to consumer preferences, in that the emission reductions would be made where they give the lowest welfare loss.

This analysis shows that the large differences that currently exist in fuel prices determine most of the dispersion of energy intensities in the developed countries; prices are partly determined by geographical reasons (resource availability, costs of providing roads paid for by fuel tax), but mainly set by government taxation policy. This poses problems for designing efficient international policy, as current differences between countries are due to policies and institutional barriers which may continue to change and are not constrained by any international agreement. The efficiency gain from a flat-rate tax will thus be undermined as it is rather like trying to build a level-playing field on top of a range of mountains! Harmonisation of energy prices (at a higher rate than at present) would provide a 'second-best' solution to the problem of reducing carbon dioxide, as this would reduce market distortions, but physical and consumer differences would be ignored. Moreover, in a more general policy analysis such a harmonisation may not be optimal as it disregards the reasons for differing national energy taxation, which include local pollution control, revenue raising, industrial policy and transport policy. Setting individual country targets and using tradable permits and/or different tax rates in each country may therefore be the most efficient solutions that are practically feasible, despite the difficulty in negotiating agreements which appear to differentiate between countries.

CONCLUSIONS

Econometric modelling of the energy sector has the advantage of measuring the relationships between economic variables empirically, which may include many interactions that have not been accounted for in a simple theoretical framework. In a second-best world, a model that relies on first-best theory may introduce large errors. Energy markets are often highly distorted and price responses vary between both countries and fuels. It has been shown that carbon taxes may have counter-intuitive results due to complexities in fuel markets, even in some cases leading to an increase in CO_2 per unit of energy. There are large differences in the responses made in the countries studied which appear difficult to reconcile with the assumptions behind economic optimality of international flat or harmonised carbon taxes.

Atkinson and Manning (1995) give an excellent review of energy elasticities in the literature, showing wide variations in values obtained according to the functional form, fuel, sector, estimation period and whether cross-country or time-series data is used. In this study, values for overall energy elasticity have been found that are considerably lower than the commonly accepted long run value of around −0.5, chiefly due the cointegration technique used, and the specification in terms of fossil fuel intensity. In the past, the main impact of the oil price shocks has been to reduce GDP in oil importing nations, due principally to trade balance effects. This leads to a decrease in energy consumption, which has been wrongly associated with the change in prices. In effect, what these models implicitly assume is that energy demand is only reduced at the expense of GDP. As a carbon tax with revenue recycled within the economy will not impose such a large GDP shock, the impact on energy demand will be smaller than external price increases. Using a model formulated in terms of energy intensity as a function of price produces a much lower elasticity as has been found in other studies such as Neuburger (1992).

The endogenous technical progress model has a novel functional form that looks at how energy intensity declines as a result of high prices, leading to irreversible improvements in energy efficiency. The long term outcome of raising energy prices is thus decomposed into a conventional demand response and induced technological progress. The former comprises both short and long run effects as determined by the elasticity and the dynamics of the equation, while innovation is characterised by a continuing decrease in energy intensity while prices are high. The model in effect endogenises the autonomous energy efficiency improvement or AEEI used in many models. As Chapter 3 demonstrated, past modelling of long term emissions has been crucially dependent upon the value chosen for the AEEI, as it directly specifies the relationship between energy use and output. In most studies its value is has been set fairly arbitrarily at ≈ 1 per cent p. a., varying for different world regions. Whether or not the value chosen accurately reflects existing trends, specifying the value exogenously excludes the possibility of technological development being affected by policy. It also means that there is no viable solution to stabilise carbon emissions in the long run, unless energy prices rise faster than economic growth for ever. This is unrealistic because at a certain price level fossil energy consumption will stabilise due to the superiority of non-polluting alternatives. Such 'backstop technologies' have often been exogenously defined in GE models (which use standard energy demand equations) in an attempt to overcome this limitation, but the assumptions required to do this further undermine the empirical relevance of these models. In contrast the endogenous trend model does allow energy use to stabilise for non-growing energy prices in an economy with steady growth, thus incorporating the idea of a 'backstop technology' inside an empirical framework with estimated dynamics and lags in price impacts.

The fact that long run direct energy elasticities are as low as −0.15 make taxes a rather ineffective means of bringing about the large reductions in energy intensity required for atmospheric CO_2 stabilisation. Inter-fuel substitution away from coal and into gas has an important role to play in the short term in some countries, but its effect is limited. In the longer run, induced technological innovation will have to play an increasingly important role, but this requires very long planning horizons. Commitment to a regime of rising carbon taxes would allow the private sector to invest in such speculative R & D, but as long as there is regulatory uncertainty private investment will be sub-optimal. One possible policy option to overcome these small energy elasticities is to recycle a proportion of carbon tax revenues into energy saving investments, in the form of public sector investment or grants/subsidies to the private sector. Institutional changes and areas such as building regulations, energy labelling, and transport policy could also play a major part, as would stimulation of energy efficiency research which would be unprofitable for private companies due to knowledge externality effects. These policies have the economic advantage of overcoming market failures which are included in the elasticity estimates given above; these include the problems lower income households have in financing energy related investments, and the perverse incentives when living in rented or public sector housing.

5

MODELLING THE MACROECONOMIC IMPACTS OF CARBON ABATEMENT

INTRODUCTION

As Chapter 3 made clear, comprehensively modelling the macroeconomic effects of carbon taxes is a complex task because fossil fuel based energy, though only a small component of the economy by cost, is used in all productive sectors and also as a direct consumption good. As energy taxes are raised, and energy use declines, there will be long term structural changes in the mix of inputs to production, types of energy goods consumed, total output of the economy and the prices of traded goods. It should also be expected that the large shifts in relative prices and sources of taxation revenue will cause short to medium term inflationary, monetary and fiscal effects. When considering the political economy of carbon taxes (as opposed to the long run economic equilibrium) the magnitude of these short term effects has important policy implications.

Analysis of the efficiency of different policy instruments to control carbon dioxide emissions has classically focused on the direct costs of investment in energy efficiency and compared this to projected increases in global consumer welfare from introducing CO_2 abatement measures. However, because energy is both a direct consumption good and an input to production, there will also be broader macroeconomic impacts from introducing a tax. Macroeconomic impacts are defined as the effects of reducing fossil energy use on gross economic output, or the productivity of other inputs, such as labour and capital. These macroeconomic impacts will vary depending on the distribution of abatement costs between countries, which in turn are a function of which policy instruments are used to co-ordinate action at the global level.

An internationally set but domestically collected and recycled tax (with no side payments between countries) results in no fiscal loss to an economy. Output losses come in reduced competitiveness in some industries and in a move to a less productive economy as investment shifts from improving labour productivity to increasing energy efficiency. As fossil fuels will continue to be used in the foreseeable future so carbon tax revenues will be substantial. If these revenues are recycled into reducing other economic distortions, such as employers' labour taxes, as opposed to being given back to households in a lump sum, then the net effect on macroeconomic output could be minimal or even positive (Barker 1994). The potential for positive output effects from recycling carbon taxes is contentious as some theoretical economic models deny such an effect is possible (Ligthart and Van der Ploeg 1994, Bovenberg and Goulder 1994). These negative results are usually driven by an initial assumption that the labour market clears at the given wage and existing mixes of taxation are roughly optimal; relaxation of these assumptions can generate models which allow recycling of tax revenue to offset completely the direct costs of abatement even without taking into account environmental benefits (Carraro and Soubeyran, 1994). Therefore, there are three components to calculating the cost of controlling CO_2 at the macroeconomic level: direct welfare costs, macroeconomic impacts and revenue recycling benefits.

This chapter describes how the structure of the existing LBS/NIESR GEM macroeconomic model (LBS 1993) has been augmented to model the long run macroeconomic impacts of carbon taxes. The first section focuses on how prices, investment, trade and the labour market respond in the short to medium term due to increased energy taxation. The second section describes the theoretical problems associated with constructing a consistent econometrically based supply side, which will accurately simulate the long run effects of an energy/carbon tax. Two different methodologies for modelling the supply side are then described in detail. The final section looks at the simulation properties of the macroeconomic model under several policy scenarios, and discusses the influence of different theoretical assumptions on the economic impacts of reducing fossil fuel use.

EGEM'S MACROECONOMIC MODELLING STRUCTURE

EGEM (Environmental Global Economic Model) consists of detailed aggregate macroeconomic models of the main world economies,[1] which includes their demand for energy, and simpler models of twelve regional groupings of the remaining world economies. A full world trade matrix connects the economies to ensure consistent use of globally traded services, manufactured goods and commodities.

In common with many other large multi-country econometric models the original model structure of GEM lacked an explicit supply side based on either a cost or a production function. Output was instead determined from a disaggregated IS curve (with inflation effects), which related production to past wages, wealth, investment and trade. Growth was driven by the accelerator and multiplier effects of investment, and by labour augmenting technical change (Harrod neutral) in such a way as to give constant returns to scale in capital and labour. These relationships constituted an *implicit* supply side, but did not have sufficient detail to look at the effects of changing energy use.

In order to model substitution effects explicitly the original structure of GEM needed to be augmented with a consistent model of the supply side of the economy, which could be estimated from observable data. The philosophy behind the construction of the model was to try to limit the restrictions placed on the econometrically estimated factor demand equations, while ensuring that basic theoretical properties, such as constant returns to scale and homogeneity, were present in the completed structure.

Output, welfare and equilibrium in EGEM

EGEM models the economy in aggregate with no sectoral breakdown of production and works solely in monetary units, as its equations are estimated from national accounts data. Therefore, there is no measure of household utility and the implied measure of global welfare is economic activity expressed as GDP/GNP. GDP is based on the standard national accounting identities, with all components except government expenditure and government investment being endogenously calculated in the model; government investment and expenditure is calculated based on historic levels adjusted for transfers to the unemployed.

For each country in the model the long run level of domestic economic activity is driven by labour force growth, exogenous technical progress and expansion in world trade. Labour markets do not a priori clear, so persistent involuntary unemployment is possible, due either to high real wage costs (which can persist due to bargaining effects) or to insufficient demand, and this is indeed a feature of model forecasts. There are no resource constraints on the growth of the productive sector, though the price of traded commodities (including oil) rises slowly with increased global consumption.

Real output in each country is equal to the sum of consumption, investment, trade in goods and services and balancing financial asset flows to the rest of the world. Output grows over time because nominal compensation to employees grows with exogenous productivity growth. In response to the rise in income, investment also rises and the effects of these increases feed into the trade and financial sectors. Below we discuss the crucial equations governing the short to medium term behaviour of EGEM: that is, the wage, price, employment, trade and investment equations. All these equations are represented by error correction models based around cointegrating relationships, and so have unique long run solutions.

Wage/price responses to energy taxes

The form of the wage equation determines how much workers are able to offset the rise in consumption prices due to energy taxation with higher real wages; if wage increases outstrip productivity growth, unemployment and inflation is likely to increase in the medium term. Disregarding long run output effects, revenue recycling through direct taxation should leave real disposable income to consumers unchanged. However, the dynamics of price adjustment can lead to increases in perceived purchasing power, and inflationary or deflationary spirals in the short to medium term.

In EGEM, wage equations are based on a 'bargaining' type model in which real wages rise with labour productivity growth and fall with unemployment. Nominal wages depend on factory gate prices (Layard *et al.* 1991), and the inflation rate has a positive long run effect on nominal wages in some countries. In the long run wages do not rise to account for increases in consumer taxes; that is, the 'wedge' is equal to indirect consumer taxation, import and energy prices. In the short run however, import prices and indirect taxation can inflate wages. The nominal wage/price system is indeterminate and prices increase proportionately to changes in labour, energy and import costs. Prices also rise to reflect demand pressures in the economy, which are measured by capacity utilisation; this is defined as the ratio of industrial production to potential output calculated from the supply side equations described below.

The form of the wage equation is critical when tax recycling through employers' labour tax contributions is modelled. The imposition of a carbon tax effectively reduces workers' real wages if it is not recycled, because the rise in the aggregate price level is uncompensated. If the tax is recycled through employers' labour tax contributions then in a perfectly competitive market, and ignoring changes in factor usage, output prices would drop proportionately to the tax increase, leaving the aggregate price level and thus real wages unchanged in the long term. However, in an imperfectly competitive market some of the savings in labour taxes may be retained by firms as extra profits, or workers may expect this to happen. Therefore, prices (or expected prices) will be permanently higher than before the tax was imposed, and there will be pressure for wage increases from the workforce with consequent effects on inflation, monetary policy, competitiveness and employment. Such expectation effects can be modelled in EGEM by an adjustment to the wage equation that allows long run changes in the wedge to be lower than the change in indirect taxes. If lower producer wages from labour tax recycling give any reductions in unemployment this will also put upward pressure on wages because the expected cost of unemployment to workers has fallen, and so their bargaining position has strengthened.

The effects of tax recycling are different for the employed and the unemployed, and this has importance for the political economy of imposing energy taxes. Employees would prefer to see taxes recycled through direct taxation, as this would directly compensate them for higher prices. If the revenue is recycled through employers' labour tax contributions the effect on real consumption wages is ambiguous; if tax reductions are passed through in prices there will be no net change in real incomes, but if some is retained by firms as profit then existing workers will suffer an absolute pay cut. On the other hand any unemployed workers who find work due to lower producer wages are unambiguously better off, whatever the change in prices.

Compared to lowering employers' labour taxes, reducing direct taxes is unlikely to produce much new employment, as real consumer spending on non-energy goods will be unchanged and leisure/work trade-offs are assumed to be small in an imperfect labour market. Therefore, with labour tax recycling the largest welfare gains are likely to accrue to initially unemployed workers who gain work due to reductions in labour taxes; there will be some benefit in this for existing employees as transfer payments will drop with unemployment. If price reductions are fully passed through to consumers, reduced transfer payments will give direct economic benefits to existing employees and so labour tax recycling will be superior to direct tax recycling for both groups. Even if firms raise their profits, labour tax recycling should lead to greater overall economic growth compared to recycling through direct taxes, as the taxation system will be less distortionary. Therefore, from an economic perspective recycling through employers' labour taxation is most likely to be the superior policy, but will only be acceptable to existing employees if the price of non-energy intense goods is expected to fall significantly as a result of tax recycling.

Employment equations

As described above, the gains and losses from tax recycling and decreases in direct taxation depend on the reaction of the labour market. As with most macroeconomic models we assume the labour market to be non-clearing and involuntary employment to exist. As household utility is not modelled there is no leisure/work trade-off which would give efficiency gains if direct labour taxes are lowered. In much of the theoretical and CGE work in this area, the demand-side changes in work patterns are the dominant determinant of the amount of labour available to the economy; this assumption leads to a rejection of the hypothesis that tax recycling strategies can increase employment and also raise output (Ligthart and Van der Ploeg 1994, Bovenberg and Goulder 1994).

The original employment equations in GEM were based on the profit maximising labour demands of an inverted production function (Whitely 1994); capital is substituted out of the equations so that employment is dependent on output, technical progress and the real wage. Constant returns to scale in labour are imposed on the equations so that employment increases proportionately to GDP, with the available labour force growing at past demographic trend rates. Employment decreases with increases in real wages, so short run wage inflation due to energy taxes increases unemployment, which in turn depresses wages to an equilibrium level. For all countries the elasticity on real wages is always greater than -1, implying that the underlying technology of production is not Cobb–Douglas, but can be represented by a value-added CES function. The employment effects of recycling carbon tax revenues through employers' labour taxes are modelled as decreases in the real wage as seen by employers, but nominal compensation received by workers is unchanged. The augmentation of the employment equations to include the effects of energy price rises on long run employment is described in detail below.

Trade equations

EGEM models world trade as flows between the G9 countries and seven other country or regional groupings. Trade is divided into visible and invisible sectors, with invisibles being further subdivided into non-factor services, returns on overseas assets and unrequited transfers. Volumes of trade are based on 1980 trade patterns and are affected by costs and market growth. Essentially visible exports from each trading block depend on demand in traditional import markets and relative labour and 'export' prices in the trading countries. 'Export' prices reflect the mix of manufactures and commodities in each country's trade; manufacturing prices are determined by imports, domestic energy and labour costs; while traded commodities are priced in world markets. Visible imports into a country are determined by domestic real incomes, export prices in traditional trading partners and relative labour costs. Invisible imports and exports are determined by growth in world income and relative consumer prices.

Increases in energy based taxes affect trade by increasing producer prices, though any increases will be offset by changes in labour costs if taxes are recycled through employers' labour tax contributions. Carbon taxes are both consumer and producer taxes, so using all their revenue to reduce employers' labour taxes effectively provides an export subsidy compared to the baseline case. The size of the export price drop will depend on the relative energy usage in households and industry. In EGEM the split of energy use between the demand and supply sides of the economy is determined by coefficients for each of the main three fuels derived from input–output tables. The proportions of each fuel used in the productive sector does not change over time, but, as the consumption of different fuels changes due to the imposition of a carbon tax, the proportion of energy costs borne by consumption and production will change. If carbon tax revenues are recycled through direct taxes it would be expected that imports would increase due to demand growth, while exports decreased due to producer price rises.

If real exchange rates are fixed, a change in relative prices between countries leads to a permanent change in each country's trade balance. EGEM is run in this mode when simulating the effect of carbon taxes, because a devaluation of currency to regain a current account balance reduces the relative welfare of consumers by lowering their international purchasing power. As EGEM contains no measure of household utility this welfare change is approximated by comparisons of GDP changes given constant real exchange rates. This method of simulation allows welfare effects to be approximated as real income losses and avoids having to model exchange rate reactions to interest rate and price changes inside EGEM, which greatly simplifies the model and improves its stability in the long term.[2]

Investment equations

Aggregate investment is divided into business, housing and government in EGEM (there are also equations for stockbuilding), with government investment being set exogenously and funded through taxation. Business investment is proportional to GDP in the long run but experiences greater shifts in the short term; it is also affected by changes in long run real interest rates. Long run interest rates are linked to short run rates, which in turn are set in order to stabilise inflation by targeting money supply growth.

This structure reflects an 'accelerator' modelling of new investment plus replacement of fully depreciated capital. Housing investment is determined by short run interest rates and consumer prices. Analogously, consumption is proportional to real disposable income with effects from inflation, interest rates and real wealth where appropriate; consumers are seen as the owners of all corporate capital and so dividend and equity income changes are modelled in this area.

The strength of inflationary effects from imposing energy taxes will depend on both the energy intensity of the country modelled and that of its principal import partners. The changes in interest rates needed to control inflation (fiscal policy is targeted at maintaining a constant government budget ratio) will depress investment and consumption, but increase real wealth, leading to complex and long lasting effects on the demand side of the economy.

Long run growth properties

Along with trade competitiveness the majority of attention in modelling the impact of carbon taxes has been to investigate the effect of energy price rises on long run growth. As the existing structure of EGEM adequately describes the short/medium term dynamics and price reactions in the economy, adding a consistent supply side allows analysis of all major macroeconomic interactions.

In the original GEM model there was no explicit aggregate production function, though the form of the employment equation can be derived from a CES production function between labour and capital. The lack of an explicit supply side meant that investment only affected aggregate annual demand directly, and not the implied capital stock. Under these assumptions all technical productivity improvements can be considered to be associated with labour usage (Harrod neutral). Real wages are driven by labour productivity, and therefore in the long term so are consumption, investment and total output. When solved analytically the long run growth rate in this type of economy is a multiple of labour productivity and demographics, as in the Harrod–Domar growth model.

The system of equations derived below augments this growth model to include explicitly the effect of capital accumulation and total factor productivity, and ensures factor demands and economic growth change in a logical way as relative prices are altered. Given the number of different approaches which have been used to model these effects a short literature review and history of other studies is given to place our work in context.

SUPPLY SIDE MODELLING, DERIVATION AND ESTIMATION

There is a major difference between modelling the results of consumers' and producers' reactions to energy taxes in an econometric model. While robust energy demand equations can be estimated from observed data, as in Chapter 4, the quantitative interpretation of these equations in terms of welfare and output is rather more difficult. This is because the underlying structures of the economy, that is, the relative preferences of consumers for different goods and the set of production possibilities open to producers, are unobserved and can only be inferred from empirical data by invoking assumptions about aggregation, rationality and functional forms.

On the consumption side an increase in the price of directly consumed energy (petrol, electricity, heating fuel, etc.) will lead to a shift in the basket of goods bought by households; consumers will drive less, heat their homes to lower temperatures or invest in double glazing. If we assume that the original consumption bundle was optimal then this shift will obviously decrease the welfare of consumers. The range of this decrease will be from zero, the consumer is indifferent to the shift, to the total consumer surplus lost by the decrease in energy consumption; that is, the integral under the energy demand curve over the price increase. As we have estimated the energy demand curve we can calculate the consumer surplus and then hypothecate a proportion of this as being 'lost' in the substitution into other goods. However, this proportion cannot be directly calculated without estimating a consumption function for all major groups of goods in the economy. There are many known market failures in consumers' use of energy, mainly because it is such a small part of consumption that the transaction costs of increasing efficiency are proportionately very high. These failures tend to make consumers' energy demand less elastic than it would be in a perfect market. Therefore, estimating a utility function using rigorous market clearing optimality assumptions will lead to an over-estimate of the utility that consumers will lose from reducing energy use.

Modelling the supply side is more theoretically supportable, because companies have proportionately lower transaction costs, greater access to information and higher incentives to rationalise their resource use. Assuming that producers optimally combine their available inputs (neglecting for the moment dynamics and accumulation processes) so as to produce a final product, the rationality condition argues that at equilibrium:

$$\partial Y/\partial E_w = \partial Y/\partial M_w = \partial Y/\partial L_w = \partial Y/\partial K_w$$

where $\partial Y/\partial E_w$ is the marginal product of a cost weighted unit of energy (e.g. a dollar's worth of energy), M denotes other material inputs, L labour input and K capital input. Therefore, at equilibrium the marginal cost weighted products of each input to production are equal, and there are no changes in input mix which would increase the efficiency of production. Increasing energy prices will move producers away from equilibrium, and so they will substitute out of energy into other inputs until this condition holds again. If an energy tax decreases energy use by $-\Delta E$ the resulting loss of output ΔY will be approximated by:

$$\Delta Y = \Delta E.(\partial Y/\partial E - \partial Y/\partial O.\partial O/\partial E)$$

where all variables are in physical quantities, O represents the new mix of other inputs (M, L & K) and the cost weighted marginal products are again equal. Energy use will only decline if the marginal product of a unit of energy is less than its after-tax cost; so, for a constant output, the total direct economic loss to individual producers must be less than the taxation level multiplied by the decrease in energy use.

When a firm faces inelastic demand the average benefit to the firm from saving a unit of energy is always less than the tax rate, as substitution into other factors is costly but never more costly than the tax. If demand is elastic then as a firm's price rises it will lose market share, thus raising the per unit cost of energy taxation to the firm; these costs will increase as the elasticity of demand falls. For a profit maximising firm with constant returns-to-scale which charges a constant mark-up on each unit of output, the optimal level of energy saving will remain constant whatever the elasticity of demand (see Appendix 5.1). If companies are maximising not simply profit but also value market share or turnover, they will save more energy than the equilibrium conditions above would suggest. In this case behaviour will depend on the expected loss in sales, which is a function of the firm's assessment of its rivals' pricing behaviour, the extent of competition from imports which may not face an energy tax and the elasticity of demand in the market.

The above arguments hold for individual companies, but the macroeconomic costs to the economy may be larger than those estimated at the microeconomic level. This is because a shock such as imposing large energy taxes will have macroeconomic externalities which are not controllable by individual producers; inflation may rise, necessitating higher interest rates and lowering investment, which will have a knock-on effect on the supply side. Output may fall further due to terms of trade effects, and reducing unemployment through substitution into labour could stimulate inflationary wage increases. Balancing these negative effects are the positive benefits from recycling revenues to remove existing taxation distortions in the economy. Along with demand-side and fiscal effects it is also possible that there are production externalities in the economy from changing patterns of relative prices. Therefore, though each company makes the rational decision to save energy, the diversion of investment away from other factors reduces (or, conversely, increases) the rate of sectoral productivity growth. Such investment, innovation or knowledge spillover effects between companies could reduce potential output by a larger amount than would be expected from microeconomic analysis. Conversely, energy taxes could stimulate new innovations which lower *all* production costs, giving a win–win solution. Such a process cannot be modelled inside a traditional general equilibrium/scarcity approach to factor allocation and pricing, and so this response is a priori rejected by most modellers.

By estimating cross-restricted demand equations for all the relevant inputs into production an underlying aggregate production structure can be inferred, and output losses from substitution calculated as above. We would expect the unit cost of saving energy, discounting demand effects, to be below the tax rate if companies are assumed to be rational profit maximisers. However, this simple picture is complicated by the potential for positive and negative externalities from energy taxes, and by the fact that estimation is over historical periods when actors had certain expectations about future energy prices (usually that they would continually rise) which drove their behaviour. It is not always relevant to extrapolate theoretical restrictions on factor substitution derived at the microeconomic level to macroeconometric estimation. As the review below shows, the theoretical concepts invoked to allow the estimation of aggregate production functions are varied, difficult to pin down and more often honoured in the breach than in the observance.

Previous econometric work on factor substitution

Much of the early research into the cost of reducing carbon dioxide emissions has been performed on computable general equilibrium (CGE) models which were calibrated either by abstracting from econometric studies or by fitting them to a particular year's data (see Chapter 3 for a review of this work). The empirical relevance of this approach rests on assumptions of optimal use of production technology, infinitely malleable capital and clearing labour markets, and these studies generally avoid shorter term dynamic, fiscal or monetary issues. As interest has grown in the precise magnitude of these macroeconomic costs, researchers have focused on basing their models more firmly in empirical data (e.g. Jorgenson *et al.* 1993, Barker and Gardiner 1994, Barker and Madsen 1992, MERGE 1992) by attempting to estimate multi-factor production functions rigorously and econometrically. However, this has proven to be fraught with difficulty, both due to problems with data (especially on capital stocks) and because the current 'technology' of production functions place great restrictions on the data (Chung 1994). It should also be remembered that there is no reason why a theoretically consistent production function should exist at any level above that of a single technology firm; assuming the existence of such a production function imposes the assumptions of homotheticity and homogeneity of production technology in every sector to be aggregated (Fisher 1965).

These problems outstanding, many authors have aimed to estimate production functions at the industry and aggregate level using Labour, Capital, Materials (or Intermediates) and Energy as primary factors of production. A useful survey by Chung (1994) of the most sophisticated work using flexible trans-log production/cost functions shows that there is general agreement that labour is a substitute for both capital and energy, but no agreement on the relationship between capital and energy.

Economic theory would tend to argue that capital and energy are substitutes if they are separable primary factors, and micro-level engineering research emphasises the need to invest capital to improve in energy efficiency. However, many studies at the aggregate level have found them to be complements, while sectoral studies on the same data showed a mixture of results (e.g. Pindyck 1979). Divergent results coming from similar data sets and analytic techniques seem to be mainly caused by differences in a priori factor separability assumptions; these reflect different researchers' opinions about the productive sector as well as limitations in data. The most common split is to separate the functions into two bundles of capital/labour and energy/materials. It should be noted that estimating a trans-log function restricts the marginal rate of technical substitution between the bundles to one, as the trans-log reduces to a Cobb–Douglas function of the bundles (which are trans-log); this type of restriction has been rejected for sectoral estimates in the USA by Jorgenson *et al.* (1987).

Intuitively the existence of historical complementarity between energy and capital seems reasonable, because if productivity growth is embodied in energy-using machines this would imply more energy use as capital use intensified. If relative prices are stable, or the factors are relatively price inelastic, these growth effects could statistically swamp price effects over the estimation period. Though historically plausible this result is of little use in modelling economic responses to carbon taxes, as 'bottom-up' and engineering studies advocate increased capital spending to save energy. Griffen *et al.* (1976) try to reconcile these results by describing the complementarity as a short run effect during which the capital–energy bundle is substituted for labour; while in the long run capital is a substitute for energy. Griffen *et al.*'s work found capital–energy substitution on cross-sectional data for the G7, but the coefficients were statistically insignificant. Chung (1987) has found statistically significant capital–energy substitution, but the elasticities were very small (0.0045) and varied markedly with different types of restriction. More advanced work on system estimation of derived demand equations, using a fully dynamic specification of a trans-log cost function, has shown the measurable elasticity of capital to energy to be insignificant when the non-stationarity of the data is removed by establishing cointegrating relationships (Allen and Urga 1995).

Many of the estimation problems discussed above could result from the timescales of data available. Traditionally, price elasticities of goods have been sub-divided into short run (1–2 years) where only variable factors of production can change, and long run (5–15 years) where fixed factors of production change. However, the above work on energy elasticities seems to point to a much longer delay in factor substitution between energy and capital, with long run changes in total capital investment unobservable twenty years after a price shock. This may mean that in the long run the absolute amount of capital investment is not elastic to energy prices, or it could just be a data measurement and timescale effect. Some researchers (e.g. Griffen *et al.* 1976) aim to overcome this by arguing that cross-country data, due to its great range of energy intensities and prices, gives estimates of very long run substitution effects. However, the validity of this approach is questionable both econometrically (Peseran and Smith 1995) and because the historic energy consumption of different countries has been as much determined by geography, climate, industrial shifts, policy choices and the availability of indigenous resources as it has by relative prices.[3]

The difficulties of finding robust and flexible estimates of multi-factor substitution elasticities should not be surprising given the relative value shares of the factors; in most advanced economies energy makes up only 5–7 per cent of GDP and an even smaller proportion of inputs to the productive sector. Given these difficulties, and the sensitivity of results to restriction specification, the extraction of energy–labour and energy–capital elasticities from econometric work to be used as 'stylised facts' in numerical general equilibrium models must be viewed with concern. These elasticities have little economic meaning as stand-alone numbers and depend on the exact form of estimated equation, or system of equations, for any statistical relevance.

Capital–energy substitution as a shift in embodied technical change

The difficulty of finding macroeconomic evidence of the capital–energy substitution, which microeconomic and engineering studies have been describing for years, gives problems for grounding any model in empirical data. One answer is just to estimate energy–labour substitution elasticities and impose a restriction of no direct effects on capital. Unfortunately, this leaves the mechanism for saving energy by increasing labour input unexplained, and restricts simulations to short time periods. With no change in the capital stock we are also basically assuming that human muscle power is a direct substitute for fossil energy in modern societies! Of course the observed substitution between energy and labour could be the result of changes in consumer preferences towards goods with lower energy intensities, and/or the destruction of energy intense industrial sectors and the growth of labour intense industries due to competitive pressures. The evidence of such industrial shifts, which would include relocation of industries such as chemicals to countries with lower energy prices, is tenuous as energy is only a minor determinant of costs except in a small sector of the economy (Smith 1994).

A different approach is to ask why such a clear micro effect is so difficult to measure at the macro level. One possible answer is that the *total* level of capital accumulation is reasonably steady relative to energy price changes, but that the *type* of capital equipment used changes from being labour productivity enhancing to being energy productivity enhancing. This is consistent with optimising behaviour by firms which have a constraint on their total investment budget relative to output, or economy wide behaviour when there is an exogenous, or price insensitive, macro-constraint on borrowing imposed by the current saving rate or government monetary policy. Therefore, a change in relative variable factor prices may redirect investment to the most profitable area, but leave aggregate levels untouched.

Without disaggregated data on the purpose of each investment this split cannot be observed at the macro level. However, the results of decreased investment on trend labour productivity should be measurable, and this could explain why robust parameter values for energy/labour substitution can be estimated. Because this substitution is not direct, but is transmitted via capital accumulation, the delays and lags on the process are much greater than if energy and labour were directly substitutable variable factors. This explanation of energy substitution as the diversion of investment flows is consistent with the energy consumption equations estimated for EGEM; these showed very small short run price elasticities with strong trend effects. These trend effects could be explained by relative price movements but not by changes in overall investment patterns.[4] The endogenous efficiency term in the energy demand equations can therefore be interpreted as a price induced switch in embodied technical progress and diffusion from labour productivity enhancing to energy saving. This process is not equivalent to using a traditional fixed capital production function which has very slow dynamics of adjustment between equilibria. This is because in these functions price shocks do not permanently affect equilibrium values once removed, and a continuously rising tax is needed to give continual reductions in the energy/GDP ratio.

In reality it is very difficult to measure how much energy efficiency is due to true innovation stimulated by price increases, and how much is due to reversible technical diffusion caused by increased investment. This means that the strong interpretation of the endogenous technical change model, which attributes all efficiency improvements to technical innovation which can be applied costlessly into the future, may only hold over simulation periods where technical diffusion can be considered irreversible, up to around forty years.[5] On the other hand, models using simple energy elasticities overstate the costs of controlling CO_2 by assuming that prices have no effect on the rate of underlying technical change in the energy sector.

Measuring embodied labour productivity

If the mechanism described above is considered a plausible explanation of energy–labour substitution, then modelling this with a constant price elasticity in a traditional production function will lead to significant errors over the medium term. These errors arise from two factors: firstly the question of dynamics and reversibility outlined above, and secondly the effect of embodied technical change on labour productivity growth.

If a significant proportion of past productivity growth has come from technical progress embodied in capital, then the cost of diverting investment from labour to energy saving capital (per unit of investment) will become higher as more energy is saved. Expressed in elasticity terms, this means that the price elasticity of substitution between energy and labour will increase over time, because more workers will be employed for every additional unit of energy saved. In a standard production function this would mean that it is less costly to save energy, because overall exogenous labour productivity is constant, and so with increased employment output will be less affected. In the endogenous technical change model it means that the shift in capital usage decreases labour productivity, and so a similar shift in employment imposes higher costs on the economy than would be calculated in a model with exogenous technical progress.

To simulate the effect of energy price rises on labour productivity two things are needed: firstly a measure of the amount of investment diverted to energy productivity from labour productivity, and secondly an estimate of the influence of capital on labour productivity. The terms describing changes in factor productivity are usually defined at the macroeconomic level where disembodied technical change implies increases in output for the same mixes of inputs, while embodied technical change implies an increase in the capital–output ratio as more is invested in new machines. However, these categories are not very helpful in understanding microeconomic behaviour as they do not identify whether technical progress is associated with new machinery or with scale and organisational effects involving no new investment.

Considering the simple case when there is no capital depreciation, if new more productive equipment is bought at *the same cost* as an older machine doing the same job, then this is clearly embodied technical progress, but at the macro level such an effect could be attributed to either total factor productivity or improvements in capital or labour. If the new machine costs more than before, but displaces labour or materials, then productivity improvement can be measured at the macro level either as a substitution of capital for labour or as embodied technical change. Depreciation further complicates the measurement of technical progress because capital is constantly being replaced by new vintages of investment, which will be more productive if technical change is embodied in physical capital. Therefore, for no increase in the measured capital stock an increase in productivity would be observed, supporting a disembodied measure of technology. The simulation properties of such a model would be in error however, because the effect of investment on output growth would be vastly understated.

As well as such identification problems, the response of markets and prices to technical change also complicates its measurement and interpretation. If technical progress increases the average productivity per unit of capital invested, then simple optimal capital market theory tells us that interest rates should rise to reflect this increase in marginal productivity. At the same time inter-temporal optimisation of savings decisions implies that the marginal productivity of capital should equal the social discount rate (adjusted for taxes on investment income); as this is fixed the capital market should clear at the same interest rate but at a higher level of saving and investment. Furthermore, if wages are linked to labour productivity (which they have been historically in Europe and Japan) they will increase with embodied technical progress. This will give a price substitution effect away from labour to capital investment, unless the rise in real wages increases the price of capital goods by a similar amount. The eventual equilibrium will be a function of both technical efficiency and the transmission of productivity changes through the labour and capital goods markets. This will eventually decide the relative costs, and therefore productive mix, of capital and labour (see Chang 1970 for theoretical discussion of two sector models and the measurement of technical progress).

In empirical studies the assignment of such effects on output and factor mix to price substitution and different modes of technical change is determined largely by the a priori assumptions of the researcher regarding functional form and the restrictions (returns to scale, separability, etc.) placed on the estimation. The actual underlying mechanics of the productive sector are unidentified, especially in single equation modelling, and because of the complex interplay between technical change, investment and markets all the factor demands would have to be estimated simultaneously to begin to tie down the underlying structural parameters. These effects are further complicated when ideas of learning-by-doing, spillover investment returns and knowledge externalities are included, so the path of technical change (or diffusion) in the economy becomes partly an endogenous process (Lucas 1988).

Given this complexity it is unsurprising that measuring the macroeconomic influence of embodied technical progress on labour productivity has a long and unresolved history in economics, forming the core of the famous 'Cambridge Capital Debate' in the 1950s and 1960s (Harcourt 1972). Much of this debate concentrated on the difficulty which lies in measuring how much productive capital there is in the economy at any one time, given the problems of measuring economic (not accounting) depreciation, the underlying dynamic of technical change and utilisation levels of different technical vintages of equipment. Without a set of often heroic assumptions no semblance of an aggregate capital stock can be constructed for any level of economic activity. Methodologies that attempt this task are diverse but Helliwell (1976), Griliches (1988), Jorgenson *et al.* (1987), Mayes and Young (1994) and O'Mahoney (1993) give good illustrations of some different approaches and empirical problems encountered by researchers. Hall (1968) derives the important aggregation restriction that, without information on the current rental values of all vintages of existing capital equipment, aggregation is only possible if the marginal productivity of capital in each vintage is assumed to be independent of labour costs. As was mentioned above, if there are properly functioning capital markets this is unlikely to be the case, thus throwing doubt on the application of simple aggregated production function approaches.

In summary, microeconomic studies imply that energy efficiency usually implies increased investment in capital goods and processes. However, finding robust estimates of capital–energy substitution at the macroeconomic level is very difficult, and this is probably due to structural rather than data collection problems. Theories of embodied technical progress give a plausible explanation of the observed response of the economy to energy price rises, but there are many identification problems in measuring these effects. One principal problem is the inability to define an accurate empirical measure of the capital stock, and this undermines both a traditional production function approach and one which considers embodied technical progress.

MODELLING THE SUPPLY SIDE EFFECTS OF ENERGY PRICE INCREASES IN EGEM

The theoretical basis for estimating factor substitution and embodied technical progress is very diverse, and can justify many different formulations of the aggregate production function with various sets of restrictions, all of which are difficult to differentiate econometrically due to identification problems. Because of this we have decided to use two different approaches to modelling the effect of energy price changes on the productive sector: a traditional estimated production function approach, and a simplified endogenous technical progress model with restricted capital investment.

Using two different methodologies allows us to compare the simulation properties of each method, and gives a first order calculation of the numerical magnitude of the theoretical differences. This will give a measure of the sensitivity of the model to different structural assumptions and allow the strengths of each approach to be assessed.

Estimating an energy dependent production function

The approach taken here is to use an enhanced version of the CES production function already implicitly embodied in GEM to estimate the derived demand for labour which, when combined with the existing energy demand equations, is dependent on energy prices. This equation can be expressed in log-linear form and so can be directly estimated, accounting for cointegration between these trended series using the Johansen procedure; thus avoiding the problems of spurious regressions and linear approximations to non-linear functions.

The general form of the production function is:

$$Y=f[L,K,t]^{\alpha}.(E.z[\Pi_i X_i,\pi_t])^{(\alpha-1)} \quad (1)$$

where Y is real output, L is labour, K is the total capital stock and E is fossil energy used in the productive sector.[6] Function f[. .] is a value-added CES production function with labour productivity driven by a deterministic time trend t. Function z[. .] is the inverse of the trend reduction in energy intensity (T_t) estimated separately using an endogenous representation of technological change in the energy sector (see Chapter 4). In this, fossil energy use is related to a stochastic time trend (π_t), output and relative energy prices. The full specification of this estimated equation is given below in equations 2a, b and c.

Measurement equation:

$$\Delta\ln(C)_t = \alpha + T_t + \beta_1\ln(C/Y)_{t-1} + \beta_2\ln(EP)_{t-1} + \Sigma_{ik}\gamma_{ik}\Delta Z_{i,t-k} \quad (2a)$$

Transition equations:

$$T_t = T_{t-1} + \pi_{t-1} + \gamma\ln(EP)_{t-1} + \Sigma_i\eta_i X_i + \varepsilon_t \quad (2b)$$

$$\pi_t = \pi_{t-1} + \nu_t \quad (2c)$$

$$(\pi_t = \pi_{t-1} \text{ in simulations})$$

where C is total fossil fuel consumption, EP is the aggregate real fuel price, ΔZ_i are lags of differences in consumption, GDP, and price, π_t is the stochastic exogenous trend, and X_i are determinants of the endogenous trend.

In estimating the energy equations, constant returns to scale in energy were imposed; therefore, the technology underlying the energy demand equation can be expressed as a Cobb–Douglas function, where increases in energy productivity are equal to the inverse of the total trend in energy intensity reduction. Changes in total trend energy intensity come from stochastic technical change π_t, other trend variables $\Pi_i X_i$ (e.g. structural economic change), or changes in energy efficiency caused by energy price rises. It is assumed that the stochastic trend captures disembodied technical progress in the energy sector, and the relative price driven trend captures all energy efficiency achieved by increased investment and price driven R&D. The proportion of total fossil fuel use occurring in the productive sector (E/C) is derived from input–output tables.

The two functions are combined into a Cobb–Douglas structure, implying separability of factors, and a unitary elasticity between the combined product of capital and labour and efficiency weighted energy use; with this structure the Allen elasticities of substitution between inputs will be constant over time. The advantage of the Cobb–Douglas form is its simplicity, and compatibility with the energy equations which use an endogenous technical progress trend. Unlike a conventional Cobb–Douglas representation the energy sector productivity trend is uniquely associated with that sector due to the form of the endogenous trend model, and as such cannot be simply aggregated together with the trend in total factor productivity.

However, for estimation of the labour demand equations a deterministic trend (δt) was used as a substitute for the endogenous trend structure ($g[.\ .]$) in order to simplify the derivation of the relationship.

The exact specification of the production function is:

$$Y = \gamma.[(L.e^{\eta t})^{-\rho} + \phi K^{-\rho}]^{-\alpha/\rho}.[E.e^{\delta t}]^{(1-\alpha)} \tag{3}$$

In this formulation all technical progress in the CES function is disembodied and has been associated with labour,[7] and the components of the efficiency trend for energy have been reduced to a single deterministic series. Assuming profit maximising wage setting, the following labour demand equation can be derived from the partial derivative of (3) with respect to labour,[8] where the capital stock has been substituted out using the production function in (3):

$$L = \alpha.(\gamma)^{\frac{\zeta}{\alpha}}.\, e^{\zeta(\eta+\beta.\frac{\delta}{\alpha})t}.Y^{\frac{(\rho/\alpha+1)}{(\rho+1)}}.E^{\frac{\zeta.\beta}{\alpha}}.RW^{\frac{\zeta}{\rho}}$$

Where: (4)

$$\zeta = -\frac{\rho}{(\rho+1)},\ \beta = 1-\alpha.$$

where RW are real wages paid to labour, including employers' labour taxes. All the series except labour supply are trended I(1) in all countries;[9] labour is only definitely I(1) in USA, Japan, Canada and France (low trend). However, labour intensity (L/Y) is I(1) in all countries, and as equation (4) is log linear in all variables the Johansen procedure was used to estimate a cointegrating long run relationship between the variables inside an error correction mechanism (Cuthbertson *et al.* 1992).

Initial estimates with no restrictions did not give useful parameter values, therefore a restriction was placed on the real output term (Y) to set its elasticity to unity. Given the structure of the function one would expect the profit maximising cost shares of labour/capital and energy to equal their respective coefficients in the Cobb–Douglas production function. As the cost share of energy in production is approximately 5–6 per cent for most countries, alpha will be nearly 1, and this approximation should not introduce too large a simulation error into the equations. With this restriction in place, the results for elasticities of labour to wages, energy and deterministic time trend that were found for the G7 countries are shown in Table 5.1.

The energy elasticities were quite low, as would be expected, and close to the cost shares of energy in each country. The exceptions to this were Japan and Italy which have very high elasticities which were robust to specification changes. In the case of Japan this may arise from the highly trended but short estimation period which was used to avoid the (mainly fiscal) effects of the 1973 oil shock distorting the long run results.

Constructing an endogenous technical progress model

In constructing the model used here to include the effect of embodied technical change and investment switching on labour productivity, we draw

Table 5.1 Estimated elasticities of labour: production function model

Country	*Wage elasticity* $(1/(1+\rho))$	*Energy elasticity* $(\beta\rho/\alpha(\rho+1))$	*Exogenous trend* (η) (%/yr)	*Estimation period*[a]
USA	−0.72	−0.0526	1.38	6702–8702
Japan	−0.66	−0.1810	4.41	7402–8504
Germany	−0.71	−0.0611	2.17	6702–8702
France	−0.24	−0.0499	2.67	7502–8702
Italy	−0.37	−0.1737	1.65	6804–8602
UK	−0.36	−0.0699	2.86	6704–8702
Canada	−0.72	−0.0262	3.03	7204–8602

a Estimation period figures denote year and quarter (e.g. 6702 is second quarter of 1967).

on recent ideas in the economics of innovation and diffusion. The model rests on the primary assumption that all the observed bias of technical change towards labour is caused by productivity which is embodied in physical capital. Using the same profit maximisation wage setting conditions as above, this implies that the extra output gained from investing in new machines is divided between workers and owners in a way that raises real wages; this is then measured at the macro level as a bias towards labour in productivity. Increases in total factor productivity are assumed to arise from innovation processes which do not change relative factor shares, or rewards, and do not change investment patterns.

The general form of the production function is:

$$Y = f[L,K_n,t]^{\alpha}.(E.z[K_e,\Pi_i X_i,\pi_t])^{(\alpha-1)} \qquad (5)$$

where Y is real output, L is labour, K_n is the non-energy capital stock, K_e is the amount of capital stock devoted to achieving greater energy efficiency and E is fossil energy used in the productive sector.

It is assumed that all price driven changes in *trend* energy use involve the diversion of productive capital from labour saving investments, to energy saving investments. Therefore, the implied energy investment K_e is a function of relative energy prices (EP) and can be calculated from the incremental, price driven, decrease in trend energy consumption in each year; given assumptions about the cost of conserving energy, equipment lifetime and the discount rate. Assuming that technological progress produces new innovations, or developments, at a rate which keeps the energy savings curve linear over time, the average cost of saving energy will be equal to half the marginal cost of energy. The cost efficient amount of investment in energy efficiency in each year can therefore be calculated as:

$$\sum_{i-1}^{n} (0.5\times EPi).\frac{1}{(1+r)^n}.[Ei-1.(1+z'(EPi-1)) - Ei.(1+z'(EPi))] \qquad (6)$$

where n is the lifetime of the investment, EP is the real aggregate price of energy, r is the discount rate and z'(EP) is the fraction of energy saved by price influences on trend energy consumption. In EGEM n is calibrated to the standard accounting lifetime for equipment used to construct national capital stock with the perpetual inventory method (fifteen years in most countries; O'Mahoney 1993) and the discount rate to the expected long run real interest rate over the next ten years. Since using only long run rates would bias the estimate of capital investment upwards, as they are low compared to usual commercial real discount rates of 10–15 per cent, an arbitrary mark-up of 5 per cent was added to account for this. This production technology amounts to a putty/clay capital model, because capital invested in energy productivity cannot be used to augment labour productivity in the future if relative prices change; therefore, choice of investment mix is fixed for the life of the capital.

As the energy demand equations have a long run unit elasticity on total output and a direct (that is, not in the endogenous trend term) elasticity ω on real energy prices, and defining A as an arbitrary constant, the following transformation holds:

$$E/Y = A.EP^{-\omega}.z[K_e(EP), \Pi_i X_i, \pi_t]^{-1}$$

$$E.z[K_e(EP), \Pi_i X_i, \pi_t] = A.Y.EP^{-\omega} \tag{7}$$

It follows from this that the function f(. .) in (5) may be estimated separately from the energy equations (E.z[. .]) as long as energy prices are exogenous, the implied cumulative energy investment K_e is small relative to the net capital stock (K_n), and the two exogenous trends in technical progress (π_t and η) can be considered independent. Though the direct effect of energy price changes on the net capital stock K_n can be taken into account using (6) there are also indirect effects on wages (which are related to productivity), which in turn will effect relative energy prices; however, given the small size of the energy sector, ignoring this effect should with luck give insignificant errors in the estimation. If this effect were considered large the whole system of derived factor demand equations, and the price equation, would have to be estimated simultaneously with appropriate cross-equation restrictions.

Following from the above assumptions and substituting for E.z[. .] from (7), the precise form of the production function is:

$$Y = \gamma[(L.e^{\eta t}.P)^{-\rho}+\phi(e^{\eta t}K_n)^{-\rho}]^{-\alpha/\rho}.[A.Y.EP^{-\omega}]^{(1-\alpha)} \tag{8}$$

$$\Rightarrow Y = \theta[(L.e^{\eta t}.P)^{-\rho}+\phi(e^{\eta t}K_n)^{-\rho}]^{-1/\rho}$$

$$\theta = \gamma^{1/\alpha}.(A.EP^{-\omega})^{1-\alpha/\alpha}$$

$$K_n = K - K_e$$

where all symbols are as in (3) and P is the proportion of labour productivity associated with technical change embodied in capital equipment, the capital stock K_n is not weighted for increases in productivity, but is calculated based on the perpetual inventory method. Inspection of (8) shows significant differences from the function given in (3). Disembodied technical change is now defined as being associated equally with labour and capital (Hicks neutral), and so the deterministic trend term relates to both labour and net capital (K_n). This defines P, the embodied labour productivity, as the bias in productivity growth associated with labour.[10]

The determinants of P are not obvious, but, to achieve the desired simulation properties of the model, P needs to increase as the capital share in the economy grows and to weight newer investments proportionately more than older ones. This assumes that optimal capital market operation estimates of relative capital productivity can be derived from expected real market interest rates (≈ future user cost of capital), given that companies expect to receive constant profit ratio on total investment costs (i.e. profit/(investment + interest charges)). Therefore, the expected return on current period investment over its lifetime can be calculated using obtainable series on gross investment and interest rates, and by assuming a standard investment lifetime (this is analogous to the process given in (6)). This expected return series was constructed, and then the baseline productivity of one unit of conventionally calculated capital stock, adjusted for energy sector investment, was defined as the per unit expected return in the first estimation period. The expected return on investment in the remaining periods was then multiplied by this ratio to give a series for the productivity weighted net capital stock PK_n, which is the amount of conventional units of capital needed to produce the calculated expected return on investment in each period. In all countries studied the unit productivity of capital calculated by this method was seen to grow over time. In simulations the growth in the embodied productivity of each unit of capital was given by a time trend fitted to the past values of the PK_n/K_n series, which is consistent with exogenous technical progress, not learning-by-doing.

Therefore, (8) combined with the energy demand equations allows changes in energy prices to affect production and employment. As energy prices rise energy use drops, implying a rise in investment in energy efficiency which can be calculated from (6); this reduces the amount of capital available for productive output ($K_n = K - K_e$), and the productivity of labour by lowering $P = h[PK_n, Y]$. Given that the derived demand for labour is modelled using the profit maximising wage conditions as in (4) above, increased energy prices will increase the demand for labour at a constant level of output by reducing its productivity. The derived demand for labour is given by the first order conditions of (8) with respect to labour (substituting for the net capital stock K_n):

$$L = \theta^{\zeta}.e^{\zeta.\eta t}.RW^{\zeta/\rho}.P^{\zeta}.Y \tag{9}$$

$$P = h[PK_n, Y]$$

$$\zeta = -\rho/(1+\rho)$$

Unlike (4) this equation is not uniquely identified because there are two terms in Y in the equation. Therefore, parameter estimation was carried out in two stages; firstly, P was substituted out from (9) using (8) and the following linear equation derived:

$$(Y-L.RW) = \phi.\theta^{-\rho}.e^{-\eta\rho t}.K_n^{-\rho}.Y^{(1+\rho)} \quad (10)$$

The economic interpretation of this relationship is straightforward, as the left-hand side is the cost share of capital in the cost of total non-energy output, and the right-hand side is the marginal productivity of capital ($\partial Y/\partial K$) multiplied by the measured capital stock. This cointegrating vector was estimated using the Johansen technique with cross-restrictions on the parameters on Y and K_n imposed. Values for the major parameters were recovered from the estimated coefficients, and the constants identified separately by defining P=1 in the first period of estimation, taking values for α and ω from Table 5.1, and solving through (8). This procedure allowed all the major structural parameters not uniquely associated with P to be derived. The second step was to estimate the parameters for P from the following cointegrating expression derived from (9):

$$(L.Y^{-1}.RW^{\zeta/\rho}.\theta^{-\zeta}.e^{-\eta.\zeta.t})^{1/\zeta} = PK_n^a/Y^b \quad (11)$$

The estimated long run elasticities of labour to each variable are given in Table 5.2.

Comparison of exogenous and endogenous model parameters

Comparing the two methods from the results in Tables 5.1 and 5.2, the elasticity of labour to real wage levels is lower in the endogenous technical change model in all countries except Italy, implying that direct substitution of labour and capital is harder. This reduction in apparent substitution into capital means that more of the decrease in labour intensity (L/Y), which occurs in all these countries over the estimation period, is attributed to technical progress (embodied and disembodied) in the endogenous technical progress model. There are positive embodied technical change terms in all countries (PK_e/Y rising over time), and as would be expected the rate of disembodied technical progress (η) is lower in the endogenous model in

Table 5.2 Estimated elasticities of labour: endogenous technical change model

Country	*Wage elasticity*	*PK_n/Y elasticity*	*Exogenous trend (η) (%/yr)*	*Estimation period*[a]
USA	−0.48	−0.36	1.23	6001–8702
Japan	−0.43	−0.74	2.21	6501–8504
Germany	−0.25	−0.76	1.15	6501–8702
France	−0.18	−0.62	0.91	6902–8702
Italy	−0.39	−0.51	1.31	7402–8602
UK	−0.23	−0.76	0.98	6203–8702
Canada	−0.35	−0.60	2.04	6601–8602

a Estimation period figures denote year and quarter (e.g. 6702 is second quarter of 1967).

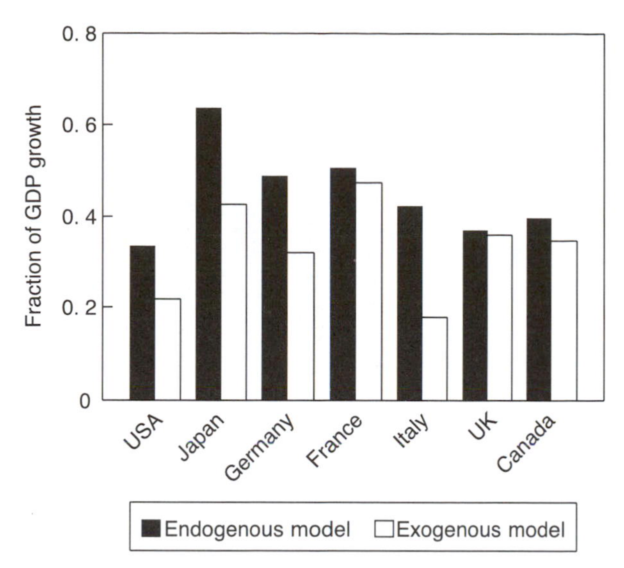

Figure 5.1 Drop in GDP growth 1995–2030 if all productivity terms fixed

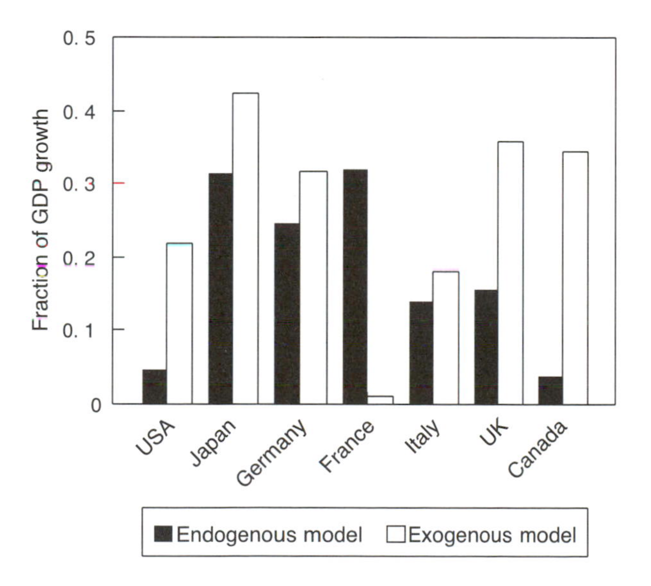

Figure 5.2 Drop in GDP growth 1995–2030 if labour productivity terms fixed

all countries. Figure 5.1 shows the drop in GDP growth in each country if the productivity terms (i.e. ηt and P or δt) are held constant. Because the production function is non-linear in factors and productivity this cannot be interpreted as the 'share' of growth attributed to technical change, but it does allow comparisons of the influence of these terms in each model. The figure shows that the productivity terms are more important in the endo-

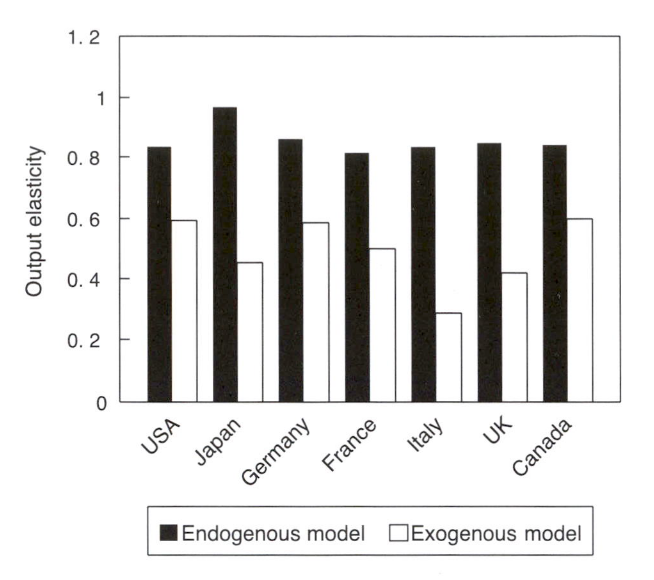

Figure 5.3 Average output elasticity of labour: endogenous and exogenous models

genous technical change model, though the difference is only marked in Japan, Germany, Italy and the USA.

Figure 5.2 shows that bias in productivity growth towards labour is much lower in the endogenous model because the disembodied productivity trend is associated with both capital and labour (except for France where there is no labour bias in the exogenous model).

However, because overall growth in the endogenous technical change model is more associated with productivity increases, rather than factor accumulation, the total marginal productivity of labour is higher. Figure 5.3 shows this effect by comparing the average calculated output elasticities of labour ($\partial Y/\partial L.L/Y$) of the two models over their estimation periods.

This means that any increased employment from recycling carbon tax revenues through employers' labour taxes should have a greater effect on output in the endogenous model, though this effect will be reduced due to the lower price elasticities of labour. Figure 5.4 shows that the endogenous model has a higher output elasticity to producer wage reductions in all countries except Germany and the USA (marginal).

Comparing the two supply side models suggests that the model with the endogenous trend in labour biased productivity will give a larger increase in output if carbon taxes are recycled through employers' labour taxes, but this will result in lower reductions in employment (discounting demand side effects). However, the relative sensitivity of the two models to energy price rises cannot be completely predicted from comparing their coefficients, but must be analysed in full simulations with the rest of the model.

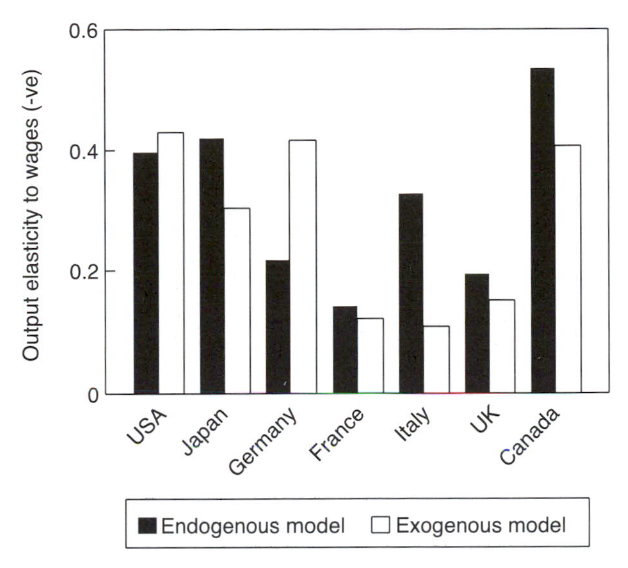

Figure 5.4 Average output elasticity to wages (constant capital)

Reduced form representation of the long run EGEM model

The two models outlined above give demand equations for labour and energy which are mutually dependent; both models differ from a CGE supply side because the energy demand equations have a price driven endogenous technical progress term.

With constant prices investment is modelled as a fixed proportion of output, and so long run growth in this model will be a function of total productivity growth. This consists of disembodied and embodied technical progress in labour and capital, labour supply growth and energy sector productivity increases. A one-off increase in energy prices will affect the *level* of output in the exogenous model, and the medium run growth rate in the embodied technical change model by reducing the amount of productive capital and thus P. The variables exogenous to each country are disembodied labour productivity, labour force supply growth and international energy prices, though these do respond to international demand. Baseline growth scenarios for all countries were produced which, unsurprisingly, showed trend growth in the industrialised countries to continue at historical rates of ≈ 2.5 per cent per annum except for Japan which showed far higher growth rates of ≈ 4.5 per cent. This rate of growth was considered unsustainable and an artefact of the rapid convergence towards Western productivity levels which occurred in the Japanese economy during the 1960s and 1970s. To produce a more reasonable baseline scenario the labour productivity growth rate (η) in Japan was reduced to 2.65 per cent per annum resulting in a total average growth rate of ≈ 3.15 per cent over the simulation period. Figure 5.5 shows the adjusted average growth rates

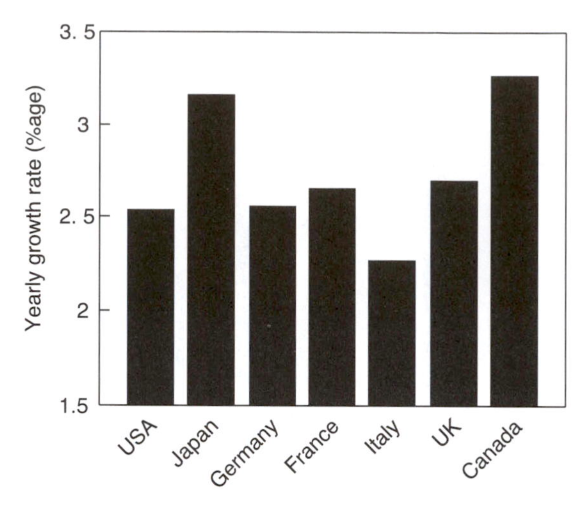

Figure 5.5 Average growth rates for adjusted exogenous model, 1994–2030

for the baseline scenario calculated using the exogenous technical change model.

In order to allow comparisons between the two supply side approaches the same baseline scenario is used for both and the percentage deviations compared; the adjustment made in Japan probably means the effects of carbon taxes will be overstated in the endogenous change model, but this is an unavoidable consequence of estimating equations based on past data. The result of these relative growth rates on GDP levels in the G7 at the end of the EGEM simulation period (2030) is shown in Figure 5.6.

It can be seen that while Europe maintains basic parity with the USA, Japan experiences significant relative growth; the USA experiences reason-

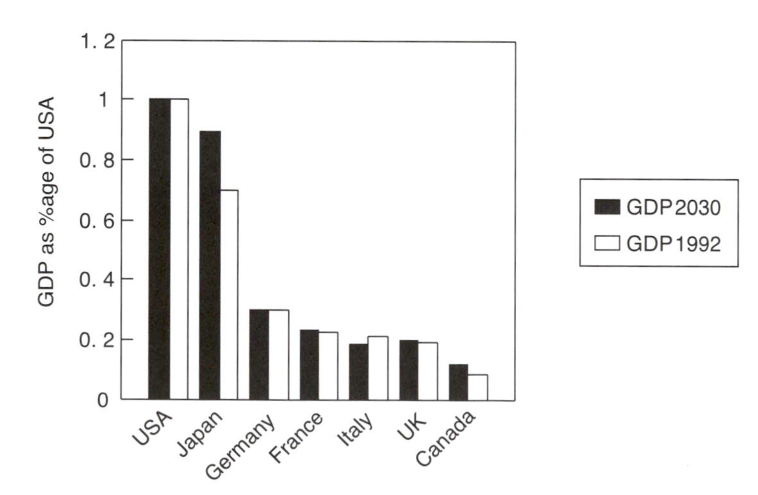

Figure 5.6 Forecast GDP as a percentage of USA GDP, 1992 and 2030

able GDP growth despite its low growth rate in labour productivity because immigration allows the available labour force to expand much more quickly than in the other developed countries. This growth scenario forms the basis for all comparative simulations.

Ignoring the details of the trade and government sectors, the reduced form set of equations for analysing the long run effect of energy price increases and tax recycling in EGEM is therefore:

$$\text{Employment (L)} = [NW/P_o, E, t, Y]$$
$$\text{Energy Use (E)} = [EP, \pi t, \Pi iXi, Y]$$
$$\text{Investment (I)} = [LR, Y]$$
$$\text{Nominal Wages (NW)} = [\partial Y/\partial L, P_o, U, INF]$$
$$\text{Output Price } (P_o) = [CU, ((L \times NW) + EP + MP)/Y_p]$$
$$\text{Consumption (C)} = [NW \times L, LR]$$
$$\text{Output (Y)} = C/P_o + I + X - M$$

where square brackets denote a functional relationship and all symbols are as previously defined, or standard accounting definitions; MP is the price of aggregate manufactured imports, LR are long run real interest rates, U is unemployment (percentage), CU is capacity utilisation and INF is the annual inflation rate.

In this system the wage equation has been altered so that long run real wages tend towards the marginal productivity of labour ($\partial Y/\partial L$), which is calculated from the production functions in (3) and (8). The original equations were estimated using average productivity measures (Y/L), but this formulation has unstable simulation properties as increasing the labour force reduces wages leading to a hiring spiral towards full employment. This is not observed in the real economic system, but is a function of aggregate demand-based modelling.

The output price (P_o) equation is based on the original estimated relationship in GEM, but the long run solution has been redefined as being equal to total input costs (that is, energy and import prices weighted by input share, and nominal labour costs) divided by potential output Y_p. Y_p, the economic output available from the mix of factors consistent with input prices, is calculated from the calibrated production functions (equations [3] and [8]) consistent with the factor demand equations. If energy prices rise, energy demand falls and correspondingly labour demand rises due to the different mechanisms outlined above for the two supply side models. Increasing labour demand increases consumption and so nominal output, but the combination of the rise in energy prices, and the shift in factor input mix reducing potential output, increases the output price; leading to an overall fall in real economic output.

As in the real economy, because all quantities are priced in monetary units, this system of equations has no unique solution for the price level P_o. In these simulations none of the equations in the model uses expected prices, and the inflation rate only enters the long run solution of the model through the wage equations of some countries. Therefore, the nominal price level could rise indefinitely at any rate without affecting long run real economic activity. In order to reflect the real economic cost of inflation the monetary side of the model is closed by an interest rate equation linked to growth in the nominal money supply. If prices rise too quickly in the short to medium term, interest rates rise to choke off demand and stabilise the rate of price level rise; this rise in interest rates constitutes a real economic cost of shocking the economy with a large change in factor prices.

In the long run, the price level P_o rises as more energy is substituted out of the economy and production becomes less efficient; this will be balanced by decreases in energy costs and wages as the marginal productivity of labour falls. This implies that raising energy prices will raise long run interest rates, lowering investment and the capital stock. Energy and capital therefore work as complements in the model. The interaction of labour and capital is more complex; a fall in real employee wages temporarily lowers inflation growth through multiplier and dynamic effects, but increased employment raises wages and inflation. Interest rates will fall if real wages are linked to productivity, showing that a fall in the marginal productivity of labour prompts substitution into capital, but with employers' labour tax recycling the fall in unemployment is likely to increase interest rates as the labour market tightens up.

SIMULATION PROPERTIES OF EGEM

To compare the general magnitude of economic effects in the model using the different supply side models and forms of tax recycling, a standard simulation was run over a thirty-five year period. In this a carbon tax was applied in 1995 at a level which stabilised aggregate G7 CO_2 emissions in EGEM at 1990 levels over the whole of the simulation period; in 1990 US dollars this amounted to a flat rate tax of $275 per tonne of carbon. This simulation is based on a renewal of the commitments in the Framework Convention on Climate Change (UNEP 1992) until 2035, and gives a 31 per cent drop in G7 CO_2 emissions by the end of the simulation period.

The results of the simulations can be split into two broad categories: firstly, the short to medium term responses of the demand side of the economy, mainly the wage, price and interest rate equations, and secondly the medium to long run changes in factor mix and production given by the supply side equations. The results of the simulations are slightly stylised because the relatively high tax is imposed in one quarter, leading to a sharper jump in prices and interest rates than would exist with a gradually introduced tax.

Recycling carbon taxes through direct taxation

If the revenues from a carbon tax are recycled through direct income taxation, keeping the government's budget to GDP ratio constant, then

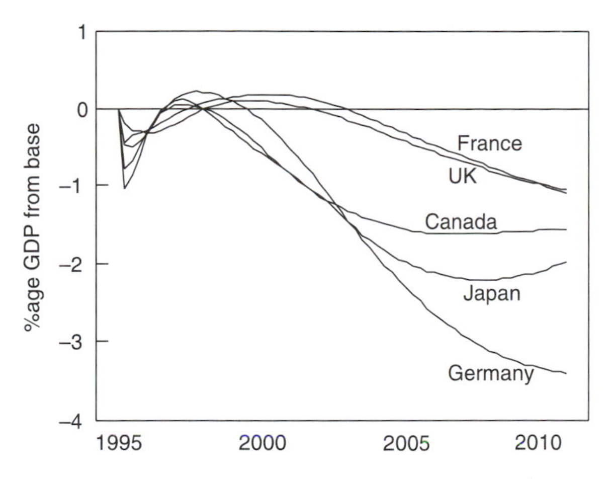

Figure 5.7 Initial GDP dynamics in five countries: income tax recycling

there are strong dynamic effects in the economy in the first 5–10 years. Employees see an almost immediate drop in income taxes (a slight delay of one year is introduced because of payment lags), but the rise in the price of goods is delayed due to the reluctance of companies immediately to pass through higher input costs. As Figure 5.7 shows this results in a small 'boom' in several countries where an initial fall in output quickly disappears, as consumption increases and imports fall due to higher export prices in other countries; all simulation results are produced using the exogenous technical change model unless otherwise specified.

This increase in consumption is unsustainable however, and leads to increased inflation as capacity utilisation increases and nominal wages try to keep up with consumer prices in the short run. Inflation leads to higher interest rates which depress investment, and in countries with a significant mortgage sector this reduces consumption as well; this initial inflationary effect of recycling through direct taxation sets up a particular 'business cycle', which is clearly shown in Figure 5.8.

Figure 5.8 plots the short run output of the economy ($Y = C+I+G-M+X$) relative to the long run sustainable output calculated from the supply side. The European Union (EU) is used as the example because its inflationary dynamics are particularly clear! Measured output from income tax recycling tends to be at a peak at the end of the simulation period, and this will affect comparisons with employers' labour tax recycling. The dynamics of economic adjustment can be seen to be of a similar magnitude to the long run output changes in the short to medium term. Over the whole simulation period, however, pure demand side effects (discounting trade balances) are neutral if undiscounted, but the effects of rising interest rates on capital accumulation and output from the supply side are non-trivial. This implies that monetary and fiscal considerations

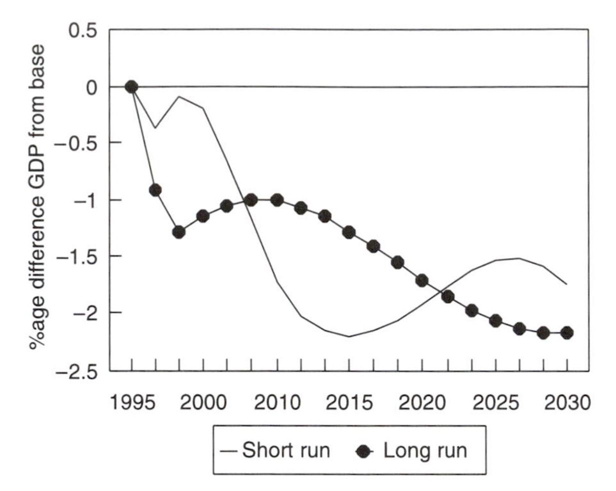

Figure 5.8 Long run and short run output for the EU: income tax recycling

will be important to politicians (especially if they have positive discount rates!), when they consider the timing and levels of any taxation scheme.

Recycling carbon taxes through employers' labour taxes

If carbon tax revenues are recycled through employers' labour taxes then the economic effects, especially in the short term, are very different to the income tax recycling case. Assuming a constant long run mark-up by firms, the reduction in labour taxes is eventually completely passed through to consumers as a price reduction. This deflationary effect is re-enforced by the initial reduction in the price of imports from other countries caused by

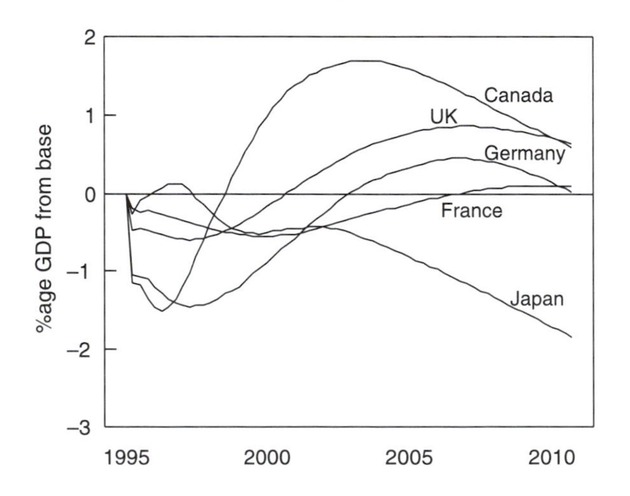

Figure 5.9 Initial GDP dynamics in five countries: employers' labour tax recycling

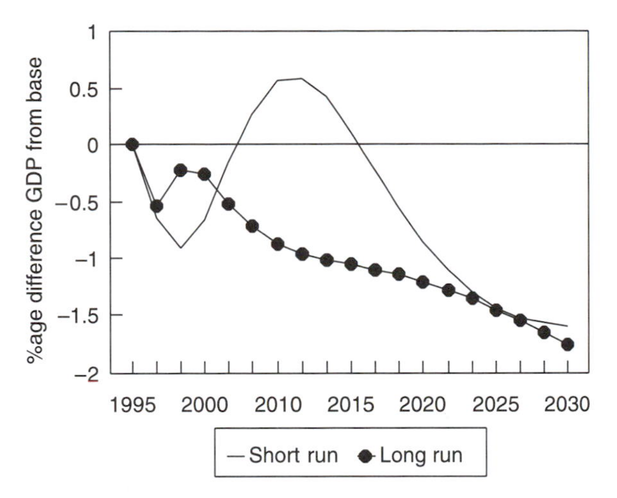

Figure 5.10 Long and short run output for the EU: employers' labour tax recycling

this form of tax recycling. As direct taxation remains unchanged, and imports rise, there is a depressing effect on GDP (Figure 5.9). The only inflationary pressures come from the response of wages to increased consumer prices, and, because in the long run the wedge does not affect wages, this is only a small, short run effect. Therefore, there is no consumption boom in the first five years and interest rate rises are correspondingly lower.

However, because decreasing labour taxes increases employment this raises real output above the income tax recycling case in the long term, and also triggers wage inflation because of the bargaining form of the wage equation. This large inflationary effect causes interest rates to rise towards the middle of the simulation period and is clearly shown in Figure 5.10.

Comparing Figures 5.8 and 5.10 shows that the business cycles of the two different forms of recycling are essentially mirror images of each other and economic comparisons will depend on the point of the cycle at which they are compared, and the discount rate used (if any). Figure 5.11 compares short run output in the G7 for the two cases. The difference in investment dynamics somewhat masks the higher long run output from labour tax recycling, but these gains are large enough to make labour tax recycling the preferred policy choice on output grounds alone.

These changes in total economic output are only slightly influenced by terms of trade effects. Higher net prices and lower demand lead to a drop in exports to the rest of the world; with labour tax recycling ≈10 per cent of the output losses come via this avenue, increasing to ≈17 per cent with income tax recycling. It should be remembered that, because exchange rates are fixed, this figure includes losses which would appear as decreases in welfare under a floating currency regime.

Figure 5.12 shows that labour tax recycling does give large employment

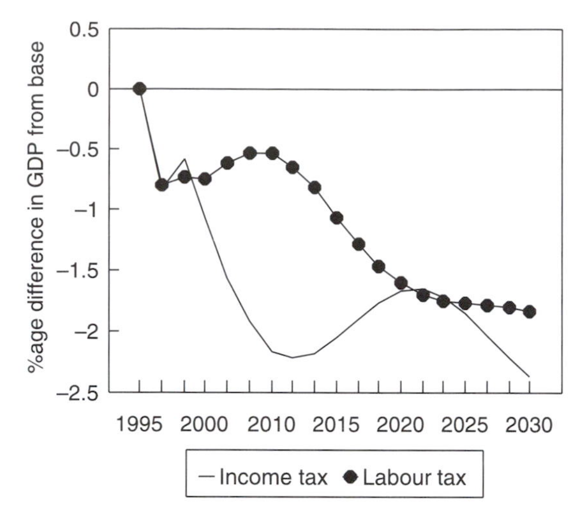

Figure 5.11 Short run output changes in the G7: recycling comparison

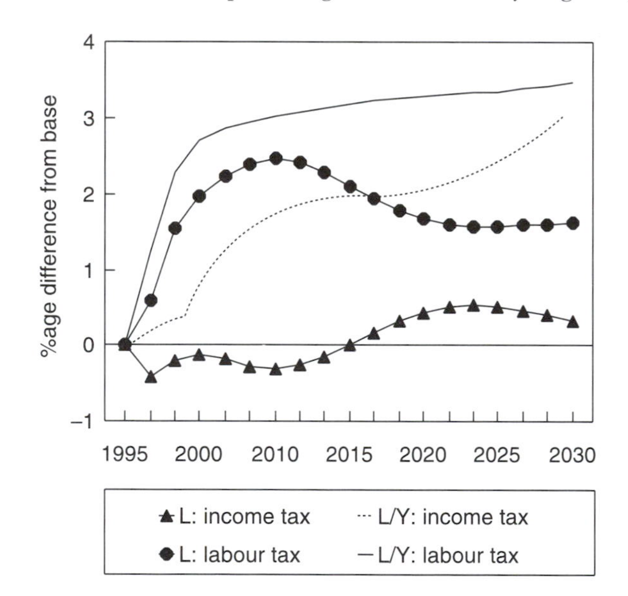

Figure 5.12 Employment (L) and labour intensity (L/Y) in the G7

gains over the base-case, despite the fall in overall output. However, the change in labour intensity in the economy (L/Y) is affected more by labour–energy substitution in the supply side than the lower producer wages caused by labour tax recycling.

Though there are significant differences between the recycling methods, especially in the first 10–20 years of the tax, it seems that the long run influence of carbon taxes on output and employment is dominated by factor substitution behaviour.

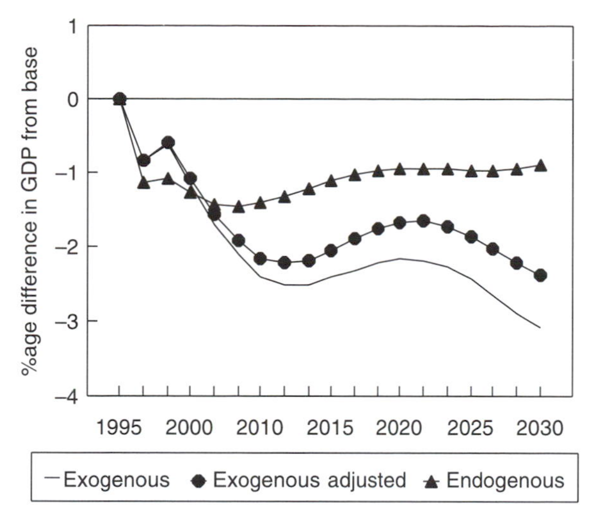

Figure 5.13 G7 output for exogenous and endogenous models: income tax recycling

Comparing endogenous and exogenous technical progress supply sides

Figure 5.13 compares output changes in both supply sides developed above for the standard simulation. The dynamics of output are similar in the two cases because these stem from the common parts of the model, but the percentage output drop for the exogenous model is trending downwards over time, whereas that for the endogenous model is relatively constant. This simulation property is caused by the price sensitive trend term in the energy demand equations, which produces a constantly increasing percentage drop in energy use and intensity over time. The exogenous model interprets this drop in energy use as being a result of substitution into labour, and so reduces the productivity of energy in the production function proportionately. In the endogenous model the productivity of *technology weighted* energy use is constant, and energy efficiency measures which are paid for at the beginning of the simulation period are assumed to be replaced at no extra cost at the end of their lives.

The exogenous model therefore overstates the macroeconomic costs of saving energy, by assuming that there is no costless technical progress in the energy sector. To account for this the exogenous model was adjusted, so that over time more and more of the reductions in energy intensity (E/Y) in the simulation come from technical progress, until at the end of the simulation all further reductions are costless. This represents an approximation to a learning-by-doing process in technological development, where the extra unit cost of new energy efficient products decreases over time as their design is refined. The output drop from this model (exogenous adjusted) is virtually untrended towards the end of

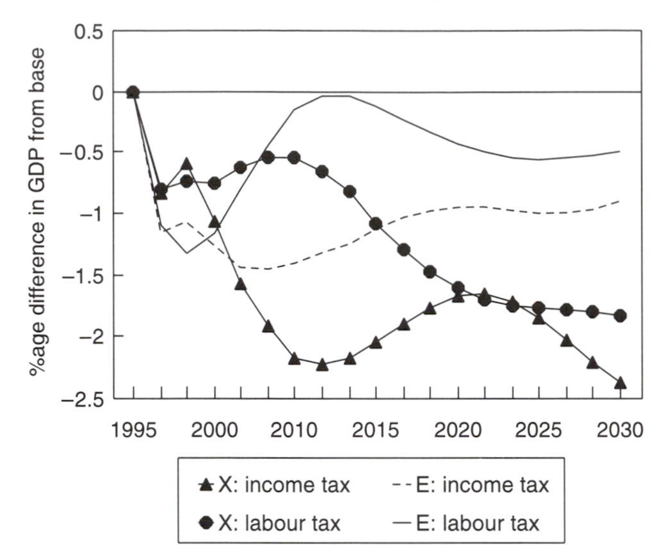

Figure 5.14 G7 output for exogenous and endogenous models: recycling comparison

the simulation period giving a more consistent comparison with the endogenous model.

Using the adjusted exogenous model (X), Figure 5.14 shows that the endogenous model (E) gives approximately half of the drop in output with recycling through income taxes and that the output gain from labour tax recycling is larger and more consistent over time. This is a reflection of the higher productivity of labour in the endogenous model, which is a result of the technological assumptions underlying the estimated equations. Despite the large differences in output, the changes in employment in Figure 5.15 are quite similar in the two models; Figure 5.16 shows that this is because of the underlying similarity of the labour–energy substitution behaviour, not the form of recycling.

In the long run the exogenous model predicts that energy taxation will lead to a more labour intense economy than the endogenous model, because energy and labour act as direct substitutes without the mediating force of capital accumulation. The amount of investment diversion needed to save energy reduces towards the end of the simulation; therefore, the productivity of labour in the endogenous model increases, giving a trend reduction in labour intensity. In the exogenous model the continual substitution of energy for labour leads to a trend increase in labour intensity over the same period. The ability of tax shifting from labour to energy to decrease unemployment permanently (at any level of output) will therefore depend on the evolution of these trends, which are clearly shown in Figure 5.16.

Though the aggregate results of the two models show some differences

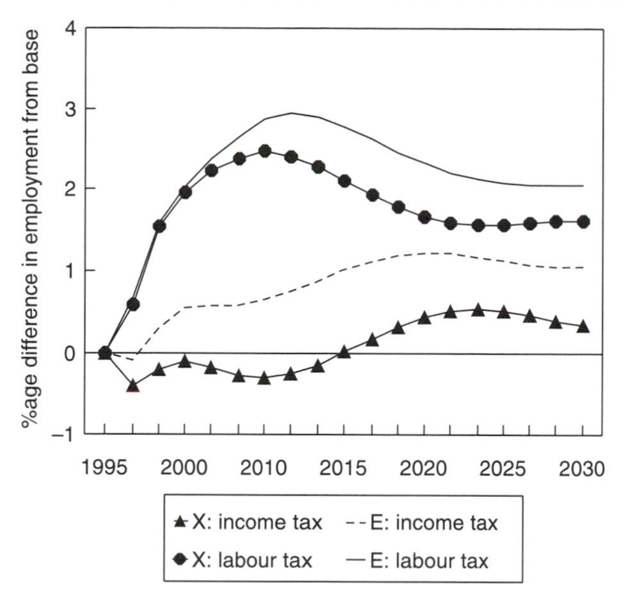

Figure 5.15 G7 employment for exogenous and endogenous models

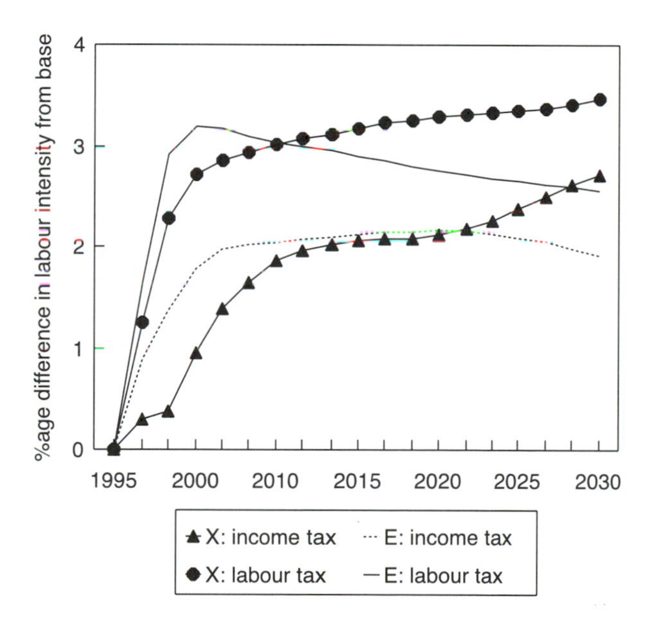

Figure 5.16 G7 labour intensity for exogenous and endogenous models

the most marked contrasts are between the regional results for output changes and employment. Table 5.3 gives the average macroeconomic cost of saving a tonne of CO_2 in each region, and for each type of tax recycling, for the stabilisation tax of $275/tC. Differences between the

Table 5.3 Average undiscounted cost of saving carbon dioxide, (US$/tonne) 1990

	Income tax recycling		*Labour tax recycling*	
Region	*Exogenous*	*Endogenous*	*Exogenous*	*Endogenous*
North America	480	302	366	184
Japan	3935	536	3473	251
European Union	778	772	360	210

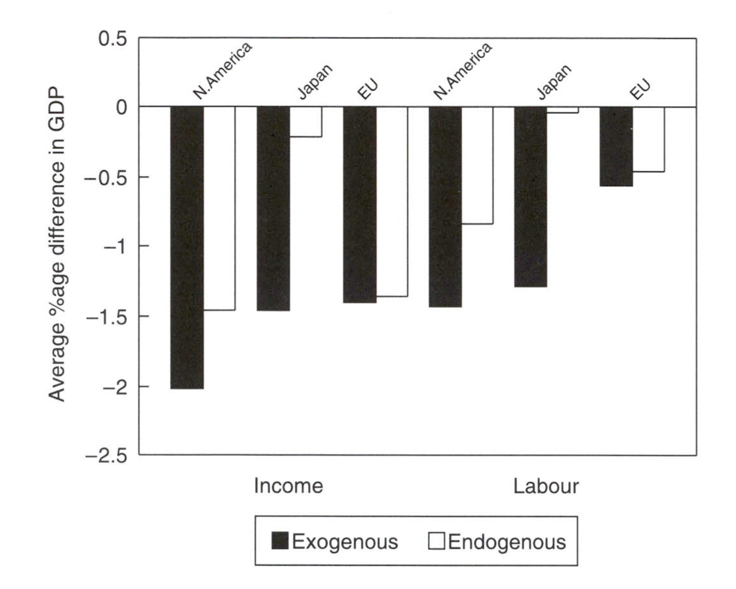

Figure 5.17 Regional output changes: labour tax and income tax recycling

models in the cost of saving carbon dioxide are also apparent in the changes in economic output shown in Figure 5.17, but the relative changes in employment are similar in the two models (Figure 5.18).

The average cost of control for the exogenous model is much higher than the tax rate. Following the logic above, this implies that either too much energy is being saved or there are strong macroeconomic externalities from imposing an energy tax. On investigation it seems that, despite the adjustment to the exogenous model, the productivity of energy in the production function is high compared to the labour being substituted for it. The elasticities for the exogenous model, given in Table 5.1, show that this is especially true in Japan and Italy, and that these are the countries contributing most to the large cost of abatement in the exogenous model. The difference in costs in Japan between the two models is of an order of magnitude which completely changes the least cost distribution of emissions reductions one would expect from a tradable permit system.

The average costs of control for the endogenous model are nearer the

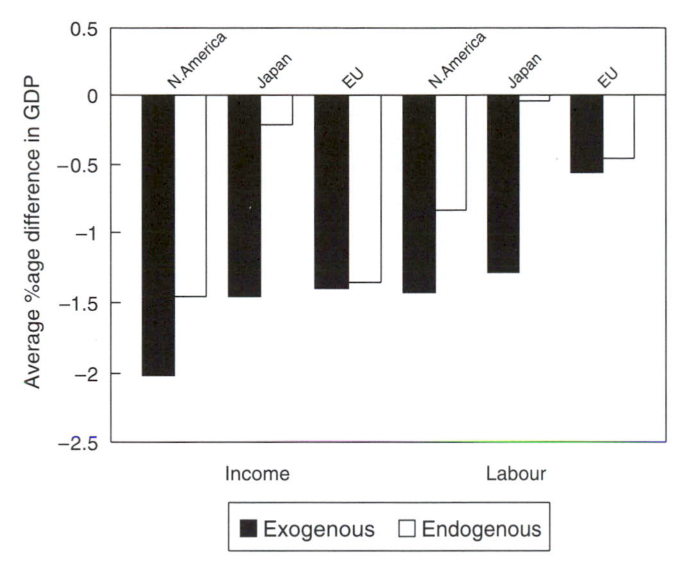

Figure 5.18 Regional employment changes: labour tax and income tax recycling

tax rate, and below it if revenues are recycled through labour taxation. Of this GDP drop approximately 10–17 per cent comes from terms of trade effects depending on recycling methods (see Chapter 9 for regional effects), and 15 per cent from decreased capital accumulation (i.e. the effect on the supply side, not the demand effect of decreased investment) from higher interest rates. However, the investment effects are much higher in Japan than the other two regions. Further simulations showed that these effects were proportionately higher at lower tax rates.

As was outlined in Chapter 2, these differences in the distribution of macroeconomic costs have very important implications for the efficiency of different policy instruments. In the exogenous technical change model the cost of saving carbon dioxide differs enormously between the regions, with the cost in Japan being particularly high whatever form of tax recycling is used. This would suggest that a tradable permits scheme, whatever the distribution of permits, will produce large macroeconomic efficiency gains for reaching stabilisation. However, these cost differences are far less in the endogenous technical change model, and virtually disappear with labour tax recycling. In this case the value of retaining tax revenues for recycling is large (especially in Japan and the EU), and an internationally levied flat rate tax, which is collected domestically, would probably be the simplest and most efficient policy instrument to achieve stabilisation.

These regional differences stem from the fundamental assumptions underlying the production technology in each model. The general case of a traditional production function approach is shown diagrammatically in Figure 5.19, where each factor or bundle of factors (energy and materials in this example) experiences an exogenous change in productivity which is

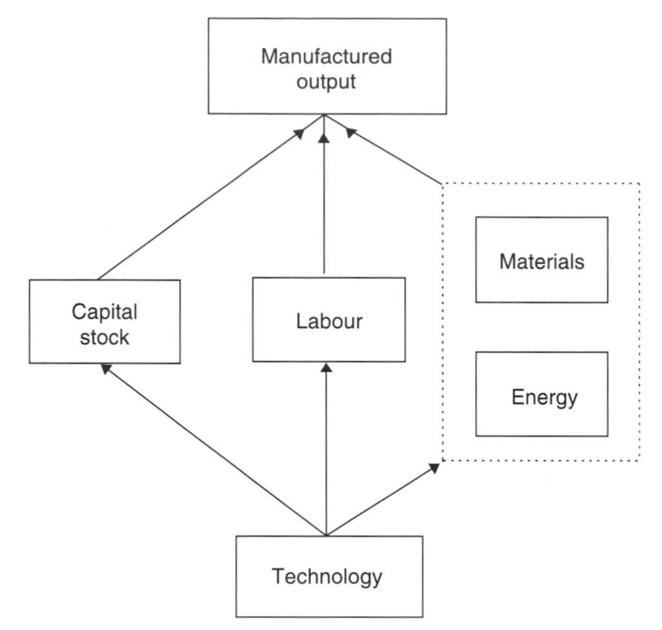

Figure 5.19 Traditional production function approach: schematic

dependent on time. The resultant technology weighted factors are then combined inside the production function to give final output. Each factor, or bundle, enters the production function separately and so, all other things being equal, the less of a factor which is used, the higher its marginal productivity in final output. Therefore, a country such as Japan, with very low energy intensity (≈ 0.25 that of the USA), experiences large output losses for each unit of energy saved.

In contrast, Figure 5.20 shows an extreme version of the endogenous technical progress model, where all productivity improvements are mediated through investment in new physical capital (of course the concept of capital could be expanded to include human capital, but this would unnecessarily complicate this example). Here the marginal productivity of each factor is determined by the cost of not investing in new machines which would increase the productivity of the other factors. If the capital stock is large relative to GDP and total investment, then investing in energy efficient machines will only marginally affect the productivity of the rest of the economy. A slow turnover of machines also means that a delay in, for example, labour efficiency investments will have little effect on the total productivity of the economy. The same effect occurs if total investment is large relative to the amount needed to save energy.

Japan is a high investment economy, and so in the endogenous technical progress model experiences relatively small output losses from saving energy; despite the fact that it already uses it very efficiently. North

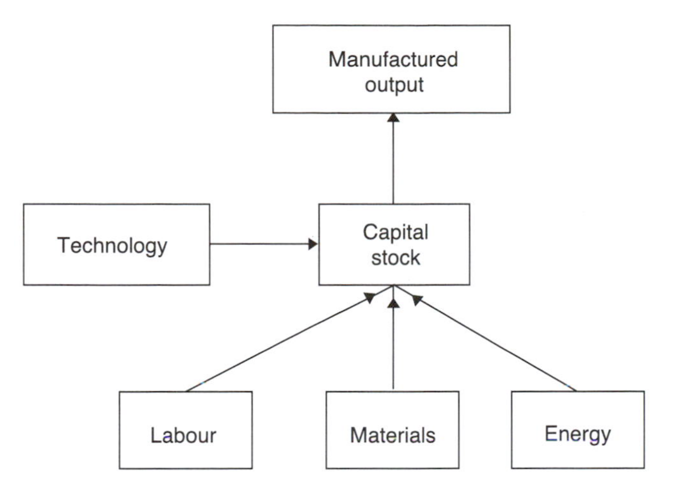

Figure 5.20 Endogenous technical progress model: schematic

America also seems to be significantly less affected by energy price rises in the endogenous model. On closer inspection, this is due to the low after-tax price of energy, implying a smaller amount of investment in energy conservation than would be the case in one of the European countries where energy is already highly priced. There is therefore both an investment and an implicit marginal productivity of energy effect in the endogenous model, though this is linked directly to energy prices rather than being mediated through the factor shares, as in the production function model.

CONCLUSIONS

In the realistic scenarios for OECD carbon dioxide abatement modelled here, the theoretical assumptions underlying the supply side of the economic model used have been seen to influence greatly the magnitude of macroeconomic costs, and their distribution between different countries. These assumptions are difficult to test and to discriminate empirically because of identification problems, but should at least be consistent with the microeconomic analysis of energy conservation. This work shows that reductions in energy use come about because of investment in new capital intensive machines. However, most empirically based production and cost function models used to date have assumed direct substitution between labour/materials and energy. Energy/capital substitution has only been included in calibrated general equilibrium models which have little numerical relevance due to their unfounded assumptions in the derivation of critical parameter values.

To investigate these effects we have estimated two supply side models for the G7 countries: one based on direct energy/labour substitution, and the other on the assumption that energy efficiency is embodied inside capital investment. Both models include some notion that price driven changes in trend energy efficiency can be applied costlessly into the future, but the effects are more explicit in the embodied efficiency model.

The greatest deviation between the two different approaches to modelling technology comes when estimating the distribution of costs between countries; this seems to be a structural property and not due to particular parameterisation. In international negotiations the absolute size of macroeconomic costs is probably less important for agreement than a relatively accurate calculation of their distribution, as this allows the construction of institutional mechanisms to spread the costs of abatement fairly across countries. In the absence of equitable burden sharing it is likely that some countries will renege on the agreement, causing long lasting environmental damage.

The production function approach assumes that countries with the lowest share of energy use will experience the highest unit costs of conservation, and generally predicts very high macroeconomic costs compared to the size of energy tax imposed. If technical change is embodied in capital this result is qualified by putting the investment needed to save energy in the context of the size and growth rate of the country's economy. High investment rates, or large labour productivity augmenting capital stocks, will mean that investment diversion to energy saving will have a smaller effect on overall output than predicted in the standard approach.

The crucial role of technological assumptions suggests that the most important direction for future research into quantifying the costs of controlling climate change is to investigate further the evolution of the technology of production and energy saving, and how these interact. Prices certainly stimulate technical innovation, and, as Chapter 4 explained, this is the *only* way to control emissions in the long term. However, innovation is not a free lunch, as R & D resources must be diverted from elsewhere in the economy.

Bottom-up engineering models can give an idea of available current and future technologies, and the amount of investment needed to save a certain amount of carbon dioxide. Econometric models, such as the one developed and estimated here, can give us an idea of the timescale of technical diffusion and evolution, and perhaps trade-offs with other sectors, but only based on past behaviour. By developing macroeconomic models which more accurately articulate the role up on productivity of R & D, investment, and vintages of investment, perhaps such microeconomic data can be used to help to calculate macroeconomic impacts. EGEM is a first step towards such models as it endogenises the process of technical change to a degree where there is no unique economic equilibrium, and the dynamics of imposing a carbon tax are all important to its macroeconomic cost in the long run. This enhancement of model simulation properties allows more definite estimates of the parameters surrounding some of the pressing political problems about the timing, and levels, of carbon reduction measures.

APPENDIX 5.1: MICROECONOMIC COSTS OF SAVING ENERGY

Assume an oligopolistic firm sells into a market which has a demand elasticity η. Its production process has constant returns-to-scale, and so its pricing policy is to charge a constant money mark-up (MU) over the average/marginal cost of production (C). This pricing policy corresponds to a desire to maintain profits as a proportion of capital costs, or share price, not as a percentage of gross turnover including variable costs.

If a carbon/energy tax is imposed at a rate of T per unit output for the original mix of inputs into the production process, the resulting prices and demands will be:

Original Price: $P_1 = C + MU$
After Tax Price: $P_2 = C + T + MU$
After Tax Demand: $Q_2 = Q_1(1 - \eta.(P_2 - P_1)/P_1)$

The producer re-optimises the input mix to production so as to reduce energy use by a proportion of $(1-\alpha)$; this increases the average cost of production, discounting taxes, to βC $(\beta > 1)$. The resulting prices and demands are:

Price after re-optimisation: $P_3 = \beta C + \alpha T + MU$
Quantity after re-optimisation: $Q_3 = Q_1.(1 - \eta.(P_3 - P_1)/P_1)$

Profit increase from re-optimisation:

$$\begin{aligned}
\Pi \text{ saved} &= -MU.(Q_2 - Q_3)\\
&= -MU.Q_1.(1 - \eta.(P_2 - P_1)\,/P_1 - (1 - \eta.(P_3 - P_1)/P_1))\\
&= -MU.Q_1.(\eta.(P_3 - P_2)/P_1)\\
&= -MU.Q_1.(\eta.(\beta C + \alpha T + MU - C - T - MU)/P_1)\\
&= -MU.Q_1.(\eta.(\beta - 1).C - \eta.(1-\alpha)T)/P_1\\
&= Q_1.MU.(\eta/P_1).((1-\alpha).T - C.(\beta - 1))
\end{aligned}$$

Therefore, for the case of inelastic demand ($\eta = 0$) profits remain constant whatever the tax level, which is also the result for a perfectly competitive free market because profits are constant with changes in variable costs. For non-zero elasticities profit will increase with substitution if $C.(\beta - 1) < (1 - \alpha).T$; that is, unit costs rise slower than energy taxes per unit of output. Profits from substitution will increase as the elasticity of demand rises, and the level of substitution α which maximises profit is defined by:

$$\partial\Pi/\partial\alpha = Q_1.MU.(\eta/P_1).(-T - C.\partial\beta/\partial\alpha) = 0$$

$$\therefore \qquad \partial\beta/\partial\alpha = -T/C$$

Therefore, the amount of substitution is not affected by the price elasticity of demand, but assuming that substitution gets harder as more energy is saved ($\partial^2\beta/\partial\alpha^2 > 0$) the optimum amount of substitution will rise as the relative size of the tax (T/C) increases.

6

CARBON ABATEMENT IN DEVELOPING COUNTRIES

A case study of India

INTRODUCTION

Though currently excluded from controls under the FCCC, in the future developing countries (DCs) such as China and India will become the major emitters of greenhouse gases, and will therefore have to consider how to control their emissions (World Bank 1992). Surveys of the literature, such as Chapter 3, on the economics of controlling greenhouse gas emissions reveal that very few models assess the impact of policies to control greenhouse gas emissions on developing economies. Some global models (for example, OECD's GREEN model and Manne and Richels' Global 2100 model) do include developing regions, but these regions are mostly modelled with or based on data for developed countries. However, in the real world DCs have characteristics such as perpetual market disequilibria, large public sectors, restricted market entry (especially in the infrastructure related sectors) and lack of perfect information, which make them very different from industrialised countries. Additionally, in developing countries, many carbon abating technologies are available but are not exploited due to non-market reasons such as the lack of availability of capital. These technological options could be very important in the determination of costs of carbon abatement in DCs. Therefore, the economic costs of abating carbon emissions in a developing country have to be evaluated using a framework developed to study its specific features.

A methodology to integrate microeconomic (that is, explicit technological options to reduce carbon emissions) and macroeconomic analyses in a single framework is developed here and applied to India. In developing this methodology problems specific to developing countries have been given special consideration. For example, short and medium term analyses are more important for economies in their developmental stages, and it is this time-frame which has been addressed rather than the very long term. An econometric approach is used instead of a computable general equilibrium model, because developing economies do not have perfect markets and equilibria in their economies. This approach is feasible because recent advancements in econometric estimation techniques, such as cointegration analysis, have made the econometric approach more appropriate for longer term analysis.

As Chapter 3 showed, when simulating future emissions of carbon dioxide, and the cost of emissions stabilisation, the most important parameter in a macro model is the rate of change of energy intensity in an economy. This determines both the baseline projected emissions and the measured value of price response. In almost all models the efficiency improvement parameter is exogenously specified, and is the same in the business-as-usual case as when policies are employed to control emissions. To overcome this limitation for India we employ a variant of the endogenous technical progress model which was used in Chapter 4 to analyse energy usage in the developed world.

The first part of this chapter analyses the energy situation in developing countries, focusing on India, and highlights the particular characteristics of energy markets that make them different to the developed world. We then discuss the components of a macroeconometric model for India. In this macro framework additional technological options which could affect the demand for carbon energy are not considered, and energy use only responds to relative prices and economic growth. However, the measured elasticities of energy demand in India are small, or even positive, due to the supply constrained history of energy systems. Therefore, to produce a useful model, a methodology is proposed for integrating the micro analysis of carbon abatement technologies within the macro framework. The feasible technological options which would reduce emissions of carbon dioxide in the Indian economy are analysed and ranked, based on their investment requirements. Finally a relationship between the level of investment and carbon savings in the economy is estimated, and this is used inside the macroeconomic model to enhance the potential for carbon abatement. To calibrate the policy simulations we construct a base-case scenario for CO_2 emissions in India up until the year 2020. Given the large uncertainties regarding the future energy demand growth, two base-cases are considered. Four policy scenarios for reducing carbon emissions in India are then examined in detail, showing the costs of different types of control target, policy instruments (taxes or investment in technologies) and tax recycling option.

This chapter only gives an outline of the modelling methodology involved. An exhaustive description of the Indian model, its estimation and assumptions, and additional policy scenarios are to be found in Gupta (1995).

ENERGY CONSUMPTION PATTERNS IN INDIA AND OTHER DEVELOPING COUNTRIES

Primary energy can be classified into commercial energy and non-commercial energy. The term commercial energy is used for coal, oil, natural gas, hydroelectricity and nuclear power. Non-commercial, or traditional, energy sources are fuel wood, animal dung and crop or agricultural waste, and include non-traded energy forms which either are collected or are by-products of agriculture and allied activities; if they are being harvested sustainably these traditional fuels produce no net carbon dioxide emissions. Total primary energy use is the sum of commercial and traditional sources of energy. In developing economies, traditional sources of primary energy are significant contributors to the total energy supply. It is estimated that these sources currently account for nearly 40–50 per cent of total primary energy consumption in India (for example, Hall 1991, Government of India 1991).

Energy consumption goes through three stages in the evolution of an economy. In stage one commercial energy consumption is low, there is heavy reliance on traditional forms of energy and per capita incomes are very low. In stage two, with economic development, the share of commercial energy increases. This is partly due to a switch from traditional energy to commercial energy, and partly due to a change in the composition of GDP towards more energy-intensive (i.e. energy used per unit of GDP increases) output and an increase in the availability of commercial energy. There is a strong preference for commercial energy because it is more convenient to use, it has a higher efficiency than traditional energy and its availability is increasing, whereas the supply of traditional energy is limited. In this stage, commercial energy intensity increases with growth in GDP and per capita income levels. In the third stage, although commercial energy consumption is high, commercial energy intensity starts to decline with development and growth in the economy. Also, the growth in commercial energy due to substitution from traditional energy ceases. The share of energy intensive goods in total output declines, and technological progress makes production more efficient and less energy intensive. In the later part of stage three, total commercial energy consumption may start to decline.

India is in stage two of this process, where the share of traditional energy sources is significant and commercial energy intensity and commercial energy consumption are increasing. With subsidised energy prices, constrained energy supply and lack of reliable time series data on traditional energy sources for India, it is difficult to say when the economy will move from stage two to stage three in its energy consumption pattern. Between 1971 and 1991, commercial energy consumption has increased at an average rate of 5.5 per cent per year, and that for fossil fuels at 5.6 per cent per annum.

Estimates of indigenous fossil fuel reserves show that coal has a reserves to production ratio of over 200 years, compared with 26 years for oil and 40 years for gas (Tata Energy Research Institute 1993), thus making it by far the most abundant primary energy source. This implies a high carbon intensity of fossil fuels in the future as demand grows and oil and gas supplies decline. Figure 6.1 shows the growth in the consumption pattern for commercial energy sources and point estimates for traditional energy

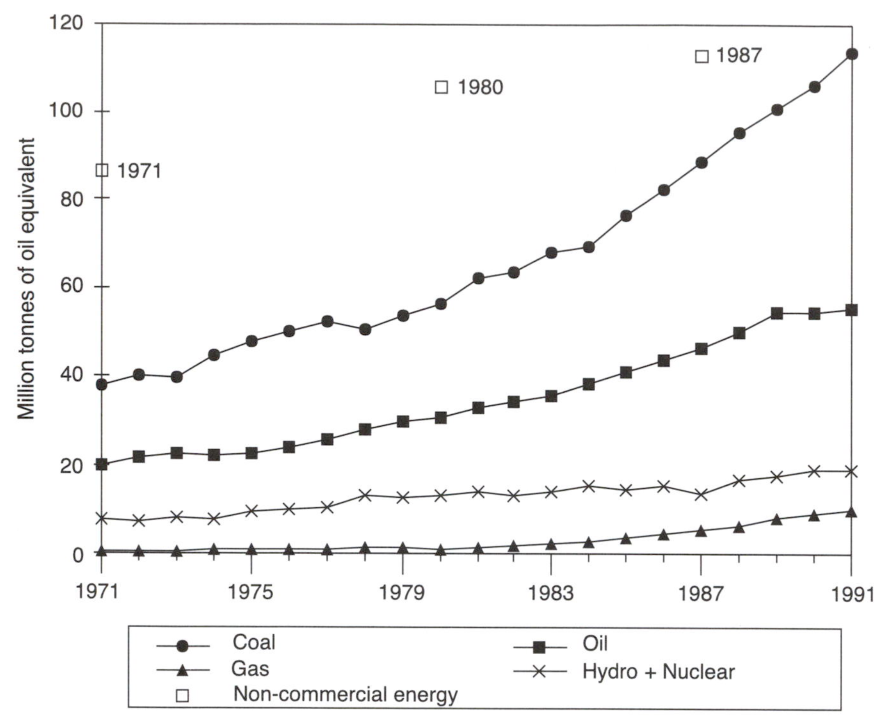

Figure 6.1 Commercial and non-commercial energy consumption in India

consumption. The growth rates for carbon energy and commercial energy are clearly increasing over time.

The share of coal has remained high over the years and accounts for more than 55 per cent of commercial energy consumption. The share of natural gas has increased and that of oil has fluctuated around 30 per cent and has declined from 1989 onwards. The relative shares of non-fossil commercial energy, that is, hydro and nuclear power, are also reducing (Table 6.1).

Table 6.1 Percentage share of different sources of primary commercial energy

Year	*Coal* (%)	*Oil* (%)	*Natural gas* (%)	*Hydropower and nuclear*[a] (%)	*Fossil fuels (Mtoe)*	*Total commercial energy (Mtoe)*
1971	57	30	1	12	58.2	65.8
1976	59	28	1	12	74.5	84.4
1981	56	30	1	12	94.2	107.9
1986	56	30	3	11	130.4	145.8
1991	57	28	5	10	179.2	198.2

Source: International Energy Agency 1993

a The following assumptions have been made for converting hydro and nuclear power in terms of primary energy equivalent: a conversion efficiency of 33 per cent and 860 kcal/kWh.

Table 6.2 Energy intensity and per capita energy consumption

Year	*Annual per capita GDP (1987 Rs)*	*Energy intensity (kgoe/'000 Rs)*			*Annual per capita energy (kgoe)*		
		Carbon energy	*Total commercial energy*	*Total energy*	*Carbon energy*	*Total commercial energy*	*Total energy*
1971	2823	36.8	41.6	96.2	103.8	117.4	271.6
1976	2931	40.5	45.9	NA[a]	118.7	134.5	NA[a]
1980	3084	41.6	47.7	97.6	128.3	147.1	301.0
1987	3691	47.6	52.2	90.4	175.7	192.8	333.8
1991	4243	48.7	53.9	NA[a]	206.8	228.8	NA[a]

Sources: World Bank 1993; International Energy Agency 1993; for non-commercial energy consumption, Government of India 1991

a NA indicates non-availability of data for non-commercial energy sources.

In Table 6.2 carbon emissions, energy, intensity and per capita energy consumption, with respect to carbon energy, commercial energy and total (commercial and traditional) energy, have been estimated. From the analysis of past energy consumption data we can conclude that both total commercial energy and carbon energy intensity in the economy are increasing and that the growth rate for carbon energy intensity is greater than that for total commercial energy. All these trends are outstripping population growth, as per capita energy consumption of carbon energy, commercial energy and total energy are also increasing rapidly, pointing to the industrialisation of the Indian economy. The obvious result of these combined trends is that *gross* carbon emissions from the Indian economy are growing, and the increasing growth rates for carbon energy and carbon energy intensity imply that this growth is accelerating. However, a decline in total energy (commercial and traditional) intensity and the decreasing rate of increase in commercial energy intensity indicate that a reduction in the rate of growth of *net* carbon emissions could be possible at a later stage.

In India, energy markets are imperfect and energy pricing decisions involve social and developmental objectives, as well as economic, political and institutional factors. This is particularly true of the coal and electricity sector. Average electricity tariffs in India are below the cost of power generation and supply. Coal prices are set below the average cost of production, but this difference is gradually reducing. Crude oil is both imported and produced domestically, and the weighted average of consumer prices for oil products is equal to, or higher than, the border prices. However, cross-subsidisation of different oil products causes several distortions. For example, kerosene, which is perceived as the poor person's fuel, is subsidised, and this leads to the adulteration of high speed diesel with cheaper kerosene (Tata Energy Research Institute 1992). To prevent this practice of adulteration the government has reduced the differential between the prices of high speed diesel and kerosene. However, this pricing policy has prompted automobile owners to retrofit their vehicles with inefficient diesel engines, further increasing the demand for diesel and kerosene which are imported at the margin.

Subsidised prices prevent consumers and producers of energy from receiving appropriate signals and so energy markets are often in disequilibrium. On the supply side, with low prices, producers do not receive an adequate incentive or generate large enough surpluses to invest in production expansion. Energy markets are thus supply constrained. This situation results in a very low responsiveness to prices; price elasticities for energy demand are very low and sometimes positive.

In summary, there are a number of geographic, social and economic factors which affect the *growth rate* of per capita energy consumption, but at low levels of income per capita income is the single most important determinant. At low income levels, as per capita incomes rise, a more than proportionate increase in per capita commercial energy consumption is highly likely (Figure 6.2). On the other hand, after a certain point energy

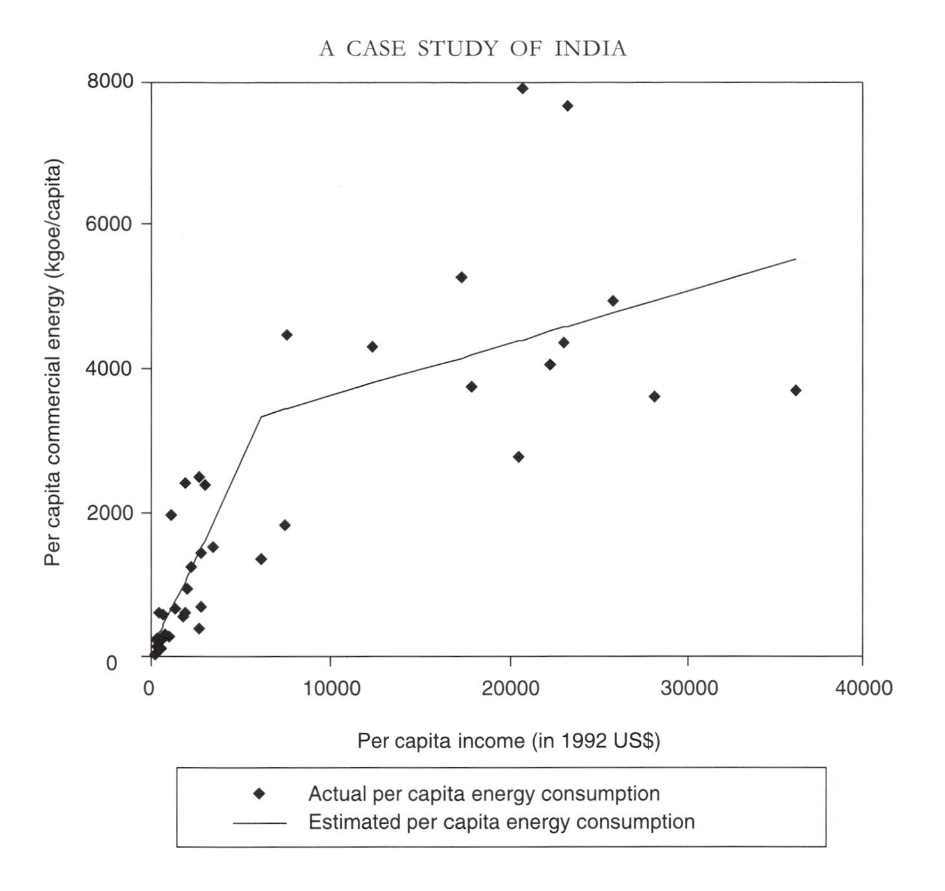

Figure 6.2 International comparison of per capita commercial energy use

intensity in an economy is negatively related to per capita income levels. Developing countries such as India and China have high energy intensities compared to developed nations, and substantial scope exists to improve the efficiency of energy use in these economies. Therefore, with increasing per caput incomes there are two opposing forces in operation: one which is likely to increase energy consumption (income effect) and the other which will reduce energy consumption (efficiency effect). The former is likely to be more dominant for India over the next two to three decades. However, in the absence of reliable data for traditional energy sources it is difficult to predict when substitution into commercial energy will stop, and commercial energy intensities would be able to decline.

MACROECONOMIC FRAMEWORK

The macroeconometric model developed and estimated for India was based on the structure suggested for developing economies by Allen *et al.* (1994). The Indian model consists of an aggregate demand side, a supply side, the government sector, the balance-of-payments accounts and an energy sector. The structure of the prototype model has been extended to include an energy sector.

The demand side consists of private consumption, total investment, government consumption expenditure and net exports. The government sector includes government revenues, government expenditure and the resultant budget deficit. The deficit determines the size of the domestic debt which together with the reserves of foreign exchange determines the stock of money supply in the economy. The supply side is represented through the specification of the price level in the economy. The supply side also includes the real wage rate, imports prices, export prices as well as the real exchange rate. The external or the balance-of-payments accounts determine the deficit (surplus) of the external account, and give the extent of accommodating flows, as well as the change in foreign exchange reserves, required to balance the account. In contrast to the past, when fixed exchange rates prevailed, the future real exchange rates are determined after a number of sensitivity runs, to determine real exchange rates which result in a reasonable balance in the external sector.

An illustrative example of the demand side functioning of the macro model can be given by tracing the consequences of an increased deficit in the government sector. The government finances the deficit by increasing the public debt and thus the money supply. This has two effects: in the short term it implies an increase in nominal wealth which increases consumption; in the medium term it increases the level of inflation, which reduces real wealth. In the short term the net effect will be an increase in apparent wealth, and so consumption and GDP will increase. This higher level of GDP results in higher imports, thereby reducing foreign exchange reserves. Higher domestic prices would imply an increase in export prices leading to a reduction in the level of exports and a further deterioration of the balance of payments. In the absence of exchange rate effects the economy stabilises through the following effects: declining foreign exchange reserves reduce the money stock in the economy and so lower expansion in the money supply. A higher level of income, due to the temporary increase in real wealth, increases the revenue receipts of the government, leading to a lower budget deficit and consequently a smaller increase in public debt and money supply. This lower level of money supply reduces inflation and lowers the balance of payments deficit through higher exports.

The supply side of the model links the price level in the economy to labour, energy and import prices; if energy prices rise it models the effect on productivity of the substitution out of energy and into other inputs. However, the maturity and structure of the Indian economy, and the lack of reliable data, precludes the estimation of a fully consistent supply side as in Chapter 5.

As most of the time-series data were non-stationary and integrated of order one, the Johansen procedure was used to estimate the long run consumption, export, investment and the price level equations in the model. The long run cointegrating vectors were then embedded inside an error correction formulation. The details of the model formulation and estimation are given in Gupta (1995). Below we describe the derivation of the energy sector in detail as this is the focus of the study.

Estimating an energy demand model for India

Energy demand in an economy depends on the level of output in the economy, the price of energy relative to the general price level, the price of other inputs and on other non-price factors such as the resource endowment, the composition of GDP, technology and government policies.

The consumption levels for total fossil carbon-based energy – coal, oil and gas – are shown in Figure 6.1. The carbon energy intensity in the economy (measured in kilograms of oil equivalent per rupee), that is, the ratio of fossil fuel consumption and GDP, is shown in Figure 6.3. Unlike the data for developed countries, both consumption levels and energy intensity are increasing over time.

In estimating the model, we only consider consumption of fossil fuels as these produce carbon dioxide emissions and account for nearly 90 per cent of the total commercial energy demand. Other primary sources of energy are accounted for by the trend factor. The energy sector model determines fossil fuel prices, aggregate fossil energy demand, and the shares of coal, oil

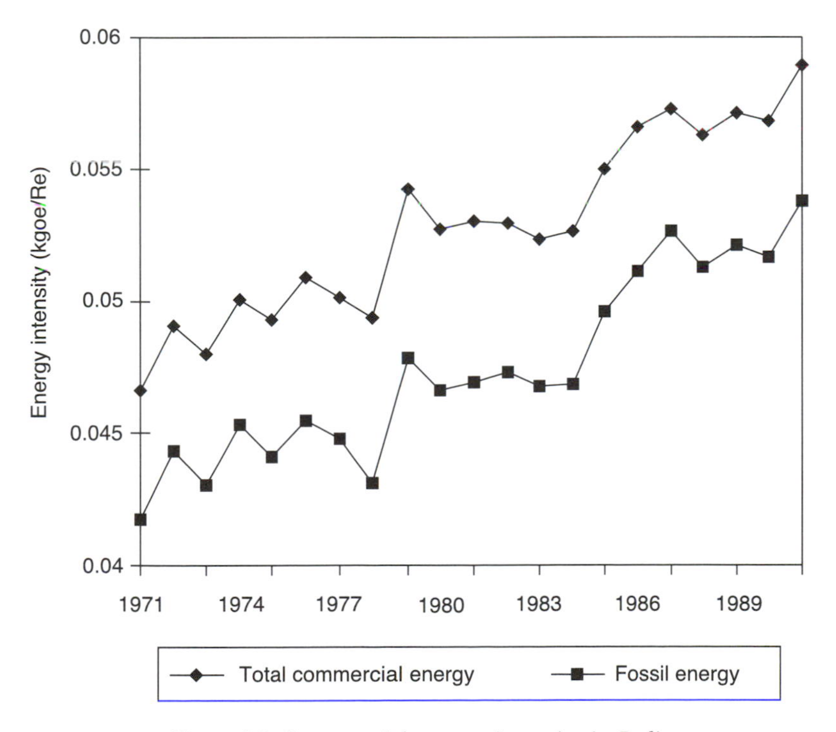

Figure 6.3 Commercial energy intensity in India

Table 6.3 Time series properties of the data: augmented Dickey Fuller (ADF) test

Variable[a]	*ADF(1)*	*ADF(2)*	*ADF(4)*
ln CEPR	−2.7	−2.7	−1.9
ln CE	1.1	0.9	2.8
ln YD	0.7	0.5	0.9
ln CEYD	−1.5	−1.6	−0.5
D ln CEPR	−2.9	−3.1	−2.5
D ln CE	−4.4	−3.7	−3.3
D ln GDP	−4.5	−3.9	−3.8
D ln CEYD	−4.8	−4.2	−4.4

a Explanation of variables appears in accompanying text.

and gas. The aggregate carbon energy prices used in the model are the average retail prices of coal, oil and natural gas, weighted by the share of each fuel in total fossil fuel consumption.

Time series data for carbon energy demand, GDP and carbon energy prices were found to be non-stationary. Table 6.3 gives the results for the stationarity tests for the relevant variables for the energy demand equation. The critical minimum value for the Augmented Dickey Fuller (ADF) statistic, above which the series is stationary, is −2.93. The ADF statistics for the natural logarithms (ln) of carbon energy (CE), carbon energy intensity (CEYD) and real price of carbon energy (CEPR) fail the test for stationarity, implying that they are not integrated of order zero (I(0)). Next, the first differences of the logarithms (D ln) of all the variables are tested for stationarity. Here the ADF statistic is greater than the critical value, implying that the first differences of the variables are stationary or that the variables themselves are integrated of order one. Therefore, a valid cointegrating relationship can exist between these variables to give a lower order of integration (I(0) in this case). Table 6.3 gives the ADF statistics for lag lengths of 1, 2 and 4. The Johansen method of estimation was used to estimate the cointegrating vector for the long run carbon energy demand.

Long run aggregate carbon energy demand is modelled using the following functional form:

$$CE_t = A.CEPR^{\alpha}{}_t.GDP_t.e^{\gamma T} \quad (1)$$

where: A = constant, CE_t = carbon energy demand, $CEPR_t$ = real weighted price of carbon fuels, T = trend, GDP_t = GDP, α = long run price elasticity of carbon energy demand, and γ = parameter relating to the trend term.

Constant returns-to-scale are assumed, so the coefficient on GDP growth is restricted to unity. This pushes all changes in energy intensity into the price and trend terms. The long run price elasticity α would normally be expected to be negative, but the largest negative value for α was estimated as −0.028, implying a very low price responsiveness. This was expected, as the energy markets are supply constrained and the supply side factors have dominated the markets for fossil energy.

The trend represents other non-price factors (for example, a change in the composition of GDP as agriculture gives way to industry), as well as technological evolution and responses to policy changes. The parameter γ is related to the factors capturing the trend in energy intensity. The value estimated for the γ coefficient was found to be positive, indicating that in the past energy intensity has increased over time. As discussed before, the economy is a developing one and the energy intensity is expected to increase. However, the increase would be at a decreasing rate. Thus, for the future, the trend term is modelled such that the increase is not at a constant but at a declining rate. The trend term T is defined as:

$$T = T(-1) + 1 - Be^{\gamma TIME} - \lambda.(\ln CEPR_{simu} - \ln CEPR_{base}) \quad (2)$$

The term $Be^{\gamma TIME}$ in equation (2) accounts for the less than constant increase in energy intensity over time. In a policy simulation, the trend term is further affected by the policies undertaken. For example, if increasing energy prices through carbon taxation resulted in an improvement in the energy using technologies, removing the energy tax would not revert the economy to the earlier path of inefficient energy consumption as in the business-as-usual scenario. The difference in the logarithms of energy prices in the simulation ($\ln CEPR_{simu}$) and the base-case ($\ln CEPR_{base}$), multiplied by $-\lambda$, the technological response coefficient, models the technological response to policy. This effect is permanent and accumulates over time, as the specification for T includes its lagged value, thereby incorporating price-induced technical progress in the model. Since it is not possible to estimate directly the technological response coefficient for India, given the imperfect energy markets and lack of other relevant information, the value for λ is based on the estimations carried out for industrialised countries in Chapter 4.

The remainder of the energy model is composed of share and price equations for the three main fossil fuels. Future shares of oil depend on the relative prices of different fossil fuels. The share of gas is a function of time and coal accounts for the residual share. Oil prices are positively related to the world price of oil and to inflation (measured by the lagged value of the consumer price index). Coal prices are dependent on oil prices and on the level of inflation. Gas prices are linked to world oil prices and a positive time trend.

MICROECONOMIC ANALYSIS OF CARBON ABATEMENT OPTIONS

The low price responsiveness of energy demand and the fast structural change in the Indian economy mean that the econometric model is not adequate to assess the potential for carbon abatement. Therefore, as a first step to producing an integrated model of carbon dioxide abatement in

India, strategies to reduce net emissions of carbon dioxide without reducing end-use services were investigated. To limit the number of choices only proven technologies which could be implemented by the turn of the century are considered. For each strategy, the realistic potential, the investment required, and the CO_2 emissions saved/fixed over the lifetime of the option are estimated.

As options with different life periods are being compared it is more meaningful to use the annualised investment cost of a project. The capital recovery factor (CRF) was used to annualise investments. The CRF is the inverse of the annuity factor and distributes the investment cost in equal proportions over the life of the asset, taking into account a discount rate. Investments were annualised using discount rates of 1, 2, 5 and 10 per cent. The lower rates of 1 and 2 per cent represent the case arguing for the social rate of time preference as the appropriate discount rate for long term environmental projects. Given the environmental implications of this study, subsequent analysis was carried out using a discount rate of 2 per cent. The ratio of annualised investment to annual carbon savings gives the specific investment cost (SIC) per unit of carbon reduction for each option:

$$SIC_i = I_i / AC_i \tag{3}$$

Where: SIC_i = specific investment cost for option i, I_i = annualised investment for option i, AC_i = average annual carbon savings for option i.

The different CO_2 abatement measures identified were ranked in ascending order of specific investment costs for CO_2 reduction. An investment–abatement curve was then generated by plotting the cumulative potential for the annual reduction in CO_2 emissions on the horizontal axis and the cumulative annualised investments on the vertical axis, as illustrated in Figure 6.4.

For India, twenty-six different options to reduce carbon emissions were considered. These are listed in Appendix 6.1. These measures cover energy supply as well as demand side efficiency. Details for each carbon abating technology can be found in Gupta (1995).

It is important to note that this investment cost curve is different from the conventionally defined cost curves. It shows the annualised *investment* per tonne of carbon saved every year and not the incremental costs. Conventionally, cost curves have shown the incremental cost of an option *vis-à-vis* the alternative strategy. These consist of the incremental costs for capital, operation and maintenance and fuel. Investment–abatement curves present the relationship between cumulative investment and cumulative carbon savings.

The reason for our approach is that conventional cost-benefit analysis for developed countries shows a number of technological options with negative or zero net costs. Despite their favourable economics, these opportunities are assumed not to have been exploited because of market imperfections or hidden transaction costs. However, in developing economies the main factor responsible for the non-implementation of these

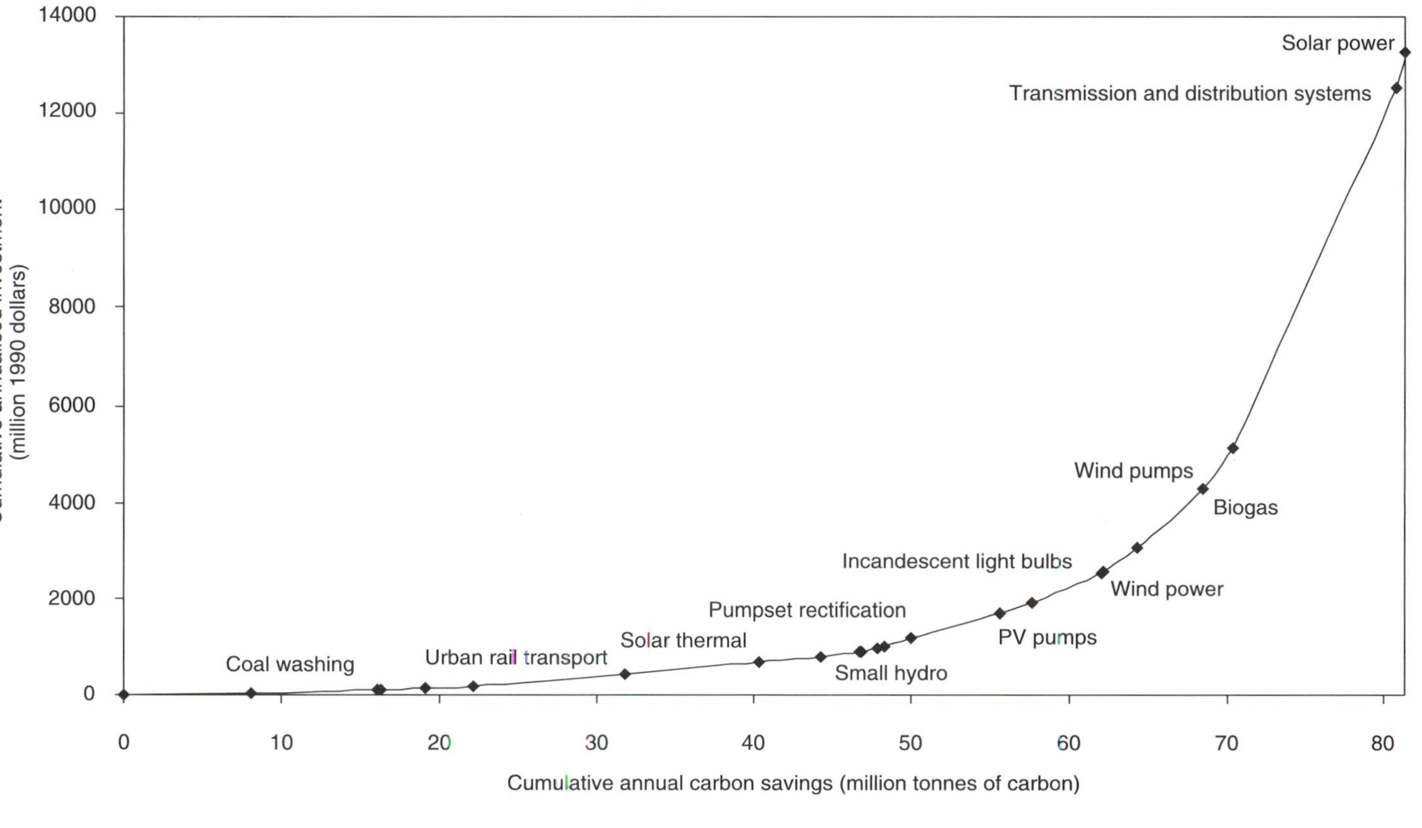

Figure 6.4 Investment–abatement curve for India

options is the acute shortage of investment capital. Therefore, to determine the diffusion of CO_2 abating technologies investment cost curves are more suitable than incremental cost curves. Also, in DCs complete and reliable data on operating and maintenance costs of the new systems are not easily available, as most technologies are imminent but not proven. As associated costs such as management costs are undervalued or ignored, reliable conventional cost curves cannot be generated for DCs.

The objective here is to assess the economy-wide potential for carbon abatement which requires a macroeconomic framework. The aggregate level of investment depends on the cost and availability of capital in the economy, and at the micro level the investment level determines the extent of technological diffusion. The integration of the two approaches can be achieved through the investment–abatement relationship which aggregates detailed micro level information, giving an appropriate link to incorporate micro level information into a macro framework.

INTEGRATION OF MICROECONOMIC AND MACROECONOMIC ANALYSIS

Two polarised approaches in modelling are currently being used to study the economics of global warming. On the one hand, there are the top-down models which oversimplify technological progress and development, and on the other, the engineering models (bottom-up) which evaluate technology in great detail but exclude all macroeconomic linkages and economic barriers which prevent a full realisation of the technological potential. Although the importance of the synthesis of these two approaches is well recognised, the actual integration has not been done for India before.

In a capital-constrained economy the level of total investment is a very important factor. In response to an abatement scenario investment in carbon mitigation options could be determined endogenously, or specified exogenously in the macro model. The carbon savings in the economy can then be estimated based on the cost of technologies considered in the micro analysis. This approach is to be contrasted to that taken for developed countries in Chapter 5, where the amount of investment in carbon abatement was inferred from the energy price response and the prevailing interest rate. This is a robust inference in industrialised countries, because markets are far more efficient and there are no capital shortages, but a similar approach is not suitable for a developing country and so the micro data supplements in this role.

To implement the above approach an investment–abatement relationship was obtained by regressing annualised investments against the annual carbon savings to produce a simple mathematical relationship for use inside the macro model. It is assumed that the policy-related investments in carbon abating technologies are operational on an average for ten years. Therefore, in the model, the 'effective' level of investment (CARINV) is the annualised investment in carbon abating technologies aggregated over the past ten years. These determine the level of carbon savings (CARSAV) based on the investment–abatement relationship derived in the micro level analysis of technologies.

The investment–abatement curve for India is relatively flat at first, but beyond a certain level of carbon abatement it is very steep. To approximate this, two different functional forms were fitted, one for the more elastic part and the second for the relatively inelastic portion. Equations (4a) and (4b) represent these functional forms for the investment–abatement curve.

$$\ln(CARSAV_t) = \text{constant} + \tau.\ln(CARINV_t) \quad (4a)$$

$$(CARSAV_t) = \text{constant} + \tau.\ln(CARINV_t) \quad (4b)$$

Total carbon savings are translated into carbon-energy saved (CESAV) by using the carbon intensity of fossil energy in the previous time period. The ratio of carbon energy savings to GDP is subtracted from the carbon energy intensity of GDP (i.e. CE/GDP) in the long run equation for energy demand in the macro model. In the base-case scenario, total carbon investment is zero, while in a policy scenario there would be a policy initiative to invest in carbon abating technologies. The modified long run relationship for fossil energy demand is:

$$\ln((CE_t/GDP_t) - (CESAV_t/GDP_t)) = \text{constant} + \alpha.\ln(CEPR_t) + \gamma T + \nu_t \quad (5)$$

Equation (5) models the long run relationship incorporating the impact of investments in carbon abating technologies. The technological response and the declining energy intensity over time are included through the trend term, T.

The integrated framework enables the overall assessment of policies such as exogenous investment in carbon abatement technologies, carbon taxes and the complete or partial recycling of these revenues into carbon abating technologies, diverting domestic investments to carbon abating technologies, and changes in international energy prices. Examples of some of these policies are modelled below.

CARBON ABATEMENT SIMULATIONS FOR INDIA

A scenario approach has been adopted, that is, cases with different policy initiatives are considered and the model is used to determine their impact on the economy and its energy-related carbon dioxide emissions. The impact of a policy simulation is measured as a deviation from the base case. Analysing the results under different 'what if' cases gives useful insights about the likely impact of different initiatives undertaken to control CO_2 emissions. The model for India gives results for the economy till the year 2020.

Defining a base-case

The average rate of growth of GDP over the data period used for estimation (1976–1990) was 4.9 per cent per annum. In the base forecast case the economy grows at an average rate of 4.1 per cent per annum from 1991 to 2020. This overall growth rate of 4.1 per cent over thirty years is not unlikely, given that the economy has grown at nearly 5 per cent in the past.

Carbon energy demand has increased at 5.9 per cent over the data period. The demand projection for carbon energy in 2020 is 738.5 Mtoe implying an average annual growth rate of 4.9 per cent for 1991–2020. The growth rate decreases in each decade (Table 6.4), and energy-related carbon emissions grow almost as fast as total carbon energy demand. The difference (0.1 percentage point) is due to the declining carbon intensity of fossil fuels. In the base-case, carbon energy intensity increases, although at a decreasing rate, till 2018 and stabilises thereafter. The overall rate of growth of carbon energy intensity, in the forecast period (1991–2020), is 0.7 per cent per annum.

Four scenarios were compared with the base run. The description for each scenario is given below. In Scenarios 1 and 2 three different cases were considered and in Scenario 4 two cases were considered.

Abatement policy scenarios

Scenario 1: low carbon tax

A carbon tax of 200 rupees (1987 prices) per tonne of carbon, in real terms, is levied on fossil fuels from the year 2000. This tax is equivalent to approximately 10 per cent of the real average price of carbon energy in the early 1990s. This scenario reflects a moderate policy to control carbon emissions.

Scenario 2: high carbon tax

A carbon tax of 600 rupees (1987 prices) per tonne of carbon, in real terms, is imposed from the year 2000. This tax is equivalent to approximately 30

Table 6.4 Growth rates for GDP, carbon energy and CO_2 in the base-case (%)

	Data period	*Forecast period*			*Forecast average*
Time periods	*1976–1990*	*1991–2000*	*2000–2010*	*2010–2020*	*1991–2020*
GDP	4.9	4.6	3.9	4.0	4.1
Carbon energy demand	5.9	5.7	4.7	4.4	4.9
Real price of energy	1.4	0.7	0.5	0.5	0.54
Nominal energy price	9.0	6.0	4.4	4.3	4.9
Carbon energy intensity	0.9	0.9	0.8	0.4	0.7
Carbon emissions	5.8	5.6	4.6	4.3	4.9

per cent of the average real price of carbon energy in the early 1990s. This scenario signifies a stronger initiative from the government to control carbon dioxide emissions.

Scenario 3: investment diversion

Part of gross total investment in the economy is diverted to carbon abating technologies. Total investments in the rest of the economy decline and the government bears the cost of the investments in carbon reducing options. Two cases are considered, one with 0.5 per cent of gross investment and the other with 2.5 per cent of gross investment being diverted to carbon abating technologies.

Scenario 4: stabilisation of emissions

Optimal control techniques are used to estimate the tax required to stabilise emissions at the 2010 level. Two different cases for utilising carbon tax revenues are considered for this scenario.

Within the four different policy scenarios there are three different ways of utilising carbon tax revenues.

Case A All carbon tax revenues add to government revenue thereby reducing the government budget deficit. Other than imposing the tax the government takes no policy initiative for controlling the emissions of CO_2. This case measures the direct as well as indirect price effects of a change in energy taxes.

Case B Fifty per cent of the carbon tax revenues add to the government revenue so that the government budget deficit is reduced. The balancing 50 per cent of the tax revenues are invested in carbon abating technologies. The results from this case show the combined effects of policy initiative taken to promote carbon abating technologies, as well as the effects of changing energy prices.

Case C All the carbon tax revenues are invested in carbon abating technologies. There is no reduction in the government budget deficit as a result of this tax. These investments are not as productive as investments in the rest of the economy. In this case control of carbon emissions is a relatively high priority issue.

In Case A, the additional revenues are not used to offset existing taxes. The macroeconometric model for India projects a persistent government budget deficit till the year 2020 (although decreasing as a percentage of the GDP). The additional carbon tax revenues just reduce this budget deficit and so there is no offsetting of other taxes as a result of the carbon tax.

Scenarios 1 and 2: carbon tax

In Scenario 1, a carbon tax of 200 rupees per tonne of carbon, in real terms (1987 prices), is levied on fossil fuels from the year 2000 onwards, in Scenario 2 this is tripled to 600 rupees. Three different cases, A, B and C, for utilisation of carbon tax revenues are considered. The results for the three cases of this scenario are given in Tables 6.5 and 6.6. Table 6.5 gives

Table 6.5 Percentage change from base-case for Scenarios 1 and 2

	Case A		*Case B*		*Case C*	
Variable	*2010*	*2020*	*2010*	*2020*	*2010*	*2020*
Real price of energy						
Scenario 1	11.1	11.1	11.0	11.2	11.0	11.3
Scenario 2	32.4	32.8	32.4	32.9	32.4	33.4
GDP						
Scenario 1	−2.3	−2.3	−2.3	−2.3	−2.3	−2.3
Scenario 2	−5.8	−6.0	−5.8	−5.8	−5.8	−5.8
Carbon energy						
Scenario 1	−3.5	−4.5	−29.2	−22.8	−31.9	−24.4
Scenario 2	−8.8	−11.5	−38.2	−31.8	−40.5	−33.3

Table 6.6 Selected results from Scenarios 1 and 2

	Case A		*Case B*		*Case C*	
Variable	*2010*	*2020*	*2010*	*2020*	*2010*	*2020*
Tax revenues (bn 1987 Rs)						
Scenario 1	87	132	64	107	62	105
Scenario 2	246	366	166	282	160	276
Total investment in carbon abatement (bn 1987 Rs)						
Scenario 1	0	0	32	53	61	104
Scenario 2	0	0	82	139	158	272
Carbon emissions (MtC)						
Base-case	452	693	452	693	452	693
Scenario 1	435	660	319	535	308	524
Scenario 2	410	609	277	469	267	459
Carbon energy intensity (kgoe/Re)						
Base-case	0.051	0.053	0.051	0.053	0.051	0.053
Scenario 1	0.050	0.051	0.037	0.042	0.035	0.041
Scenario 2	0.049	0.049	0.033	0.038	0.032	0.037
Total carbon savings (MtC)						
Scenario 1	16.7	32.2	132.7	157.2	143.7	168.2
Scenario 2	41.7	83.6	174.7	223.6	184.7	233.6

the percentage change over the base-case and Table 6.6 gives the absolute values for some of the important variables. The focus is on variables where the policy instruments are expected to have an impact.

For the lower tax the real price of carbon energy increases by approximately 11 per cent. This increases inflation in the economy above the average annual rate of inflation in the base case of 4 per cent. In this scenario the average annual rate of inflation increases to 4.5 per cent in Case A, to 4.6 per cent in Case B and to 4.7 per cent in Case C. The average loss in GDP for the last ten years or the second half of the taxation period is approximately 2.3 per cent in all three cases.

Part of the decrease in carbon energy demand is due to the decrease in GDP. The remainder for Case A is due to the direct and indirect price effects, but Figure 6.5 shows that these effects are very small. The much larger decrease in carbon energy demand for Case B and Case C comes from the investment in abatement technologies which is obviously more effective than the price mechanism alone. The carbon tax revenues decrease from Case A to Case B to Case C (Table 6.6), in line with the decrease in the demand for carbon energy.

In Scenario 2 the level of carbon taxation is three times that in Scenario 1, and the results are similar in general form if different in magnitude. GDP losses are higher at 5.8 per cent over the base-case and the carbon energy savings are somewhat larger than in Scenario 1 (Figure 6.6). Carbon energy

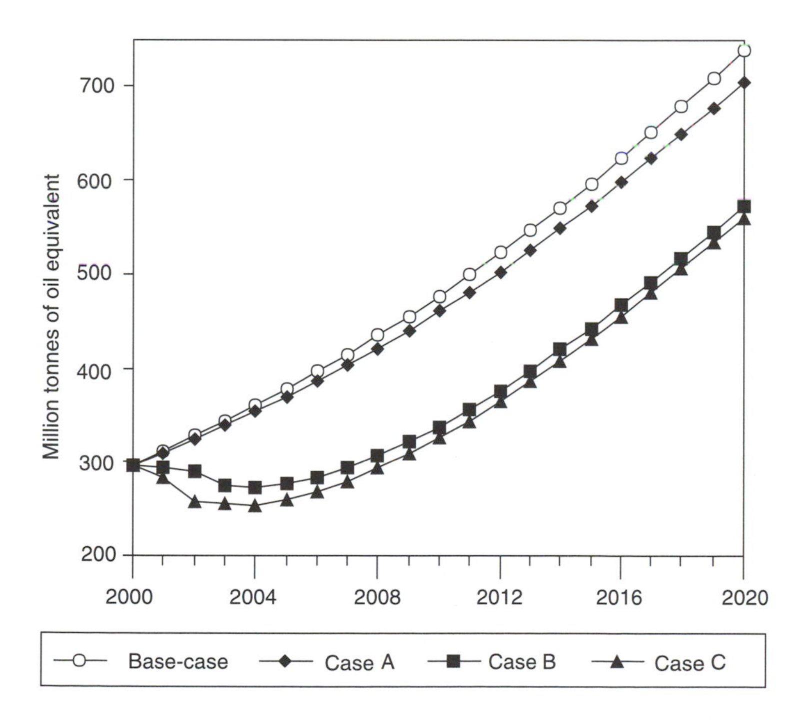

Figure 6.5 Carbon abatement in Scenario 1

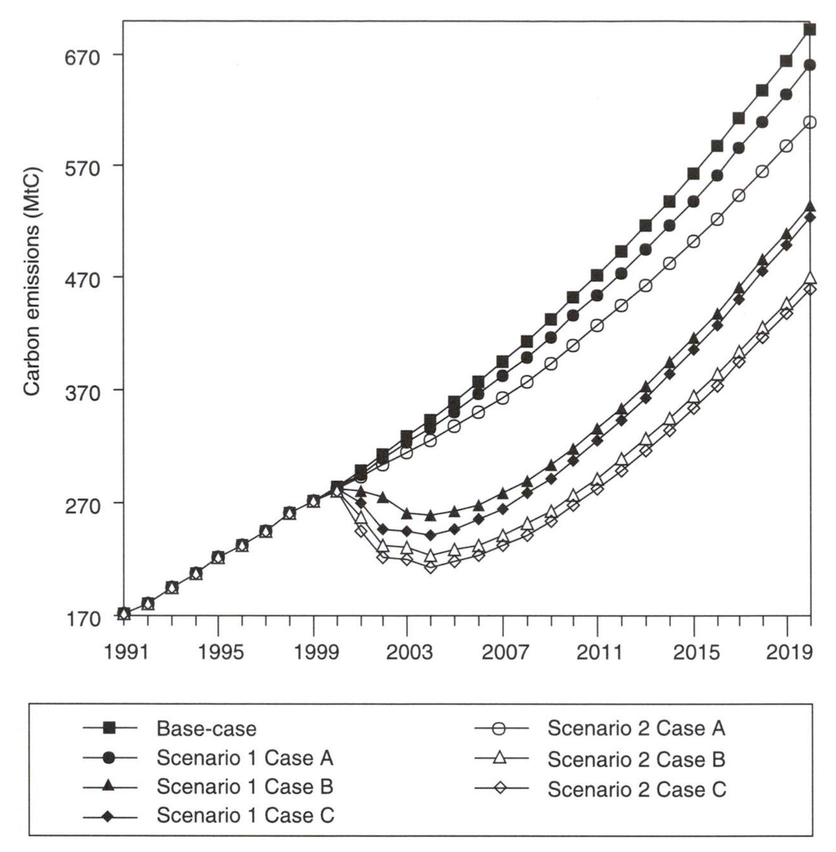

Figure 6.6 Carbon abatement in Scenarios 1 and 2

intensity of GDP decreases compared with the base-case and for the year 2020 is lower by 6.3 per cent, 38 per cent and 41.1 per cent for Case A, B and C respectively.

These results illustrate the limited marginal effectiveness of investments in carbon abating technologies. An increase of investments in carbon abating technologies, by 2.7 times (over those in Scenario 1 Case C) only increases carbon savings over the base-case by an additional 9 per cent, part of which is due to a lower level of GDP.

The following two observations emerge from the results:

- In Case A, the per centage of carbon savings increases over time, that is, a higher percentage of carbon is saved in 2020 in comparison with 2010. In cases B and C, a higher percentage over the base-case is saved in 2010 compared with 2020 (Table 6.5).
- Carbon savings increase from 157.2 million tonnes of carbon (MtC) in Case B, to 168.2 (MtC) in Case C, for the year 2020. Additional carbon

savings of only 11 MtC (6.8 per cent more than in Case B) result, despite a doubling of investment in carbon abating technologies (Table 6.6).

The non-linear relationship between investments in carbon abating technologies and the carbon saved (Figure 6.4) can explain these two observations. In the initial stages, it is possible to bring about large savings in carbon with relatively small investments in carbon abating technologies but further investments bring about less than proportionate increases in carbon savings. Increasing investment in carbon abating technologies, for Case B and C, bring about larger absolute savings, but are a smaller percentage of the growing base-case carbon emissions.

Compared to just relying on the price mechanism, investing the carbon tax revenues in carbon abating technologies increases the carbon savings substantially for the same loss in GDP. When 50 per cent of the revenues are invested in carbon abatement options, the carbon emission reductions are 25.7 per cent (Scenario 1 Case B). In Case C of Scenario 1 all the carbon tax revenues are invested in carbon abating technologies. Increasing investments in carbon abating technologies by nearly 100 per cent over Case B increases carbon savings by an additional 2–2.3 per cent. This reflects the limited capacity of micro-level investments in achieving additional carbon savings.

We can conclude from the results of Scenarios 1 and 2 that, given the low price elasticity for energy demand in the economy, the price mechanism is quite an inefficient way to abate carbon, with an 11 per cent reduction in energy related carbon emissions resulting in a loss of 6 per cent of GDP over the base-case. Policies for promoting investment in carbon abating technology can increase the carbon savings significantly. The limitation is that further investments in carbon abating technology do not result in proportionate increases in carbon savings.

Scenario 3: investment diversion

The objective of this scenario is to study the impact of diverting investments into carbon abating technologies from the other sectors of the economy. These investments in carbon abating technologies will be financed by, or subsidised by, the government, thereby increasing government expenditure. In Scenario 3.1 it is assumed that 0.5 per cent of the total investments in the economy are diverted towards carbon abating technologies, and in Scenario 3.2 it is assumed that 2.5 per cent of the gross investments are similarly directed. Table 6.7 gives the percentage change in the relevant variables over the base-case, and Table 6.8 gives some absolute values for the two cases.

In Scenario 3.1, the price level in the economy increases by 2.1 per cent and 3.3 per cent over the base-case for 2010 and 2020, respectively. Government spending is more than in the base-case due to the additional expenditure of investing in carbon abating technologies. This increases the

Table 6.7 Percentage change from base-case for Scenario 3

	Scenario 3.1		*Scenario 3.2*	
Variable	*2010*	*2020*	*2010*	*2020*
Real price of carbon energy	0.0	0.1	0.2	1.4
Nominal price of carbon energy	2.2	3.4	10.3	17.0
Coal price	2.1	3.3	10.0	16.3
Oil price	2.3	3.6	10.9	17.9
CPI	2.1	3.3	10.1	16.6
GDP	−0.4	−0.4	−1.7	−1.8
Carbon energy	−21.7	−15.8	−30.7	−21.4
Carbon emissions	−21.7	−15.9	−30.7	−21.4

Table 6.8 Scenario 3: selected variables

	Scenario 3.1		*Scenario 3.2*	
Variable	*2010*	*2020*	*2010*	*2020*
Total investment in carbon abatement (in bn 1987 Rs)	12.02	18.6	59.0	91.9
Carbon emissions (m tonnes of carbon)	353.8	582.8	313.1	544.4
Total carbon savings over base-case (m tonnes of carbon)	97.9	109.8	138.6	148.2
Carbon energy intensity of GDP (kgoe/Re)	0.0397	0.0444	0.0356	0.0420

government budget deficit. Money supply increases to finance the larger deficit, giving a higher price level in the economy. For Scenario 3.2, the increase in the CPI is five times the increase in Scenario 3.1. This is expected as the investments in carbon abating technologies increase by a factor of five, from 0.5 per cent to 2.5 per cent of the gross total investments in the economy.

The loss in GDP is 0.4 per cent and 1.8 per cent (in 2020) for Scenario 3.1 and 3.2, respectively (Table 6.7). The lower GDP is partly due to a lower level of investment, and partly due to lower consumption expenditure and exports in the economy. Although the aggregate investment in the economy does not change, the investments diverted into carbon abating technologies are less productive and hence affect output. The price level is higher, due to the larger government budget deficit and lower gross investments. The increase in the price level affects real wealth negatively, lowering the level of consumption.

Carbon emissions (Table 6.8) reduce as a result of the operation of the carbon abating technologies, and a small contribution is due to the lower level of GDP (0.4 per cent and 1.8 per cent for Scenario 3.1 and 3.2, respectively). The real price of carbon energy is nearly the same as in the base case for Scenario 3.1, but higher for Scenario 3.2. The change in the CPI in Scenario 3.2 affects the real aggregate price of carbon energy by a small percentage. Therefore, the total carbon savings in Scenario 3.2 include some carbon savings effected due to higher carbon energy prices. These are very small, given the small change in aggregate real price of carbon energy and the very low price elasticity for carbon energy demand. The shares of different fossil fuels show no substantial change, as the relative prices of coal and oil do not change by much over the base-case. The absolute carbon saved is higher in 2020 than in 2010 for both the cases, but in percentage terms the carbon savings over the base-case decline over time. As in the other scenarios, this is explained by the non-linear relationship between investment in carbon abating technologies and carbon savings. Over time, carbon savings are increasing, but the rate of increase is less than the growth of carbon emissions in the base-case, therefore carbon savings decline as a percentage of base-case carbon emissions.

Diverting investments into carbon abating technologies from the rest of the economy is a less expensive method of reducing carbon emissions, compared with a policy of carbon/energy taxation. Diverting 0.5 per cent of the gross investments saves 18.3 per cent of carbon emissions and the loss in GDP is 0.4 per cent (Scenario 3 Case 1) and diverting 2.5 per cent of the investments saves 25.2 per cent of carbon and the loss in GDP is 1.8 per cent (Scenario 3 Case 2). Though the aggregate level of investment in the economy does not change, the loss in GDP can be explained as a result of diversion of investments in less productive sectors.

However, given the lack of detail in the supply side of the macro model, especially the role of investment in productivity and growth, these GDP losses may be understated. Still, they would have to increase substantially to match those from the energy taxation in Scenarios 1 and 2.

Scenario 4: emissions stabilisation

Under the Framework Convention on Climate Change (FCCC), industrialised countries are encouraged to constrain their emissions at 1990 levels by the turn of the century. No commitment is required from the developing countries as yet. However, for stabilisation of global carbon emissions and their subsequent reduction to give sustainable atmospheric concentrations, developing countries will have to meet some targets in the next century. To investigate such scenarios optimal control methods are used to determine the level of carbon tax which would stabilise carbon emissions at 2010 levels over the simulation period.

Optimal control techniques are used to determine a tax regime which would stabilise carbon emissions beyond 2010, at the base-case 2010 level of emissions. The state variable is the level of carbon emissions and the control variable, or the policy instrument, to achieve the desired values for the state variable is carbon tax. The solution period starts before 2010 as it is likely that abatement policies would be implemented prior to 2010, in order to control emissions beyond 2010. For example, the policy makers may impose a small carbon tax before 2010, and use the carbon tax revenues to finance investment in carbon abating technologies, to meet the target for carbon emissions beyond 2010. The solution period for the optimal control runs was 2007 to 2020 and was divided into two sub-periods of equal length. The difference between the stabilisation target and the base-case will increase over time as the base-case emissions are increasing over time. The carbon tax would have to increase over time, to keep carbon emissions below a given target value. Dividing the solution period into sub-periods allows a higher tax rate to be imposed in the later years. The span of the sub-period should not be very small, because frequent policy changes are inconvenient and expensive to administer and difficult to model.

Carbon emissions in the last quarter of 2010 are 115 million tonnes of carbon. This is the target value for carbon emissions in the objective function for the second sub-period in the optimal control runs. In this scenario, Case A and Case C were considered to see how the economy meets the objective of controlling its emissions with and without investments in carbon abating technologies. Different values are used for two parameters: the technological response coefficient, λ; and the price elasticity for carbon energy demand, α. The sensitivity analysis for price elasticity was done to evaluate the costs of controlling carbon emissions at higher (than the estimated) price elasticities. The price elasticities for energy demand used in different optimal control runs were: the estimated value of -0.028, -0.15 based on the average of the long run price elasticities of demand for carbon energy estimated for nine OECD countries (Boone *et al.* 1995), and a high value of -0.5. Runs with higher price elasticity represent an economy with more price responsive energy markets.

It is not possible to estimate the technological response parameter for India, given imperfect energy markets and lack of other relevant information. The value for λ in this study is based on estimations carried out for nine industrialised countries in Chapter 4. The technological response parameters for OECD countries were estimated using Kalman filter techniques (Smith *et al.* 1995) measuring the permanent technological response to price changes. The initial value of λ, determined from the Kalman filter estimates for OECD countries, is 0.856 (Smith 1994). In one set of runs, λ is increased from 0.856 to 1.5 to represent a greater technological response to a change in energy prices. A higher *rate* of technological response is likely in a developing country than in the industrialised countries, given the initially lower level of technological diffusion in a developing economy.

The cost of a carbon tax is measured as the average loss in GDP in the last seven years of the solution period, that is, from 2014 to 2020. The costs in the first sub-period can be considered to be transitional costs. The results are given in Table 6.9. For the estimated values for price elasticity (-0.028)

Table 6.9 Results for stabilisation of emissions

Run number	*Price elasticity of energy demand*	*Technological response coefficient*	*Case*	*Carbon tax rates (R/tC)* 2007–2013	2014–2020	*%age reduction in carbon emissions*	*Average loss in GDP (%)*
1	−0.028	0.856	A	688	9903	23.3	13.2
2	−0.028	1.5	A	878	5857	24.0	11.9
3	−0.028	0.856	C	0	1304	22.5	3.4
4	−0.028	1.5	C	0	1132	22.3	3.0
5	−0.15	0.856	A	165	4477	23.4	8.2
6	−0.15	1.5	A	319	3452	23.8	8.0
7	−0.15	0.856	C	0	826	22.9	2.3
8	−0.15	1.5	C	0	755	22.8	2.2
9	−0.50	0.856	A	0	1534	23.9	3.8
10	−0.50	1.5	A	38	1414	24.0	3.8
11	−0.50	0.856	C	0	416	23.3	1.3
12	−0.50	1.5	C	0	402	23.3	1.2

and the original value (0.856) for the technological response coefficient and Case A (with no investment in carbon abating technologies), the solution of a carbon tax of 9903 rupees per tonne (R/tC) increases the average price of carbon energy by over 400 per cent (Run 1), which is very high and rather unrealistic. In the high price elasticity (−0.15), high technological response coefficient (1.5), Case A (Run 6, Table 6.9), the average price of carbon energy increases by 150 per cent with a carbon tax of 3452 R/tC. Runs with a higher price elasticity of −0.5 were made (Runs 9 and 10), which resulted in a 61–66 per cent increase in the price of carbon energy and a 3.8 per cent loss in GDP. Therefore, the range of GDP losses for stabilisation using energy taxes with no technology investment is 3.8 per cent to 13.2 per cent depending on the assumptions used (Table 6.9).

With investment of all carbon tax revenues in carbon abating technologies (Case C), the estimated price elasticity and the original technological response coefficient (Run 3, Table 6.9), stabilisation involves a 56 per cent increase in average carbon energy prices (a carbon tax of 1304 R/tC) and the average loss in GDP is 3.4 per cent. With a higher price elasticity (−0.15) and a higher technological response coefficient (of 1.5), the average carbon energy price increases by 33 per cent and the average loss in GDP is 2.2 per cent (Run 8). A further increase in the elasticity to −0.5 (Run 11, Table 6.9) lowers the tax rate, and the carbon energy price increases by only 18 per cent and the loss in GDP is only 1.3 per cent. Therefore, with investment in carbon abating technologies the costs of carbon abatement are lower, the average loss in GDP being in the range of 1.2 to 3.4 per cent.

Figure 6.7 shows that the carbon emissions almost stabilise, even though the different parameter assumptions result in quite a large variation in base emissions before 2010. For Case A runs, the rates of tax are very high which result in a big loss in GDP. Case C runs have lower tax rates (the highest tax rate increases energy prices by about 56 per cent) and lower loss in GDP. For Case C, taxation in the second sub-period is sufficient to achieve the target and the tax rate is zero in the first sub-period.

The results from these runs indicate that, to stabilise carbon emissions, the carbon tax by itself would impose high costs on the economy. Irrespective of the values for price elasticity and for the technological response coefficient, the costs for stabilising carbon emissions are always lower in Case C runs, that is, when additional investments are undertaken in carbon abating technologies.

Optimal control analysis illustrates the interplay between the price elasticity and the tax rate required to stabilise emissions at the 2010 level. As the responsiveness to price changes increases, lower taxes are required. A higher value for the technological response coefficient (1.5) requires a lower rate of tax for a given level of price elasticity. The model quantifies the difference in tax rates for different values of the technological response coefficient, λ and α. For example, in Case A, for $a = -0.028$, tax rates reduce by 40 per cent for the higher value of λ. The technological response

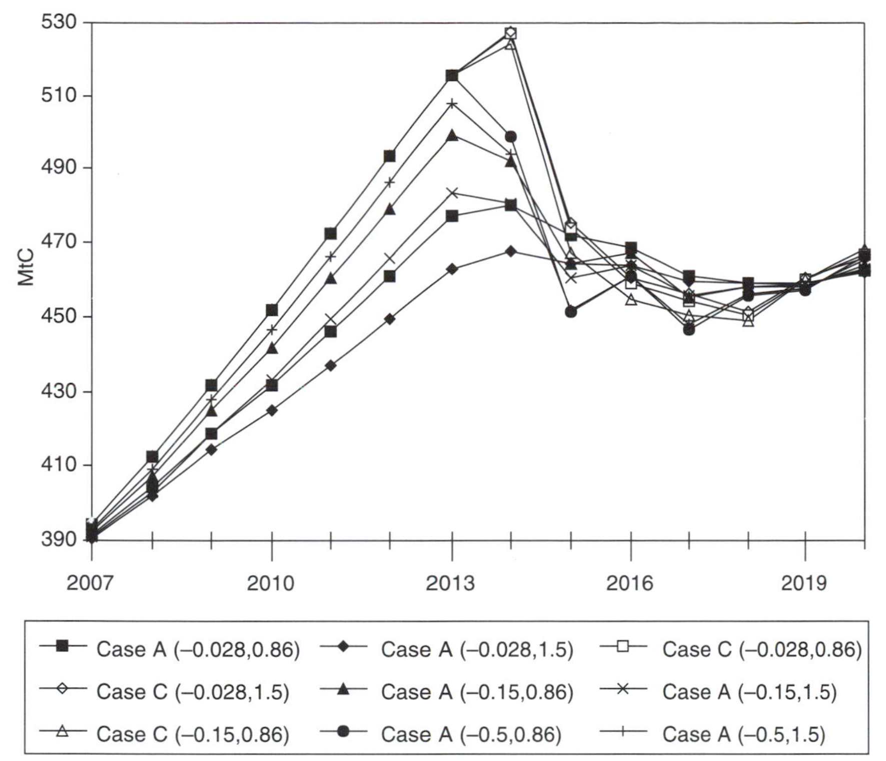

Figure 6.7 Stabilisation of emissions for different policy and parameter assumptions

coefficient is less important at higher price elasticities, for example, for Case A, for a = -0.5, the tax rate declines by only 8 per cent for the higher value of $\lambda = 1.5$.

Figure 6.8 summarises the range of stabilisation taxes for the different values of price elasticity of carbon energy demand and technological response parameter. It graphically shows that for India the largest differences in results come from the structure of the model – that is, the presence or absence of explicit carbon saving investments – rather than from variations in the estimates of economic parameters.

CONCLUSIONS

Given the importance of developing country emissions in the future it is vital that their energy markets are understood and the potential for future emissions and abatement quantified. It is likely that restrictions will have to be imposed on these countries before they achieve a mature industrialised economy. These may be subsidised by the North or be binding restrictions with no compensation. The main features of developing country energy markets are increasing energy intensity, low or positive price elasticities, lack

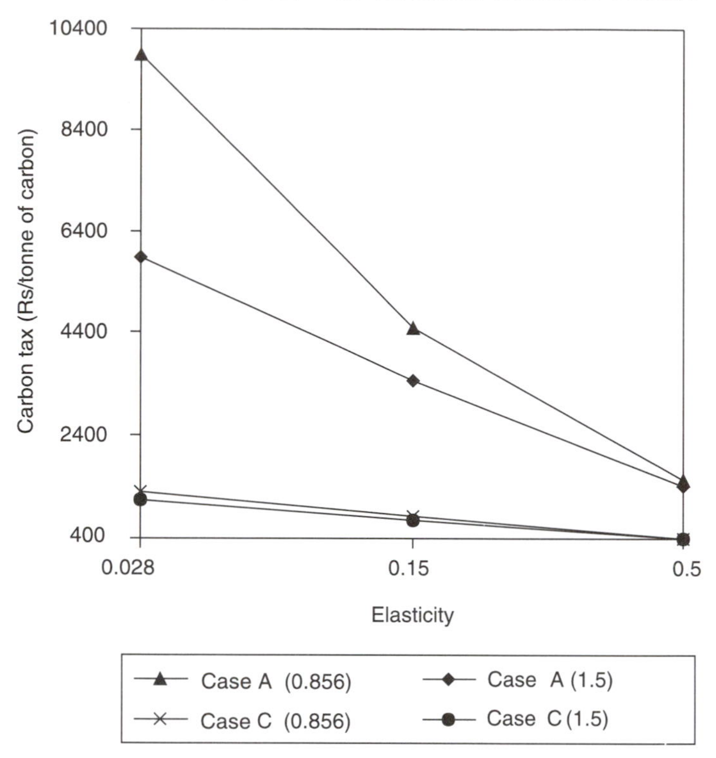

Figure 6.8 Taxes needed to stabilise emissions at 2010 levels under different assumptions

of investment and supply constraints resulting from large scale market failures, or massive government intervention in price setting and infrastructure. This intervention may support other policy goals, such as equity, but it must also be compatible with environmentally efficient resource use, something which has not been observed in the past.

By combining macroeconomic and microeconomic approaches we have constructed a model to quantify the cost of carbon abatement for India. Though relying on empirical estimation for the majority of parameters, the weakness of data and massive structural change in the economy undermines some econometric measurements, which had to be taken from studies on developed economies. Nevertheless this represents a robust empirical methodology to model policy options in the short to medium term.

The model simulations have shown that for developing countries, which have high energy use growth and very low price responsiveness, carbon taxes by themselves are a very expensive way of reducing carbon emissions. Investing the carbon tax revenues or diverting investments from other sectors of the economy into carbon abating technologies results in substantially cheaper carbon savings. However, beyond a certain level, micro level investments have limited marginal effectiveness in achieving additional carbon savings. This is due to the non-linear relationship between carbon abatement and the level of investment in carbon abating technologies.

The optimal control analysis is useful as it quantifies the magnitude of the carbon tax required to stabilise emissions at 2010 levels, in an environment of uncertainty for the key parameters. Stabilisation of carbon emissions using carbon taxes only could result in a loss of GDP of up to 13 per cent. Under more optimistic assumptions, and with re-investing of the carbon tax revenues into carbon abatement options, the loss would only be in the range of 1–3 per cent. As the responsiveness to price changes increases, lower taxes are required, and over this time range the technological response coefficient is less important at higher price elasticities.

The above scenarios are not intended to be prescriptive but rather to be illustrative. A real policy to control carbon energy demand would be a combination of a number of the different approaches considered in the above scenarios. For example, a concrete policy to control carbon emissions could include a small carbon or energy tax, some international transfer of resources for investment in carbon abating technologies with a matching domestic contribution. One fact does seem to be apparent, however: large scale technological transfer and supplementary investment funds are likely to be needed if developing countries such as India are going to be able to control emissions substantially without severely hurting their economies.

APPENDIX 6.1

Table 6.10 Potential and cost of various CO_2 emissions reduction options for India

Technological options	*Potential annual CO_2 abatement (MtC)*	*Investment cost (M$US)*	*Annualised investment ($ million)* 1%	2%	5%	10%	*Specific investment cost ($/tC)* 1%	2%	5%	10%
Electricity generation										
Coal washing	8.11	554.3	25.2	28.4	39.3	61.1	3.1	3.5	4.9	7.5
Retrofit Gas Combined Cycle	3.92	2285.5	103.8	117.1	162.2	251.8	26.5	29.8	41.3	64.2
Reduction in T&D losses	10.43	121000.0	6703.4	7400.0	9704.2	14217.0	642.9	709.0	930.4	1362.7
Industrial sector										
Improved housekeeping	2.8	364.8	38.5	40.6	47.24	59.4	13.8	14.5	16.9	21.2
Energy-efficient equipment	8.0	1094.5	60.7	66.9	87.83	128.6	7.6	8.4	11.0	16.1
Upgradation of technology	2.5	1824.1	82.8	93.4	129.42	201.0	33.1	37.4	51.8	80.4
Transport sector										
Increasing bus fleet	0.98	775.3	81.9	86.3	100.4	126.2	83.8	88.3	102.8	129.1
Metro rail system	3.05	1530.0	46.6	55.9	89.2	156.5	15.3	18.3	29.2	51.3
Enhanced rail freight	4.2	30860.0	1049.4	1234.5	1884.7	3199.0	249.9	293.9	448.7	761.9
Domestic sector										
Improved firewood stove	0.12	78.3	21.3	21.7	23.0	25.2	191.2	195.5	208.6	231.1
Tube fluorescent lighting	1.98	1890.0	199.6	210.4	244.8	307.6	100.8	106.3	123.6	155.4
Compact fluorescent lighting	4.37	4490.0	586.8	612.9	694.7	841.6	134.3	140.3	159.0	192.6

Agricultural sector										
Pumpset rectification	5.62	4867.1	513.9	541.8	630.3	792.1	91.4	96.4	112.2	140.9
Deployment of renewable energy technologies										
Biogas plants	1.84	7382.9	779.5	821.9	956.1	1201.0	423.6	446.7	519.6	653.0
Solar thermal systems	9.6	2108.6	222.6	234.7	273.1	343.0	23.2	24.5	28.4	35.8
Electricity from renewables										
Biomass	8.59	3257.1	234.9	253.5	313.8	428.2	27.3	29.5	36.5	49.8
Wind	2.15	8571.4	475.0	524.2	687.8	1006.0	220.9	243.8	319.9	468.3
Small hydro	1.7	3428.6	132.8	153.1	223.0	363.7	78.2	90.1	131.2	213.9
Sewage sludge	0.0715	86.5	4.8	5.3	6.9	10.2	67.0	74.0	97.0	142.0
Distillery effluent	0.2005	33.7	1.9	2.1	2.7	4.0	9.3	10.3	13.5	19.8
Municipal solid waste	0.458	369.1	39.0	41.1	47.8	60.1	85.1	89.7	104.4	131.2
PV pumps	0.043	108.0	4.9	5.5	7.7	11.9	114.0	128.7	178.2	276.7
Windpumps	0.017	128.6	7.1	7.9	10.3	15.1	419.1	462.5	606.9	888.4
Solar energy	0.573	11428.6	633.3	698.9	917.1	1342.0	1105.0	1219.0	1600.0	2342.0
Afforestation	136.5	4000.0	4000.0				29.3			

Part III

THE INTERNATIONAL ECONOMICS OF CLIMATE CHANGE

7

OPTIMAL CLIMATE CHANGE POLICY

Theory and practical relevance

INTRODUCTION: OPTIMISATION AND THE FCCC

In Chapter 2 we argued that explicit cost/benefit trade-offs are not currently written into the Framework Convention on Climate Change (FCCC), because current information about the impacts of potential climate change, and their economic valuation, is so uncertain. Under such large future uncertainties the 'precautionary principle'[1] was intended to drive international action, and this is compatible with emissions and/or concentration targets in the future, but not with welfare optimising taxes set to a notional marginal damage cost of CO_2 emissions. In this chapter we expand on this simplification of the arguments surrounding optimisation, placing it inside both a political and a theoretical economic context.

The drafting of this part of the FCCC was very deliberate, and resulted from highly adversarial political debate over the approach that should be taken. The idea of optimising the level of carbon dioxide abatement may be a standard *economic* viewpoint on the global warming problem, but it has been deliberately marginalised in the current negotiation process. As well as the problems of uncertainty, the other important reasons for this involve the ethics of inter-generational equity and the international distribution of future damages, and the sharing of the current abatement burden (often called intra-generational equity).

It is often argued that the precautionary principle implies a form of inter-generational equity where the earth's essential natural resources and ecosystems must be passed unchanged, either in specifics or in aggregate value,[2] from generation to generation (Weiss 1993). In this way all irreversible damage is avoided and we make no, or few, assumptions about the constituents of future 'needs'. Such conditions will often be incompatible with the idea of an acceptable level of damage which can be bequeathed to the future by our current actions (Rothenberg 1993a). Thus any optimal set of actions will also have to comply with a *sustainability* criterion, which ensures that a particular form of inter-generational equity is preserved. The second major objection is that the developing countries, who proportionately are likely to suffer the largest greenhouse damages (Fankhauser 1994a), observe that the majority of past greenhouse gas emissions have come from the developed countries. Therefore, they expect the developed countries to pay for controlling future emissions, in a way that avoids any extra damage to developing economies and their ecosystems.

Both these positions have strong ethical justifications, the first being grounded in a version of Rawlsian justice (though applied differently than in the original analysis; Rawls 1971), where the value of a policy is measured by its effect on the least well-off generation, and so at the minimum there is no trade-off between the well-being of different generations. The second argument is best expressed in terms of property rights: the developed countries have already used up their portion of the assimilative capacity (that is, the increase in global CO_2 concentrations which would produce no measurable damage) of the atmosphere and world ecosystems, and in the process have built up physical capital stocks which ensure their current wealth. Therefore, they have no right to use the developing countries' legitimate share of global assimilative capacity in order to reduce their own costs of abatement. Logically the developing countries will only have to consider a cost/benefit analysis when their emissions start to exceed their share of atmospheric assimilative capability. Of course, this second argument is theoretical in the sense that the developed countries' past emissions cannot be undone, and they will damage the developing countries anyway. However, this notional division of emission rights is a powerful argument against developed countries basing their actions on an optimisation of current abatement costs and damages, with no reference to the provenance of past emissions.

Even though welfare optimisation is not explicitly represented in the current negotiation process, it is likely that each country, especially in the developed world, will have some informal, or additive, model of costs and benefits which will help to guide its decisions. This fact is graphically illustrated by the obstructive stance taken by the OPEC countries at the recent Berlin meeting of the Council of Parties of the FCCC. Therefore, a more detailed look at the components of a hypothetical optimisation process will illuminate some of the motivations behind the current behaviour of the Parties, as well as exposing the limitations of using the concept of optimisation naively inside this type of multi-faceted negotiation.

In this chapter we analyse in detail the problems surrounding the assessment of optimal greenhouse gas abatement inside an economic framework which is purely utilitarian, pointing out where this paradigm conflicts with ideas of sustainability and inter-generational equity. We conclude that using numerical optimisation as a prescriptive policy tool is fatally flawed given current information and methodologies. However, these ideas have a useful role to play in guiding future research into climate effects, and in the descriptive analysis of how countries will interact inside global policy negotiations.

OPTIMISATION IN PRINCIPLE AND PRACTICE

The idea that greenhouse gas emissions can be 'optimised' into the future is politically and ethically contentious because of its implications for inter- and intra-generational equity. It is also straightforward to show that naive application of economic instruments, such as Pigouvian taxes set to an estimate of the marginal damage cost of carbon, will not guarantee stabilisation of CO_2 concentrations in the atmosphere even when this would be the optimal strategy. By implication this means that the world will head down an unsustainable growth path if climate sensitive environmental assets are essential to human well-being.

Existing studies which have aimed to calculate optimal control policies have ignored the issues of sustainability, damage non-convexity and intergenerational equity by sticking to tractable analytical and numerical methods. Their methodology in pricing the environment has also been flawed, as the evolution of prices and greenhouse damage have been largely subsumed into parametric assumptions, not detailed analysis of market and welfare responses. Even if an optimal growth path exists which is sustainable, it will not be reached by using Pigouvian taxation unless prices are continually adjusted to reflect changes in scarcity, productive potential and welfare gains from environmental assets as pollution concentrations increase, and future generations become richer.

Deterministic, or expected value, approaches to optimisation have obscured the policy relevance of continuing high uncertainty in the future levels of climate damage, especially the potential for extreme and catastrophic events. The probability values used in these exercises are highly subjective, and do not reflect the true dynamics of uncertainty involved in a changing climate. These flaws have lead to many participants in the political process surrounding climate change dismissing the idea, and not just the results, of cost/benefit analysis, and the resulting politicisation of approaches to the study of abatement policy by economists.

Existing studies of optimal carbon abatement

To find the optimal level of carbon abatement over time the simplest treatment is to define a cost function for abatement and an expected damage function for the *stock* of carbon dioxide and other gases emitted, and then to minimise their discounted sum over an infinite time horizon; subject to the dynamics of pollutant accumulation and decay in the atmosphere (for example, Falk and Mendelsohn 1993). Given that there is a bounded solution to this problem, that is, discount rates are high enough to make the integral finite, the optimal policy is to equate the marginal cost of abatement with the expected, discounted marginal damage cost in each period.

The structure of this type of objective function has been criticised as unrealistic in two ways: firstly, the rate of emissions accumulation usually has no effect on damage costs, but we know that ecosystems and social systems take time to adapt and faster warming will mean higher costs; secondly, the marginal costs of abatement are considered to rise continually with abatement levels, and this does not take into account the possibility of induced technical change and the development of backstop technologies as described in Chapters 3 and 4. The potential for adapting to climate change through investments in sea defences and so on is not considered.

Despite these objections several numerical studies have attempted to calculate a theoretical global welfare optimum using this type of structure. In modelling work the optimal conditions are usually represented by equating an international tax with the discounted marginal damage cost per tonne of CO_2. This is of course an abstraction from reality because the tax varies constantly over time, which is only appropriate for an infinitely malleable world, or one where actors have perfect expectations. It is also assumed that all marginal damage functions are monotonic increasing, as non-convexity could lead to the model's choosing locally optimal, but globally sub-optimal, conclusions (Starrett and Zeckhauser 1992).

Using this framework the seminal work by Nordhaus (1991a) concludes that the optimum policy now is to do little, as the costs of abatement exceed the costs of future damage once discounted at any realistic rate. This controversial conclusion has been criticised from many points, notably that it does not adequately allow for uncertainty, that abatement costs are over-estimated (Ekins 1994), that damage costs are unrealistically extrapolated from USA data to world-wide (Fankhauser 1993), and that it does not account for longer term effects past doubling atmospheric CO_2 concentrations (Cline 1992). In fact in the original work Nordhaus did consider three damage costs scenarios, and came up with three optimal levels of CO_2-equivalent abatement: virtually no action, cuts of approximately 17 per cent in emissions and cuts of around 50 per cent plus afforestation and CFC phase-out, for the three scenarios respectively. In a more recent paper, Nordhaus (1993) extends the analysis by using a dynamic treatment, rather than a static sum for the arbitrary point of CO_2 doubling, and finds that the optimum is a very small reduction, with a 20 per cent cut appearing extremely costly. Given additional uncertainties in parameters other than damage costs this leaves his conclusions rather weak: from 'do nothing' to 'do as much we possibly can'. It is this type of result that has prompted the current interest in integrated assessment models which aim to calculate optimal policies under ranges of different parameters.

However, simulating thousands of different scenarios using different parametric combinations does not help to answer the fundamental structural questions underlying the idea of optimisation, it just gives an estimate of the possible magnitude of their influence. From the above discussion it can be seen that there are four main areas of contention which have to be considered:

- The interaction of optimal strategies with ideas of global sustainability.
- The constituents of the damage costs and welfare losses and how they vary over time.
- The role of time preference, intra- and inter-generational equity in decision making.
- The treatment of uncertainty, irreversibility and learning in designing response strategies.

In the near term we must recognise that current information is imperfect, but that there is the possibility of learning more in the future. In this context it is possible to look at how countries and coalitions will choose a profile of abatement over time in the presence of such an information structure. Should we abate now as a pre-emptive strategy, or are there better ways to guard against the risk of climate change such as adaptation? What information would we need to have now to justify strong abatement measures, and what information would justify a wait-and-see approach? Is this information available, and how reliable is it?

Additionally, such analysis must aim to describe how new information, and uncertainty, will affect international behaviour and the potential for worthwhile co-operation, rather than offering set solutions for optimal policy. This less ambitious analytical task has the potential to give reliable advice about the design of institutions which will bolster international co-operation, while leaving the debate over priorities and targets to clarify in the presence of better information.

In formulating these strategies, decision makers will need to understand, both qualitatively and quantitatively, the problems with defining optimal policies and how they may be resolved. This will allow them to decide at what stage it will become meaningful, and theoretically correct, to use the tools of cost/benefit analysis to make global decisions. The following sections therefore examine each of these issues in detail to try to make explicit the fundamental problems with a simple optimising approach, and the usefulness of the theoretical and quantitative results associated with it. For the rest of this chapter we only consider how to avoid, rather than adapt to, climate change in order to simplify the discussion. Of course, investment in adaptation will have to be justified under the same rules as that for abatement, except that the consequences for environmental sustainability are rather different.

OPTIMALITY AND SUSTAINABILITY

The aim of the FCCC is to stabilise greenhouse gas concentrations in the atmosphere in a way that protects ecosystems and allows sustainable economic growth (Article 2, UNEP 1992); where the concept of sustainable development is usually defined after the Brundtland report (WCED 1987) as: 'development that meets the needs of the present without compromising the ability of future generations to meet their own needs'.

In formal economic terms this is often interpreted as implying non-decreasing per capita welfare growth into the future (Toman *et al.* 1994), with welfare including both material and non-material uses of the environment. The optimising framework does not imply this form of sustainability however, because with a large enough discount rate the welfare of future generations will have little or no value to current actors (see Appendix 7.1). In this case an optimal decision path could involve zero welfare for future generations (see Baranzini and Bourguignon 1994 for an interesting treatment of this problem).

In the short to medium term the sustainability constraint is unlikely to bite as there is ample scope for development in the world economy and natural mineral resources (albeit polluting ones) are projected to be available 100–200 years in the future. Therefore, a more relevant condition is finding the level of environmental damage which is consistent with maximising growth in human *well-being*, not just income, into the future. Reducing the availability of environmental goods reduces welfare growth in the long run, but so will increasing the proportion of the man-made economy devoted to abating greenhouse gases.

Environmental resources are finite, while man-made wealth is effectively infinite given advances in technology. This implies that the long run physical stock of global environmental assets (or conversely, the atmospheric concentration of CO_2) must remain constant, while the man-made economy continues to grow at a steady rate. In the case of climate change there are three different characteristic outcomes that are possible for such an optimal, sustainable growth path; given that we start from a stock of environmental assets bequeathed to us by past generations who did not understand the consequences of their polluting activity (see Appendix 7.2 for formal model conditions).

Technological Utopia The initial concentration of CO_2 in the atmosphere is considered to be above its optimum level and the development of cheap 'green' technologies allows the complete decoupling of economic activity from environmental degradation. With these backstop technologies in place CO_2 concentrations will stabilise at their naturally sustainable level and the level of climate sensitive environmental assets is at a maximum (see Michel 1993 for an example).

Environmental Destruction In the limit increasing concentrations of CO_2 cease to affect future well-being, as the damage caused by each successive unit of pollution continues to fall and/or there are ready man-made substitutes for environmental functions. Given very expensive abatement technology it would be optimal not to control emissions in the long run, but to let the climate sensitive environmental stock fall to zero.

Environmental Trade-Off The climate sensitive environment is essential to production in the long run, as it is more productive than its man-made substitutes (if there are any). However, abatement technology is expensive and the optimal sustainable level of environmental goods is below current levels.

Note that each of the above three scenarios is optimal to decision makers; the choice between them rests on how much the environment is valued, how much climate change will damage the environment and how much it costs to prevent emissions of GHGs. Later in the chapter we consider how these choices and trade-offs are taken, and the problems with doing this, but here we assume that these are true optimal paths.

Figure 7.1 shows a 3-D schematic of these different optimal growth paths: the growth rate in material goods in each scenario is measured along the x-axis and the size of the environmental stock (N) at any point in time along the y-axis, time progresses into the page along the z-axis. Every scenario starts from an initial environmental endowment N_0 and then, depending on the different costs and benefits of abatement, evolves along an optimal growth path into the future. The total growth rate of well-being in the economy is a function of the environmental stock and the growth rate of material goods, as shown in Figure 7.2.

The sustainable growth rate *for material goods* is the same (G_{max}) in both the Environmental Destruction and Technological Utopia cases, implying that no premium is being placed on carbon-free technologies in the base-case, and that the productivity improvements from technical innovation in energy efficiency do not improve overall productivity. The growth rate in the Environmental Trade-Off case is lower than the base-case, because environmental goods are being substituted for production. If climate change is mainly a production externality, rather than a consumption one, and it is subduing production at time zero then this may not be the

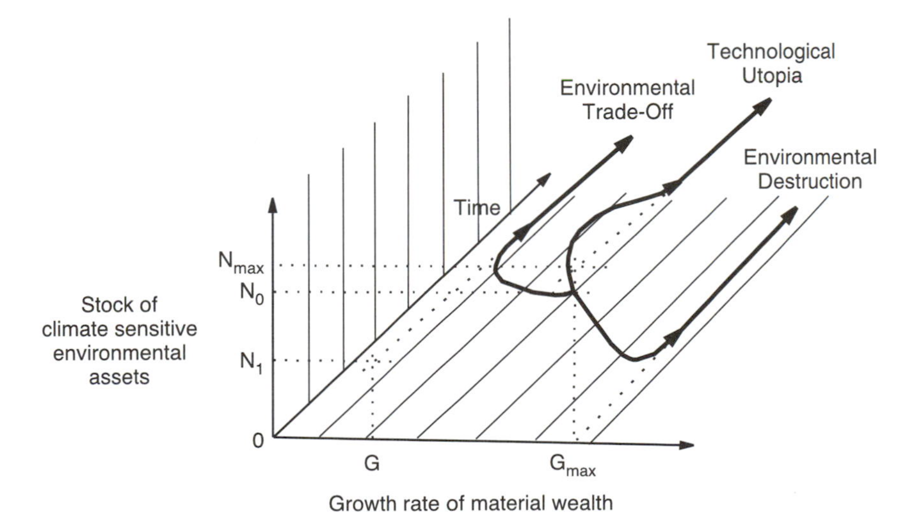

Figure 7.1 Sustainable growth path schematic

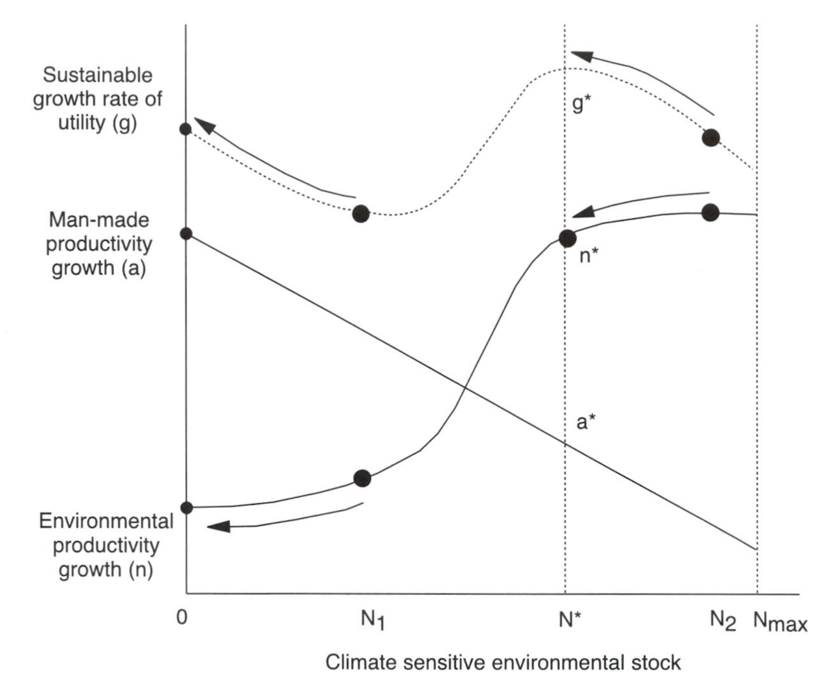

Figure 7.2 Schematic of sustainable growth equilibria and transmission paths

case, as controlling emissions could raise the sustainable growth rate. Given the magnitude of impacts attributed to the greenhouse effect at the present time this does not seem to be the case.

The dynamics are of course diagrammatic, but the transition path for the Environmental Destruction case should be interpreted as involving no emissions control, while the dynamics of the Environmental Trade-Off case will be decided by the particular costs and evolution of abatement technology. The growth rate in the Technological Utopia case falls on the transition path as new technologies are developed and implemented, but it returns to G_{max} after a period of learning and improvement.

In Figure 7.1 the equilibrium level of environmental assets, or the atmospheric CO_2 concentration, for the Environmental Trade-Off case is shown as N_1, which is well below the initial stock and a significant reduction from the unpolluted case at N_{max}. However, this choice of equilibrium point is not a purely economic trade-off with a smooth set of incremental choices; the stock nature of the environment and the irreversibility of climate damage means that under certain conditions choices are limited to discrete paths, either to abate or not to abate.

Conditions for environmental equilibrium

If the natural ability of the environment to remove CO_2 from the atmosphere is constant whatever the level of CO_2 concentration, then there will be a unique level of maximum yearly global emissions consistent with stabilisation of the stock of environmental goods. In this case the economy is limited to one particular mix of polluting and abating technology, and by implication one equilibrium growth rate with a positive environmental stock. Disregarding transition costs, in this case there is no incremental trade-off of environmental quality with long run growth rates, just the discrete choice between maintaining any positive amount of environmental stock or having no emissions control at all.

If the natural removal rate of CO_2 from the atmosphere *increases* with higher concentrations, and thus lower stocks of environmental assets, then several sustainable emission rates will exist. In the absence of uncertainty the optimum long run emissions path will depend on the cost of abating emissions relative to the productivity of the environment. If the environment is not critical to welfare then the optimal growth path could imply no control of emissions. If the natural removal rate of CO_2 from the atmosphere *decreases* with higher concentrations, then the choice of optimal growth path is again a discrete choice of either maintaining the current stock (which results in incremental improvement) or having no controls at all on emissions.

Up to this point we have considered the potential for sustainable long run growth paths with stabilised pollution concentrations, or in extreme cases zero carbon sensitive environmental amenities. To date none of the published optimisation exercises has concluded that stabilisation of atmospheric CO_2 concentrations is optimal, and so their long run growth paths are characterised by the Environmental Destruction case. There are two possible interpretations of these conclusions: either that the environmental destruction caused by climate change will not *essentially* affect human well-being or, conversely, that an unsustainable growth path is not optimal, but that these exercises are actually optimising behaviour along the transition path to a new equilibrium, not over the full future costs of global warming. There are two main reasons why the second result could occur: non-convergence to the optimal growth path for some initial environmental stock sizes, and the lack of appropriate shadow price dynamics for environmental goods.

Transition to the steady state

If the cost of transition to the sustainable path is considered, then the size of the initial stock of the environment becomes critical in defining whether the optimal sustainable growth path can be reached. For example, in Figure 7.2 the relative productivity of the environment at different stock sizes (N) is plotted against the productivity of man-made capital on that sustainable path. At lower equilibrium levels of N the amount of abatement needed to sustain a constant environmental stock is smaller and so the productivity of the man-made factors is higher. Each pair of points (a, N) represents a sustainable growth path, and their sum (g) shows the total productivity of each mix, and thus the sustainable growth rate in human well-being. In this example (a^*, n^*) is the maximum sustainable growth path, with value g^*. However, if the initial environmental stock is at N_1, which is smaller than N^*, it will not seem optimal to individual actors responding to a Pigouvian tax on emissions to invest in abatement technology in order to reach the optimal path. This is because the fastest increase in *total* growth rates at N_1 is achieved by reducing N, even though this leads to a local optimum at $N = 0$ with a growth rate below the global optimum at N^*. If the initial environment is too clean (N_2) then the optimal growth path will always be reached because reducing environmental quality saves abatement costs and increases the total growth rate until equilibrium occurs at N^*.

Given an initially depleted environment, consideration of local marginal productivity rates can therefore be a misleading guide to decisions, because the productive potential of the environment will change (most likely in a non-convex manner as in Figure 7.2) with non-marginal amounts of preservation or destruction; the assumptions underlying the usual optimisation criteria will be violated and naive use of taxation instruments will lead to sub-optimal solutions.

Even if the damage curves are shaped such that an incremental marginalist approach, such as using Pigouvian taxation, will lead to the correct growth path, a conventional numerical optimisation exercise will not do so if it fails to take into account correctly the effect of quantity changes on environmental productivity. In Figure 7.2 incremental decisions starting from an initial environmental stock at N_2 result in a globally optimum solution converging to N^*, but only if the value of environmental goods is constantly updated to reflect their growing productivity as the stock falls. However, if environmental productivity in the future is assumed to remain constant, the optimum at N^* will not be 'seen', but increased man-made productivity at lower abatement levels will be, and so maximisation of welfare will occur at $N = 0$.

In summary, all numerical optimal control exercises to date have disregarded the issue of sustainability which is enshrined in the FCCC, and so have calculated optimal solution paths which do not stabilise CO_2 concentrations in the future. Unless those parts of the environment sensitive to climate change have a finite effect on human well-being, or there is a finite amount of global warming for any concentration level above a certain point, this continual increase in emissions is an unsustainable strategy.

When optimal sustainable growth paths exist with a controlled CO_2 concentration level it is not obvious that using taxes based on marginal damage costs, either in numerical simulation or the real world, will enable the world economy to reach these desired paths. Given the structure of the problem it is likely that several competing local maxima exist and the convergence to a specific solution will depend on both the initial size and productivity of the environmental stock and how the shadow prices used to value these assets evolve over time.

WELFARE, DAMAGE COSTS AND ECONOMIC GROWTH

Accurate calculation of the value of the environment is critical to making correct decisions about greenhouse gas control. It is not just a linear problem; the wrong value estimates could mean the difference between sustainable and unsustainable growth in the future. Valuation is not an easy task because the homogeneous category of 'environmental assets', which was used above for analytic simplicity, is in the real world a set of complex, interlinked and heterogeneous effects, as increased greenhouse gas concentrations perturb the whole of the global climate and its associated natural systems. Given the assumption that we are able to predict accurately the physical impacts of climate change, which is a far from solved problem of the natural sciences, there still remains the economic problem of interpreting the importance of these changes.

For the purposes of economic analysis we are interested in how much people, now or in the future, would value not experiencing the change in their material and psychological well-being brought about by climate change. Measuring, or predicting, the magnitude of each effect will therefore depend on which sphere of human activity it enters, and a useful functional disaggregation is into three categories of damage.

Marketable impacts on production Climate change directly impedes the production of goods and services which are traded in the marketplace and thus have measurable prices associated with them. Effects include the cost of increased flood/storm damage (including extra insurance and mitigation measures), increases in the availability and price of agricultural, fishery and forestry products and construction expenditures associated with changes in building types and air conditioning loads. The costs of these changes will directly affect the decision making of firms and households, who will alter their consumption and input patterns in response.

Non-market impacts on production These are effects which reduce the availability of public goods, which impact on productivity, but which for various reasons have no immediate price response associated with them; for example, increased mortality and morbidity rates and reductions in ecosystem services, such as wetlands which purify water supplies. These costs affect production but there are no mechanisms for price signals to inform firms of these costs, unless companies directly pay health insurance costs.

Direct welfare impacts These are changes in human well-being brought about directly by climate change. These include the welfare costs of mortality and morbidity increases, loss of unique and valuable ecosystems which fail to adapt to the new climate and the enforced migration and dislocation of agricultural societies whose resource base has been undermined.

These three categories are formed around how each impact may be measured, and many physical changes will have costs in each category; for example, human health and wetland destruction. In the context of optimisation it is vital to understand how each cost category is calculated, because those which are hardest to measure may have the highest potential welfare impact (e.g. parasitic disease rate increases) and an 'optimal' policy based around partial costing information would propose too little abatement.

Marketable impacts can be directly estimated from price data, but projection of future costs will depend on the evolution of these prices as the climate changes and the economy grows. The value of non-market inputs can be measured as their shadow price into production; that is, the value of their marginal contribution to economic productivity. Though this is a theoretically simple procedure it is usually very difficult to do in practice, and projections of these prices encounter the same difficulties as for marketable inputs. Direct welfare effects are probably the hardest to measure, mainly because the theoretical justification for aggregating and projecting psychological 'utility' ratings is suspect, people's preferences are not directly comparable and preferences will change over time (especially in response to processes such as climatic change). The standard measuring technique is to ask people about their willingness-to-pay (WTP) to avoid the welfare loss, or their willingness-to-accept (WTA) compensation for experiencing the loss. In formal terms the WTP is the compensating variation for a change in quantity, while WTA equals the equivalent variation.

Price dynamics with climate change

Given the critical nature of transition dynamics for the achievement, or not, of sustainability, it is important to look more closely at how we can measure these different types of value as the climate changes. If the environment were to contribute a steady proportion of future output, that is, if the economy were to grow in scale but remain unchanged in composition, then we would expect the damage costs per marginal tonne of carbon to rise linearly with GDP. If the optimal sustainable growth path were to lie at a higher level of CO_2 concentration than at present we would expect marginal damage costs to rise at a higher rate than GDP until concentrations were stabilised.

This assumption of proportionality in prices and GDP, which is used in many of the optimisation studies, is an over-simplification because the prices of different climate sensitive environmental goods will evolve heterogeneously over time and with changes in CO_2 concentrations.

Costs linked to GDP These are generally goods which enter into production, have no man-made substitutes, and cannot be made more productive by human activities; for example, ecosystem services such as water filtration by wetlands and flood protection.

Costs which rise faster than GDP These are public goods which enter directly into consumers' utility functions (for example, biodiversity preservation and health conditions) for which man-made goods are poor substitutes, and which will be valued more as wealth increases. The presence and magnitude of this 'superior good' quality can be measured by looking at the difference between the WTP and WTA valuations of a particular resource, though empirical evidence is at the moment mixed (Kristrom and Riera 1994). Hanemann (1991) has shown that, given quantity changes in public goods without perfect private substitutes, WTA >>WTP, even in the absence of large income effects. The intuition behind this is that consumers, asked for their WTP for a good, respond with a figure *inside* their current budget constraint; while the WTA measures the importance of that good relative to an increase in wealth, which is analogous to the position of a future consumer at a higher GDP level.

Costs decoupled from GDP Many renewable resource sectors, such as agriculture and forestry, take up lower proportions of GDP as countries develop, because the man-made economy becomes more reliant on its own past capital stock for growth. Even though global population growth implies higher demand, and thus lower marginal productivity and higher prices in the future, it is unlikely that this will outweigh the effects of industrialisation. However, because these sectors have few substitutes their costs will be greatly affected by changes in productivity due to climate change.

Transition costs It may be the case that the major costs of the greenhouse effect are transitional, in adaptation to a new climate, albeit over a long time scale. In other words a climate which is 4°C warmer is not seriously less hospitable or less productive but is just unsuited to our existing infrastructure and knowledge base. However, if adapting this infrastructure were to include not only human capital and experience but also the evolution of natural systems, the adaptation timescale would be long enough to make the distinction between adaptation and absolute costs rather immaterial.

The key issue in these four valuations is the degree to which the environmental amenities removed by climate change are essential to human wellbeing. Are there man-made substitutes which will provide the same service at a lower cost than stopping GHG emissions (for example, raising coastal defences to protect against sea level rise and increased storm activity)? If there are no substitutes for an environmental asset, for example a lost species, then damage is irreversible and the welfare value is the psychological impact of this loss. This may be a large or a small value, and irreversibility does not a priori impose a high value on an asset; otherwise we would value not destroying the smallpox virus! However, irreversibility does reduce the options of future generations, thus violating some definitions of sustainability mentioned above. Unfortunately, we cannot judge the preferences of these future citizens and so must make proxy decisions for them about how they would trade-off material and environmental assets. In this situation there may be a value to the future of preserving options, which exceeds the direct utility from a good, and the economics of this is examined in detail below when we consider the effects of uncertainty on optimisation.

Given our current state of knowledge about greenhouse damages it is difficult to say which effects will dominate, unless we assume that emissions stabilisation will not occur soon, and so we are already committed to significant environmental deterioration in the future. In this case it is likely that the value of all environmental goods, as their quantity decreases, will rise over time at a rate faster then the growth of GDP.

Despite this obvious potential for price increases, most existing studies have calculated damage costs as the value of goods or services lost at current relative prices. A commonly cited paradox is that agriculture accounts for around 3 per cent of world GDP and if the climate were to change so as to prevent agriculture altogether this would result accordingly in the loss of 3 per cent of GDP! It is evident that the actual losses would be greater, and that this is not being taken into account in the simple marginal analysis. If agricultural productivity were reduced world-wide, and given that food has inelastic demand, the size of actual damage costs would be better represented by the increased cost of growing the same yield under new conditions; that is, the cost of pesticides, fertilisers, irrigation, shade houses, air-conditioned animal houses, etc., plus transitional adaptation costs (changing crops, bringing new land under cultivation). As there are tightening constraints on agricultural land due to urbanisation and population growth the supply of land in the 'no climate change' case is also likely to become more inelastic, leading to even greater increases in prices than would be calculated from current conditions. Of course, if this rise in prices is due to an increase in resource rents it merely involves a transfer of wealth from consumers to producers, but if it involves a decrease in productivity everybody loses out (see Appendix 7.3 for a sample calculation). This type of analysis may apply to many areas, but is most relevant where supply and/ or demand is inelastic, which may include impacts on 'necessities' such as food and water, or where supply is limited by natural resources such as forestry or farmland lost to sea level rise.

Secondary benefits

As well as the path of greenhouse damage costs over time and with CO_2 concentration, a related practical issue is accounting for the *secondary benefits* of greenhouse gas abatement scenarios in terms of other public goods such as reduced acid precipitation, lower traffic congestion and reduced local pollution (CO, NO_x and VOCs). The partial nature of climate change optimisation studies has meant that these effects have been ignored, thus overstating the cost of abating CO_2 and making the recommendation of a transition to a sustainable growth path less likely. This source of error is exacerbated by the fact that local air pollution damages and congestion costs will also grow faster than the rate of GDP unless emissions reach a steady level.

Estimates of the size of these benefits have been shown to be larger than the macroeconomic costs caused by relatively low levels of carbon taxes for the UK (Barker 1993), but a survey of other studies in Fankhauser (1995) shows a marked dispersion in estimates for the USA, Europe and Asia with costs differing by three orders of magnitude. This is not surprising given the difficulty of such calculations, but it underlines the problems with the results of optimal growth models which have not addressed these issues.

DISCOUNTING, INTRA- AND INTER-GENERATIONAL EQUITY

Apart from price dynamics, the other key feature of the economics of transition to a long run growth path is that global temperature rises slowly, while greenhouse gas abatement costs are immediate. We are presented with a trade-off between future environmental damage and current costs, which will define the lives of our immediate successors because of the practical irreversibility of both climate change (emissions are naturally removed to the deep oceans after ≈ 100–200 years) and the damage caused by climate change.[3] If we do not value our offspring's welfare then we could just pollute and enjoy the wealth derived from using up fossil reserves; on the other hand we could drastically cut our standard of living to prevent further increases in future damage. Of course, there may be large advantages in starting abatement now, especially if future energy use is determined mainly by infrastructure design (e.g. urban densities) which will be hard to change if climate change costs appear more serious in the future. Additionally, if the majority of energy saving is expected to come from new technical innovation, rather than life-style changes and diffusion of existing technology, then initiation of R & D now could bring large, if hard to quantify, benefits.

Decision makers do not have access to an optimal growth model of the world economy into the distant future and so cannot see the consequences of moving on to an unsustainable growth path. Traditional economic logic tells them that all changes to the economy are marginal, and so investments in pollution control should be made by comparing the potential return in the material goods sector with the future savings in environmental damage costs. The practical procedure for doing this comparison is to discount the future damage costs caused by a present unit of emissions over some very long period (for example 200 years) in order to find the present value damage per tonne. If the computation of future prices for environmental damage is done correctly, and in the absence of transition non-convexities, uncertainty and irreversibility, then setting the discount rate to the social rate of time preference will theoretically produce an optimal and sustainable growth path when modelling a single, infinitely lived, representative consumer. However, because of free-riding it is unlikely that such incremental decisions will result in a sustainable solution when made by successive overlapping generations of individuals, unless there is explicit use of a criterion for inter-generational equity (Howarth and Norgaard 1993). Because they do not guarantee sustainable outcomes, normal exponential discount rates have been criticised for not accounting for the true attitudes of current generations to their descendants, or ethical obligations between generations (Rothenberg 1993a).

When comparing the financial value of future income (rather than its utility), the social discount rate can be decomposed into the pure rate of time preference plus a factor representing the decreasing marginal utility of income:

$$d_t = \upsilon + \sigma_I . \delta(Y/P)/\delta t$$

where υ is the pure rate of time preference, Y/P is GDP per capita, and σ_I is the income elasticity of marginal utility. Values used for υ are in the range 0–3 per cent, with little agreement as to the correct value. The second term implies that future values should be discounted further at the rate at which increased income results in lower utility per unit of income. Usually, a value of $\sigma_I = 1$ is taken to imply a logarithmic utility curve. For OECD countries, with stable populations and steady economic growth, this might increase the discount rate by 2–3 per cent. However, in developing countries, where population is rising faster than GDP, declining per capita incomes imply a discount rate lower than the pure rate of time preference.

For the representative consumer case, on a first best optimal growth path the social discount rate will equal the real riskless rate of return on capital; however, because of imperfections in capital markets and the presence of investment taxes, this is not the case for observed data (see Luckert and Adamowicz 1993 for a discussion of measurement issues). In the optimal control studies attempted to date, a value of between 1–3 per cent has been commonly used to represent the total social discount rate; the ceiling for this value is benchmarked by the real rate of return on government bonds, which can be interpreted as risk free investment return for public goods.

This approach is flawed however because it assumes that climate change will not alter the marginal productivity of capital, from which we derive our alternative scenario of no abatement and the social rate of time preference. If damage from climate change is non-marginal, as is likely in developing countries, the actual riskless rate of return on capital will fall. This will have two effects: either less capital will be invested, slowing growth rates – an extra uncounted cost, or, if capital markets are 'sticky', the applicable discount rate will drop. Both of these effects will increase the cost of climate change and will argue for more abatement now, but the use of models in which climate change does not directly affect productivity masks this macroeconomic effect.

The attraction to policy makers of using incremental cost/benefit criteria is that it gives a conceptually simple and strong policy conclusion about how to produce optimal levels of global warming: just correctly measure damage costs and the social discount rate, and carbon taxes will do the rest. However, the problems of measurement, free-riding and inter-generational equity mean that many important issues affecting the optimal amount of current abatement can be hidden in the selection of this one parameter.

In response to these concerns different ways of valuing the future have been proposed, such as having discount rates evolving exponentially in the first generation but thereafter giving a positive, constant value to each future generation's welfare (Rothenberg 1993b). In practice the problem with such innovative approaches has been quantitatively defining the parameters, as there are so many possible combinations. Even when using simple discount rates, sensitivity analysis can be unenlightening as it merely gives a range of optimal policies from low to high abatement, depending on the values used. Therefore, rather than using the discount rate to subsume all these different effects by trying to find an objectively 'correct' figure, a more fruitful avenue is probably to calculate the implications on optimal policy of using different discount rates, and then estimating the efficiency costs associated with any one option. These costs can then be weighed against the uncertainty and irreversibility present in damage estimates to see if they offer cost-effective insurance, or are a reasonable price to pay for inter-generational equity. In this way policies are determined by discussion about preferred outcomes, rather than by disputes over correct methodologies, which clarifies the decision making process.

Welfare aggregation and climate change negotiations

As well as the theoretical and practical problems involved, the aggregate incremental cost/benefit approach is not compatible with the political decision making framework surrounding global warming, because it forces the imposition of one homogeneous valuation of the future on to the global population. In a public choice context the majority rules, and this majority may well have a different time preference to the minority. For example, poor countries tend to discount at a higher rate than rich ones; one rationale for this being that their mortality rates are higher and so the increased probability of dying makes the future less valuable. Therefore, assessment of *global* emissions reductions and cost sharing is problematic, because of the ambiguity present in choosing appropriate discount rates. Even in a national context it has been argued that older people care less for the future than younger people, and so if they form a majority of the electorate a sustainable growth path, which in the aggregate would be optimal, may never be chosen (Kennedy 1994).

Therefore, it is uncertain whether using aggregate approaches to find the optimal level of global warming damage will result in a solution which is politically feasible; in the sense that it commands the support of a majority of countries, or a majority of greenhouse emitters. However, if the solution is not politically feasible, or representative, it is of little use to the global community except as an abstract accounting exercise or an illustration of the outcome of one group's set of preferences.[4]

A simplistic approach to constructing a politically useful valuation analysis would be to measure the discounted damage costs in each country, using local preferences, and then sum the results at the global level. However, this raises the ethical problems of intra- and inter- generational equity explained in the introduction, because preferences and income are linked. Poor countries *cannot* value climate change damage higher than they currently state because they have restricted budgets with which to buy public goods. Prescriptions for abatement based around descriptions of current preferences will therefore internalise the status quo of income distribution and power. The contentiousness of this approach has already been exposed in the use of the 'statistical value of a life' by some climate researchers; these studies have valued the citizens of developed countries much higher than those in the developing world because they are able to pay much more to avoid risk. This is a true description of the world, but obviously has severe implications for equity if it is taken as some kind of abstract truth, and will not attract the democratic support of developing countries in the international arena.

Defining the objective function for a hypothetical 'global optimisation' of emissions reduction is therefore fraught with political and ethical concerns. Under the current FCCC structure, which defines two groups with relatively homogeneous obligations and preferences – North and South – the simplest form of global optimisation would result from:

$$max\{\phi_N.[D_N(C_N) + N.G(C_N)] + \phi_S.[D_S(C_N) + S.G(C_N)] + C_N\}$$

where C_N is the cost of abatement funded by the North, $D_N(C_N)$ and $D_S(C_N)$ are the domestic marketable benefits of abatement to the North and South respectively (e.g. gains in agricultural productivity), $G(C_N)$ are the global non-material benefits of abatement (e.g. mortality decreases, preservation of unique ecosystems) shared in proportions N and S ($N + S = 1$) between the developed and developing countries, and ϕ_N, ϕ_S are the weights on future benefits determined by discount rates and subjective expectations in the two regions.

If the South's benefits are small (D_S, S and/or ϕ_S small), the benefit equation reduces to:

$$\text{Global Benefits} = \phi_N.[D_N(C_N) + N.G(C_N)] - C_N$$

and so the global optimum is merely given by the North's unilaterally optimising its own utility:

$$[dD_N/dC_N + N.dG/dC_N] = 1/\phi_N$$

Unsurprisingly, if the South's environmental benefits from abatement are small, as stated in the FCCC,[5] and even if all abatement is funded by the

North with no altruistic intent, the outcome will approximate to perfect co-operation and intra-generational equity will be satisfied.

However, this result depends on the aggregation structure of the model and the nature of environmental utility. The aggregate global welfare function above is constructed by adding the North and the South's *monetary* valuations of climate change damage together, not their utility functions. This is unimportant if the majority of gains come from material benefits traded in global markets (D_N and D_S), as their value will be unaffected by the income level of the country. However, if non-market benefits (G) predominate, which either cannot be physically assigned to any one country or depend on income levels, then aggregation must take place at the utility level if true co-operation is to be observed. This implies that each citizen's preferences are given equal weight regardless of income level and is functionally (but not mathematically) equivalent to a democratic choice system where everybody in the world has a single vote. If utility can be aggregated in this way (neglecting Arrow's impossibility theorem!) then, if all citizens *equally* value global gains from abatement, they will all give an equal proportion of their income to preserve it (differences in time preference between North and South will be irrelevant in this aggregation if all values are discounted at the same rate as future income). Under these assumptions the two different forms of aggregating non-marketable abatement benefits can be written as:

$$\text{Globally Aggregated Utility} = G(C_N) = \rho.(Y_N+Y_S)^{\delta}(P_N+P_S)^{1-\delta}.V(C_N)$$

$$\text{Summed Financial Benefits} = S.G(C_N) + N.G(C_N) = \rho.(Y_N{}^{\delta}.P_N{}^{1-\delta} + Y_S{}^{\delta}P_S{}^{1-\delta}).V(C_N)$$

where $P_{N,S}$ and $Y_{N,S}$ are the populations and total incomes of the North and South respectively and $\rho.V(C_N)$ is the proportionate valuation of emissions abatement. If abatement benefits are valued as a constant proportion of future income ($\delta = 1$), the aggregations are identical and unilateral action by the North will be close to global co-operative action only if $Y_N >> Y_S$.

If the environment is a superior good ($\delta > 1$), then the North will value climate change damage proportionally more than the South and global aggregation by utility will give a smaller financial value than a simple summation of North and South's separate valuations. This is because global aggregation uses average world per capita income to calculate the financial value, whereas in the simple summation the high per capita incomes of the North increase the total WTP for abatement. Figure 7.3 shows this effect by plotting the ratio of summed benefits to globally aggregated utility (Global Benefit Ratio) for different values of δ.

This sheds light on the argument that the North should raise its current abatement commitments, in order to recognise the value of abatement to developing countries, who cannot afford to pay for it. If these abatement commitments were calculated by co-operative optimisation (however *ad hoc*), this argument will result in higher abatement only if global environmental damage is an inferior good, per capita utility losses in the developing world are higher than in the North, or the developed world measures the financial valuation of developing country utility based on the assumption that everybody has the per capita income of the developed countries. However, the latter case is inconsistent with maintaining a global budget constraint when assessing public spending priorities, because one could decide to spend more than total world income on abatement, while being logically consistent with one's initial assumptions! If the North is acting uncooperatively by optimising just its own utility, then the argument for tighter emissions controls is ambiguous, and depends on income inequality and the income elasticity of environmental goods.

An estimate of the size of these effects is given in Figure 7.3 where the North's unilateral financial benefit (N) is plotted as a ratio of globally aggregated utility, for two different measures of global income distribution: firstly, calculated in dollars when the ratio of per capita income in North and South is ≈ 20 (World Bank 1993), and secondly measured at purchasing power parity (PPP) when this ratio becomes ≈ 11 (IMF 1994). Given the large per capita income differences present in the 'dollars' case, 80 per cent of the financial valuation of greenhouse damage lies in the North ($\delta = 1$), and only a moderate degree of superiority ($\delta = 1.15$) is needed to make the unilateral benefits of the North equal to globally aggregated utility. If purchasing power parity is used, the North's share at $\delta = 1$ is only 66

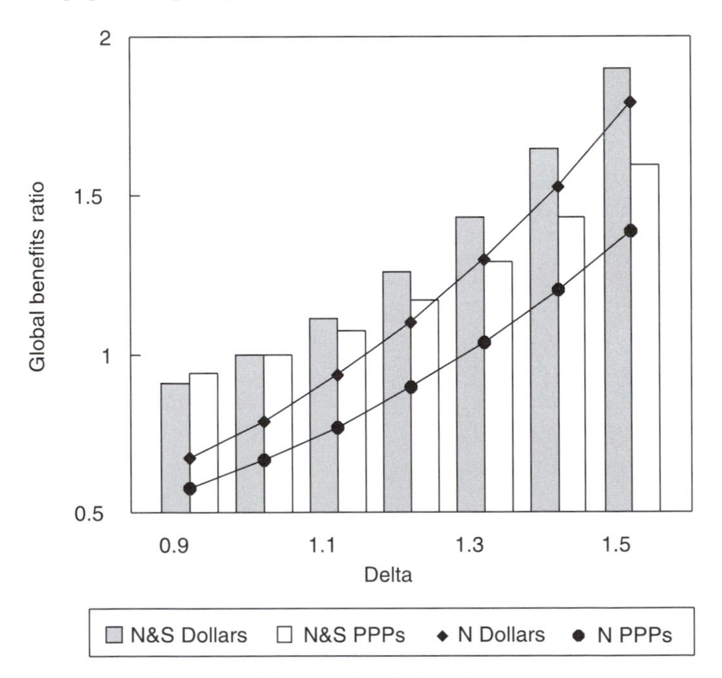

Figure 7.3 Ratio of North's (N) and global (N&S) benefits to a utility aggregation

per cent of financial value, and a significantly larger degree of superiority is needed ($\delta = 1.28$) to equal globally aggregated benefits. Therefore, given the current inequality of incomes it is possible that unilateral action by the North could equal, or exceed, a more 'democratically' calculated optimum if the environment is a superior good.

So, in a non-cooperative optimisation where the North does not take damage to developing countries into account when setting abatement levels, intra-generational equity could still be preserved. Additionally, abatement could be increased above the North's unilateral optimum level if it acted not only from self interest, but also from a sense of 'backward indebtedness' for having benefited from previous emissions, and this principle is hinted at inside the FCCC.[6] However, this optimistic result will not hold if per capita utility losses in the South are significantly greater than those in the North, due to climate change affecting subsistence crops and so on. In fact it is probable that climate change is a superior good in industrialised countries, and an inferior good in low income countries. Middle income industrialising countries will therefore value climate damage the least, as they will not suffer subsistence damage, but do not have enough money to value ecological assets, such as species loss, very highly.

Similar problems of intra-generational equity exist if the lack of perceived benefits in the South is not based on small damage values (D_S and S small), but on the perception of the *future value* of benefits (ϕ_S). If ϕ_S is small (i.e. the discount rate is large) this biases against future, conceivably richer, generations which may have regrets that conservation was not increased, especially if the environment becomes a superior good. The optimal strategy in this case would be to borrow against the future to fund conservation. Of course in the real world of government budget constraints such investment for the future is unlikely whatever the return, and so total conservation is likely to be sub-optimal if discount rates are large.

Summary

The text of the FCCC does not define the type of economic co-operation that countries should undertake, but we have assumed that the developed countries are likely to take non-cooperative action as long as they are the only ones bound by abatement obligations. This implies that the South should pay for its own benefits if any remain after the North has finished abatement (i.e. the marginal damage of emissions has not decreased because of abatement), which in the short run could result in the South's compensating the North for carrying out abatement. This result is the local consequence of the assumption that the North has no altruism, and that the South bears the same responsibilities as the North but just has fewer financial resources.

In the short to medium term while large income inequalities persist, the superior good quality of environmental and health effects may counteract this non-cooperative behaviour, but this does not satisfy concerns for explicit intra-generational equity. The income inequality between North and South will largely determine both the prevailing discount rates and the financial valuation of global climate change in different areas. Therefore, global economic decisions about global warming must be considered in parallel with aid and trade policy if any sort of rational economic optimum is to be reached, though this has not been prominent in developed country thinking so far.

In the medium to longer term, it is likely that the utility loss to the developing world from climate change will be proportionately larger than that to developed countries. This is because their economies are more dependent on climate sensitive production, and the opportunity for health degradation is higher. These problems will impinge most on the non-cash sectors of the economy, and so will have a low tradable financial value. It is therefore unsurprising that, as current valuation studies based on *market* prices have not taken this into account, they have been seen as under-valuing the effects of climate change in the developing world. If this is the case we cannot rely on the superior good nature of environmental damage to make up for all this value, if the North chooses to optimise emissions based only on its own impacts. Non-cooperative behaviour will therefore lead to emissions which exceed optimal levels calculated from either a global welfare function or some type of global voting procedure, and future damage will be higher than desired by groups in the developing world. Therefore, the issues of valuation, discount rates, intra- and inter-generational equity are closely intertwined, and no economic assessment of optimal policy can be made without prior determination of the ethical and political parameters which will define the form, nature and parameterisation of the objective function.

UNCERTAINTY, IRREVERSIBILITY, LEARNING AND STRATEGIC OPTIMISATION

In the discussions above we have examined the practical problems of measuring future costs and damages with any degree of accuracy, and the potential problems associated with using the wrong numbers, but not how to proceed under persistent future uncertainty. In reality the costs of abatement are relatively well known, but the magnitude of potential damage from climate change is very uncertain and must be treated as a stochastic, or random, variable.

There are two ways of incorporating uncertainty into the calculation of an optimal abatement policy. The first is to use the techniques of stochastic optimisation, which balance the relative probability of low and high damage when setting current and future abatement levels. Secondly, if policy makers value certainty in and of itself, that is, they are risk averse, this psychological fact will make them adopt strategies which minimise the variance of future damages. Such a strategy is known as a *hedging* strategy. Optimising against stochastic variables produces the best current strategy considering all the information we know about the future, that is, the probability distribution of future damages; a hedging strategy aims to increase the certainty of any future outcome, whatever its expected value.

It is important to differentiate the semantics of these issues: the presence of uncertainty as a quality of information implies a distribution of potential damage scenarios, each with its own probability; whereas the phrase 'more uncertain' implies that one outcome has become less likely, and thus the *mean* of the distribution of damages has changed. In the political debate some parties (notably industry, e.g. ERT 1994) argue that the uncertainty surrounding climate change damage implies that we should postpone abatement measures, but environmentalists argue that this uncertainty should lead to greater current abatement. As we shall see these misunderstandings come about due to the different implicit frameworks being used to address these problems, the factors included in the optimisation process, and how shifts in knowledge change, or preserve, the mean of the damage distribution.

The simplest way to introduce uncertainty into calculations of optimal current abatement levels, which has been utilised by researchers such as Manne and Richels (1992), involves attaching subjective probabilities to potential damage scenarios and then multiplying these figures together to produce an expected damage cost per tonne of carbon; alternatively a set of preferred parameters defining the evolution of future damage are picked and a 'best guess' scenario modelled. Optimal abatement levels are then calculated in a deterministic way using this expected net present value, or best guess, as the future damage cost of emissions.

To date the most thorough calculation of the expected value of damage costs appears in Fankhauser (1995), though an interesting attempt to analyse the uncertainty and dynamics of linkages in the physical side of the climate system has been done by Filar and Zapert (1995). Fankhauser's results are reproduced in Figures 7.4(a) and 7.4(b), which show a distribution of damage costs for a doubling of CO_2 concentrations in the atmosphere, derived by assigning a probability distribution to the parameters used in the calculations. It should be noted that Tucci (1995) argues that this method will not produce the true confidence intervals around the damage distribution, because the covariance of the parameters has not been considered, and future work in this area will have to include this effect. The distribution is skewed, and bounded above zero, because the potential for a highly adverse or catastrophic impact from climate change is much higher than the potential for a benign or favourable scenario.

As well as skewness in the distribution, Figures 7.4(a) and 7.4(b) also show that the standard deviation of the distribution increases over time (from 14.3 to 19.0 in this case), demonstrating how the magnitude of costs grows with increased global output. A more realistic case would have the confidence intervals surrounding the parameters also growing with time, as forecasts become less certain the further into the future they go. This effect is considered by Tucci (1995), and implies that the variance of the damage

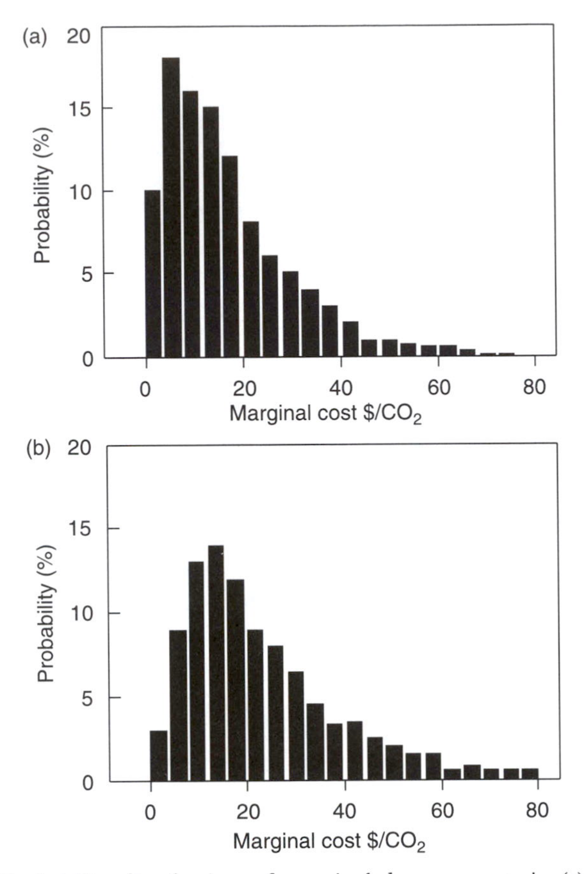

Figure 7.4 Probability distribution of marginal damage costs in (a) 1991–2000, (b) 2021–2030

function grows faster than the mean; as will be seen below this relative increase in uncertainty has potentially strong effects on current abatement policies.

As a general result, using the expected value of a variable inside a deterministic simulation only gives the correct optimal strategy if the system being optimised is linear. If the system is non-linear, then finding an optimum using the mean of the stochastic parameters will not give the same answer as a stochastic optimisation which generates the true distribution of outcomes;[7] in fact the two values will be offset by terms including the variance and skewness of the distribution (see Hall and Stephenson 1990, and Johansson 1993 for formal treatments). This is unsurprising, as when trying to optimise an uncertain future it would seem natural to include in the calculation all the non-stochastic features of that future which are known in the present (that is, the complete parameterisation of the probability distribution), and not just the mean of the distribution.

In the climate change problem we are trying to use a control variable, anthropogenic emissions of GHGs, to optimise the difference between two endogenous variables in the system – damage costs and costs of abatement (*minimise* [Damages + Costs]). Even assuming costs are known with certainty, the relationship between emissions and greenhouse damages is usually modelled as being both uncertain and non-linear, so the expected value approach is strictly inapplicable. Using expected damage values constitutes a *partial* determination of the stochastic behaviour of the system, as endogenous changes due to the optimisation process will not affect expected damages.

For example, consider the role of the expected emissions path in optimisation: to calculate the future marginal cost of an extra unit of emissions now, the future atmospheric concentration levels of CO_2 must be known, because as concentrations increase the additional radiative forcing caused by each unit of CO_2 decreases. If future damage is high, abatement will also be high, thus decreasing concentrations and increasing the future marginal effect of a unit of GHGs emitted now. In a full stochastic optimisation each damage scenario is modelled, along with the associated control reaction, and the resulting equilibrium damage cost is used to calculate the expected future damage. In a partial optimisation the baseline emission concentration remains the same, whatever the damage scenario, and so the strategy will be inconsistent with future actions when the true state of the world is revealed. Therefore, in a full optimisation the optimal abatement strategy will hold for all future states of the world defined by the probability distribution; in a partial optimisation the current strategy may be different if the future true state of the world is revealed, or if it remains uncertain, which means the strategy is not optimal (see Appendix 7.4 for a mathematical explanation).

Whatever the model structure, using a 'best guess' approach will not give the correct result because it involves using the means of the underlying parameter values, and thus does not take into account any of the non-linearities of the system. To date, either the 'best guess' approach has been taken or simple expected values have been calculated from extreme 'high' and 'low' estimates of damage costs, which again ignores the non-linearities of the damage distribution.

The practical importance of these effects on deciding current policies depends on how closely the damage costs and the other determinants of the model (for example, the macroeconomic and welfare costs of abatement) are connected. If marginal damage costs are relatively unconnected with concentration levels, abatement costs are only a minor constituent of GDP, and damages are mostly welfare costs, then using the partial approach may give a good approximation to the optimum. If there are non-trivial links between the productive and climate sensitive sectors, then a full stochastic optimisation of the model must be attempted, and this is one of the rationales behind the construction of the integrated assessment models mentioned in Chapter 3. Additionally, as the variance of damage costs grows faster than the mean value in the future, consideration of noncentral cases will become more and more important. Even in short to medium term studies the gases emitted will cause damage for 100–200 years, so partial optimisation using the mean of the damage distribution seems too likely to produce erroneous results.

Irreversibility, uncertainty and learning

Although in theory using a full system optimisation removes the problems of combining non-linearity and uncertainty, our knowledge of the processes involved is often not strong enough to attempt such a sophisticated treatment. Some attempts have begun to use numerical stochastic optimisation techniques which include learning and irreversibility (see Manne and Olsen 1994), but these are not yet based on realistic simulations and use only small stylised models. Because of this much of the literature in this area has avoided quantitative analysis, but it aims to give qualitative insights about how various factors should alter the optimal level of abatement calculated from a *partial optimisation* model. In this way the partial optimisation approach, flawed though it is, is seen as a baseline for optimisation under uncertainty, and its limitations are explored as special cases. The three important structural properties of climate change which have been considered are: changes in the range of future uncertainty, the irreversibility of GHG emissions and the ability to learn more about damages in the future (Ulph and Ulph 1994a, Kolstad 1993). In our analysis we add the dynamics of abatement costs to this list. The reason for studying these cases theoretically is that, in the absence of a full optimisation, it is important to know whether new information indicating, for example, greater irreversibility in damage costs than previously anticipated implies a raising or a lowering of current abatement levels.

For a single decision maker, the strategy defined by partial optimisation will, by definition, only depend on the mean of the expected damages, not mean-preserving changes in its variance. This will be true if damages from emissions are either reversible or irreversible. By concentrating on the most likely scenario partial optimisation ignores the potential downside (or upside!) of extreme cases, and the dynamics of control associated with them. In contrast, the full optimisation calculates the optimal response to each set of potential future damages, and so a mean-preserving change in the variance of damages can have an important influence on current emissions.

For instance, dynamics in abatement/absorption (for example, time is needed to invest in infrastructure or to perform R & D into energy saving technology) imply that a larger variance in climate damages will affect the current optimal abatement strategy. An increased damage variance involves higher extreme damages, and the corresponding adoption of more aggressive abatement policies. If these involve a significant shift in investment then keeping our options open now by, for example, not building cities dependent on high car use will on balance save abatement costs in the future. Therefore, in this case an increase in the expected damage variance will raise the present level of abatement. The amount of pre-emptive abatement will depend on the severity of the constraint; that is, how fast damage costs evolve compared to investment and research dynamics (see Appendix 7.4). In fact, current abatement may fall in the presence of increasing uncertainty if abatement measures are more costly to reverse should damages turn out to be trivial than they are to construct should damages be severe – it is a problem that needs quantitative measurement.

The importance of considering this effect is shown by the fact that the structure of EGEM, and many of the models reviewed in Chapter 3, assumes that it takes time to put abatement measures in place (even if the base technology is available), and the faster abatement is carried out the higher the unit costs are. If optimum abatement policies were calculated on these models using a partial optimisation approach, only the dynamics of responding to the expected damage scenario, and not the potential extremes, would be considered; thus ignoring much of the simulation detail of these models.

Another dynamic influence is that emitted CO_2 causes damage until the atmosphere removes it naturally, and much of this damage is itself irreversible. Irreversibility reduces the options for actors in the future and so provides a constraint which may raise the expected cost of emissions above that in the reversible case, thus prompting more current abatement. As in any constrained optimisation, irreversibility will only matter if the constraint actually bites in the future; this case being termed *effective* irreversibility by Ulph and Ulph (1994a). Effectiveness implies that there is some future state of the world where we would want to have negative emissions, a condition we can test by using a deterministic model such as EGEM. Note that here we are considering the effects of irreversible emissions, not damages. As mentioned above in the section on pricing, the potential for irreversible damage may imply that future generations are willing to inherit less material wealth from us, in return for greater environmental options. When performing optimisation on emissions this information should be included in the *pricing* of future damages, otherwise current abatement will be too low.

Figure 7.5 gives a graphical example of emission irreversibility effects for a single future period. Each value of the marginal damage distribution D has a probability π_i associated with it, and the extreme per centiles of the marginal damage distribution are D_H = 95 percentile, D_L = 5 percentile. The average value $D_A = \Sigma\pi_i.D_i(e)$ exists for every emission level (e), and the expected value $D_E = \Sigma\pi_i.D_i(e^*)$ is the probability weighted sum of the marginal damage costs at each *optimal* level of emissions e*. Partial optimisation equates current marginal economic benefits of emissions (B(e)) to the discounted average damage value D_A, and full optimisation equates it to the discounted expected value D_E.

Considering the calculation of a full optimisation, when damages are at the high level (D_H) optimum abatement will be negative at e_H. The

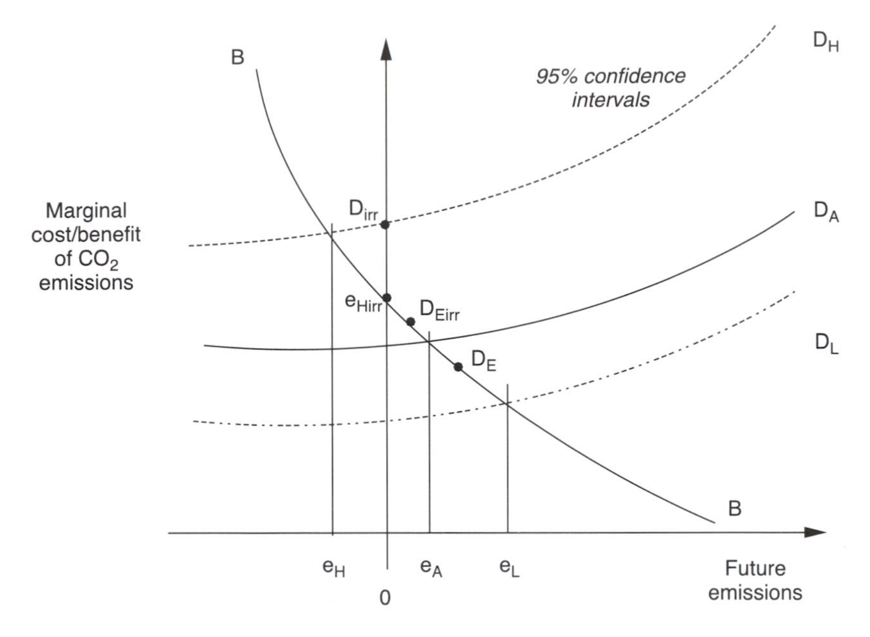

Figure 7.5 Optimisation and irreversibility

irreversibility of emissions means that the best action which can be carried out is to cease emitting, but the marginal damage cost of this scenario will be higher (at D_{irr}) than in the reversible case with no constraint. Therefore, the expected marginal damage from climate change seen in the present (D_{Eirr}) will be higher than in the case of reversible emissions (D_E), and so more abatement will be carried out sooner (see Appendix 7.4 for full example). However, if (unlike the case in Figure 7.5) marginal damage costs fell with rising concentrations, current emissions would *rise* in response to increasing irreversibility and uncertainty, because raising future baseline concentrations reduces the cost of the emissions constraint in the future.

The logic of this non-intuitive result is best explained by a similar, but more graphic, example. Imagine we are calculating how tight to set the regulations for oil discharges from ships in a certain area of water, say an estuary. The damage caused by oil spills is cumulative, and after a certain level the marginal damage caused by each unit of oil decreases because ecosystems have already been seriously damaged (a 'crowding' effect). Given the finite, if small, probability of a serious freak oil accident happening in the future (the probability of which is unaffected by the setting of current regulations – this is a vital condition), the expected damage caused by a future extra unit of 'normal' oil pollution will *fall* because there is some probability that it will enter a sea which is already heavily polluted, and thus its marginal contribution will be very small. Therefore, the chance of a large freak accident will lead us to apply looser standards to regular oil discharges from shipping than if there were no possibility of major irreversible damage. It is apparent that this result only applies under several special conditions, such as when the action being considered does not affect the potential for future damage (i.e. the probability of different levels of *marginal* climate change damage is independent of past abatement levels) or when current actions do not affect the future costs of cleaning up damage.

In a partial optimisation study the irreversibility constraint will only bite if the average damage cost (D_A) implies negative emissions, and if irreversibility only bites at D_H this will not affect current emissions. Therefore, compared to a full optimisation, partial optimisation will underestimate the amount of abatement needed to provide optimal protection against future irreversibility, or vice versa if marginal damages decrease with concentration. For cases where marginal damage costs rise with GHG concentrations, and the irreversibility constraint bites, a larger variance in the damage function increases the probability of the constraint occurring, and so its value. Therefore, along with the dynamics of abatement costs, irreversibility provides another reason why a mean-preserving spread in the variance of damages can increase current abatement levels above those found by a partial optimisation study.

The final factor influencing optimisation results is that information about climate damage may improve in the future due to scientific research and observation. Learning involves reducing the uncertainty surrounding predictions of future damage, as well as potentially changing the expected value of damages. With a single decision maker, the potential for learning always gives at least as good a result as in the no-learning case because the strategy set of the decision maker has expanded, but still includes the ability not to act on the information. The difference in expected outcomes between learning and not learning defines the *value of information* to the decision maker, that is, how much they will spend to obtain it, which is calculated by researchers such as Manne and Richels (1992). From the above results, we can see that if we spend money to lower the variance of future damages (assuming the mean is constant), this will imply higher emissions in the present when marginal damage costs rise with concentration and irreversibility constraints apply, and vice versa for falling marginal damage costs. However, mean preserving learning is a rather stylised situation, and so gives only a partial insight into the effect of new information on optimal strategies.

A common methodology for dealing with this combination of conditions – irreversibility of damage, growing future uncertainty and the potential for learning – is investment theory (Dixit and Pindyck 1993). This technique analyses the value of being able to postpone an irreversible investment until the true size of a significant variable (usually a price) has been revealed. If the price is too low the investment will not go ahead, and therefore being able to wait creates an 'option value' for being able to avoid commitment to irreversible action; this is analogous to the constraint in Figure 7.5. The option value can be expressed as a single value, or as an increase to the 'hurdle' rate of return which an investment must give above the prevailing riskless interest rate. Using this nomenclature, if the irreversibility constraint bites in the future then the material benefits of polluting, a *de facto* 'investment' in a low abatement/high economic growth strategy, should be discounted at a higher rate than the damage costs saved by immediate abatement.

Dixit and Pindyck (1993) apply a stylised model of investment uncertainty to the climate change case by comparing the pay-off between two scenarios – one with no abatement and one with no emissions. In this example, both emissions and abatement costs are irreversible; indeed the sunk costs of abatement are considered more irreversible than damage costs. This is because emissions slowly decay in the atmosphere, but if policy makers decide to do anything they are committed to perpetual future abatement at that level. Under these assumptions the irreversibility of a commitment to abatement drives the results, and greater uncertainty over future damage costs motivates greater *delay* in implementing GHG controls. This example complicates the effects of uncertainty on current abatement policy given above, as this will be affected not only by the shape of the marginal damage curve but also by the relative reversibility of abatement and climate damage. If abatement is more reversible than damage, and the irreversibility constraint bites, then increased uncertainty will reduce current period emissions, and visa-versa (see Kolstad 1993 for a fuller analysis). Usually, abatement costs are clearly more temporary than damage costs,[8] arguing against Dixit and Pindyck's results and for a strategy that increases current abatement along with uncertainty.

Unfortunately, the specific mathematical tools of investment and option accounting are inapplicable in realistic quantitative models of climate change policy, because empirically based functional forms are not soluble. For tractability the evolution of the damage function variance must be linear, and the best forecast of future damage must be able to be based *solely* on the current level of damage;[9] this will not be true for climate change because of the lags in the process, and because the probability of damage will depend on the rate of past emissions production, as well as the level. The intuition behind the inapplicability of option accounting is that it assumes uncertainty is always completely resolved in the future, that is, the price is observed and, knowing this true state, an optimal forecast of future conditions can then be made. In the case of climate change, learning and uncertainty are linked more subtly, as the lags in the process mean we will never be able to observe the effect of current emissions; therefore, we will always have to make decisions based on uncertain methods of prediction.

If learning is assumed to be passive, as described above, our understanding of the world improves whatever the *abatement* actions undertaken, though it may involve some minor spending on research. On the other hand, learning may be more endogenous to our actions, because if we carry on emitting gases we will definitely learn more about future damage, but if we stabilise concentrations quickly our knowledge will remain more uncertain. Therefore, if commitment to a programme of future abatement removes the ability to learn, abatement becomes *de facto* more irreversible than climate damage, and increasing uncertainty about damages will decrease the optimal amount of current abatement. This ability to commit forever to an abatement policy which allows no learning is what drives the Dixit and Pindyck result. In this case, a form of quasi-active learning could be more appropriate, where we wait to see the effect of climate change and then institute abatement programmes based on this new knowledge. It has even been suggested that full active learning be carried out – that is, emissions of GHGs should be markedly *increased* now so we can observe the response of the climate system, improve our knowledge and set future abatement polices more accurately!

Of course, the real learning process is likely to be somewhere between the two extremes, but considering stylised cases helps to clarify the influences of uncertainty, learning and irreversibility on present behaviour. When faced with uncertainty over damages, risk-neutral policy makers will increase current abatement above the level calculated in a partial optimisation, if they anticipate a dynamic constraint on building abatement capacity in the future or perceive climate damage both to be growing faster with increased concentrations and to be *effectively* irreversible in the future. Current abatement can fall in the face of uncertainty if marginal damages fall with concentrations, or abatement measures are more irreversible than climate damage. These results will hold as long as we do not expect to learn more about climate damage in the future. If the uncertainty around future damage can be reduced by research, then anticipation of this knowledge will alter the current optimal strategy in the opposite direction to increasing future uncertainty. If learning comes only if we pollute, and so stabilisation of concentrations is an irreversible policy commitment, then the optimal strategy will involve higher current emissions than with passive learning.

Uncertainty and sustainability

Obviously, the interaction of these factors is highly complex and will depend on their specific functional forms, so few generalisations can be made. Therefore, the influence of uncertainty on current abatement policy is an empirical question. If climate change damage is so low that on our current emissions path we would not want to have negative emissions, even in the worst case scenario, then irreversibility of emissions is irrelevant for policy making. If abatement measures can be easily put into action as damage costs evolve, then the dynamics of abatement can also be discarded, and if damages decrease quickly with concentrations then current abatement may well be lower than previously estimated in partial studies.

However, it is up to researchers, if they are to use partial optimisation techniques, to prove that this is the case and not to assume it a priori as has been done in the past. At the moment our knowledge of the critical features of future climate damage does not credibly support policy formed from such sophisticated optimisation treatments. We have no idea of the simple magnitude of damages, let alone their second and third differentials, upon which the above reasoning depends. It is significant that these high order features of the climate change problem are the same critical factors which defined the different sustainability scenarios examined earlier.

Going back to Figure 7.2 the optimal strategy achieved by using marginal instruments, such as carbon taxes, was seen to be dependent on the size of the original environmental stock, or conversely the concentration of GHGs in the atmosphere. If original concentrations are high (at N_1) then much of the possible damage has already been done to the environment, and increased abatement will seem sub-optimal because marginal damages have fallen, and will continue to drop, at these increased concentrations. Therefore, policy makers following an incremental optimisation will chose the Environmental Destruction scenario, with the climate sensitive environmental stock falling to zero. This is the same logic that drives current emissions higher in the face of an irreversibility constraint when future marginal damage costs fall with concentrations. However, if the damage function is inflected, as in Figure 7.2, initial high levels of emissions have *forced* the choice of this option by reducing the environmental stock to N_1. As in Figure 7.2 the globally optimal solution would be to reduce concentrations (increase the environmental stock – as long as damage is reversible) and move to N^*. The irreversibility of emissions means that this is not possible, and so the next best solution is to take the no-abatement path leading to the achievement of a locally optimum, but a globally sub-optimal, solution.

If marginal damage increases with concentrations the initial environmental stock is probably high (i.e. N_2 in Figure 7.2), and over-emitting now will be far more costly in the future as we will have to impose draconian abatement programmes. Therefore, uncertainty and irreversibility tend to promote more abatement, leading to the Environmental Trade-Off or Technological Utopia scenarios.

Given our current level of ignorance about the climate system, it is likely that marginal damage costs will increase with GHG concentrations at the high damage end of any probability distribution, and decrease with concentrations at the low damage end of the distribution. Therefore, the damage distributions used to calculate optimal policies cover a range of values which include all possible sustainable outcomes. If this is the case then ignoring the influence of extreme values, as is done in partial optimisation, could well lead to a non-trivial shift between a sustainable and unsustainable emissions path. Coupled with the existence of inflected damage functions, linked to past levels of emissions, this fact removes the validity of optimal policies calculated by such simple models. To be truly optimal full stochastic optimisation must be used, where every future emissions and damage path is explicitly calculated, and the system outcomes are weighed against each other.

In the short to medium term it is likely that decisions will be taken under the assumption of rising marginal damage costs (positing a threshold effect for concentrations on damages), and so greater uncertainty about impacts should prompt more current abatement. This goes against the opinions of many commentators who argue that uncertainty should prompt inaction (ERT 1994). However, the main determinant of abatement timetables will probably be the relative irreversibility of abatement measures and possible damages, and whether pre-emptive abatement will reduce the costs of waiting to make decisions when information improves. The processes behind learning about climate change damage, and the ability of governments to make long term commitments, will also be non-empirical influences on any practical strategy. Concerns about the effect of subtle shifts in marginal damage costs will probably not influence policy for many decades to come, but must be considered when considering policy instruments such as carbon taxes which aim to produce optimal results by the 'invisible hand'.

Learning and strategic interactions

While the definition of optimal global strategies under uncertainty is important, it has rather obscured the need for more systematic analysis of the influence of climate change uncertainty on strategic behaviour by countries. A perfect strategy will be of no use if countries cannot agree to put it into action.

It can be argued that uncertainty as to precise impacts is a good thing, because individual countries will not be able to predict their national gains and losses, and so will be forced to take a co-operative, risk-averse stance. This is comparable to a 'Rawlsian Veil' of uncertainty, which leads self-interested actors to formulate a just economic and legal system, because no-one knows at what level in the system they will be in the future (Rawls 1971). Once the precise consequences of climate change become apparent it is quite likely that some richer countries who are major emitters (e.g. Russia, Canada and the USA) will experience only minimum harm, or even gain, from a warming climate. The subsequent reformulation of the climate change negotiation game as an uncooperative negotiation, where not only the distribution of the abatement burden but also the size of total abatement becomes a contentious issue, can lead to higher levels of climate damage than in the case where precise impacts were uncertain. Of course, if there is large uncertainty over whether the climate will warm (which seems unlikely given current research) then learning could increase abatement; it will all depend on the precise nature of the information provided.

These issues of learning and strategic interactions have been modelled both analytically and numerically by Ulph and Ulph (1994b) in a two-period, two-player model with discrete uncertainty. They decompose the effects of reversibility, learning and strategic interactions from each other by considering a number of different cases and solution regimes (open-loop Nash and Feedback-Nash).

Ulph and Ulph begin their analysis in an international situation where identical countries want to lower emissions, but may have differently correlated future damage costs; that is, learning may reveal high costs for one country and low costs for the other. Compared to a single decision maker, the role of learning is complicated with countries gaining differentially depending on the type of strategic action undertaken. If countries both co-operate and learn then they always do better relative to the no-learning case because they are effectively a single decision maker, but if damages are perfectly correlated the gains from co-operation are slightly lower in the learning case. This is just because learning raises utility in the uncooperative scenario, thus narrowing the differential between the two cases.

If damages are completely uncorrelated, the introduction of learning causes a fall in utility if the countries do not co-operate. The reason for this is that with no-learning each country abates until its *expected* damages equal its own marginal costs, without taking into account changes in the other country's emissions, and so abatement is equal in each country. However, with learning one country finds out that it will suffer low damages and thus raises its emissions, while the other faces high damages and so lowers its emissions. These effects re-enforce each other as each aims for its ideal aggregate emissions level, and so at equilibrium marginal abatement costs are very different in each country. Because in the no-learning case marginal abatement costs were identical, the ability to learn has increased the cost of non-cooperation.

Though these results come from a simple model they will extend to more realistic scenarios, and they clearly show that when countries do not co-operate some countries will be worse off when learning is possible. As in most games where actors have different risk profiles, or face different future uncertainties, there are usually gains to be made in co-operating to hedge collective uncertainty, rather than competing to minimise individual risks. This is especially true if the differences in damage variance between countries mean that some will prefer lower initial emissions than others (perhaps due to irreversibility effects etc.). Those countries preferring to 'wait-and-see' can therefore force the other players to shoulder more of the initial abatement burden, perhaps by negotiating a favourable allocation of tradable permits.

As learning can lower utility in a non-cooperative game, this increases the benefits of constructing an international co-ordination regime for emissions reductions. Benefits will arise even if perfect co-operation, in the sense of complete internalisation of global damage costs, is not undertaken. However, given the enhanced likelihood of countries being defined as winners and losers when learning is possible, the stability of international agreements has probably been undermined as there will be greater incentives for some countries to free-ride in the future. The issues of learning, uncertainty and degrees of international co-operation are therefore closely intertwined and can be usefully analysed inside the framework of optimising behaviour.

'Hard' uncertainty, risk perception and 'no-regrets' strategies

Using expected values, stochastic optimisation or option values involves assigning reliable probabilities to the potential consequences of emitting different levels of greenhouse gases. The usefulness of this approach in policy analysis depends on the provenance of the probability measures, and how these are combined with damage costs across the whole spectrum of possible future events.

In the future it is possible that 'objective' probabilities of different degrees of climate change will be derived from frequency observations of numerous past climatic responses to CO_2 forcing, as can be obtained from ice core samples. However, to date all the probability distributions used have been derived from subjective opinions of 'experts' in the field of climate modelling; using greater or lesser degrees of sophistication in extracting and processing the results. These opinions have been based on very little hard data or experience, and so must be treated with extreme caution. The mapping of climate change on to material damage to natural and human systems is equally fraught with uncertainty, and the unknowns in the two areas combine to produce highly dispersed estimates of effects. In the light of this the potential for future climate change damage is in many ways better characterised by 'hard', or non-ergodic, uncertainty – where no meaningful probabilistic distribution can be assigned to it – than by 'soft' uncertainty where accurate distributions can be easily computed (Vercelli 1994).

The hard uncertainty surrounding predictions of climate change damage is also increased by the potential for catastrophic outcomes, such as changes in tidal patterns, shifting of the monsoon patterns, or sudden sea level rise due to the melting of the Antarctic ice shelf. The large non-linearities present in the climate system mean that an interpretation of these extreme events as being of low probability has no basis in observed science, it is merely a reflection of our naive mental models of natural systems. However, this is usually how catastrophic events are seen in the damage literature; for example, Figures 7.4(a) and 7.4(b), which have very low probabilities at the high damage cost end. When the large costs of extreme events are multiplied by their subjectively very low probabilities they tend to contribute little to the expected value of climate change, and so do not significantly affect the results of partial optimisations. This simple approach is not representative of the broader risk assessment literature which commonly records that, in the presence of hard uncertainty,[10] the public are observed to value small probability/high cost events several orders of magnitude higher than would be predicted by a simple multiplicative expected value approach. While some commentators have seen this non-linear use of probability as an irrationality, which should be discounted in rational policy analysis, others interpret it as legitimate risk aversion towards irreversible catastrophic events (Crouch and Wilson 1987, Ehrenfeld 1992).

Most researchers implicitly assume that decision makers are risk-neutral when they optimise against future uncertain outcomes. However, in the presence of hard uncertainty it is very likely that rational decision makers will take a risk-averse attitude; where the mere fact of uncertainty reduces the value of an option below its calculated expected value.[11] Including this hedging effect in a optimal strategy will give an even higher weight to the variance of future damage, so decision makers will tend to want to abate more now if this will lower the range of future risks. In linear decision theory the difference between the strategy chosen and the expected value is often called the *risk premium*, as it reflects the attitude towards uncertainty of the decision maker. It is obvious that the existence of damage irreversibilities, climate non-linearities and the potential for catastrophe mentioned above are often conflated with the effects of hard uncertainty and the psychological risk aversion of decision makers, thus rather confusing the different components of an optimal strategy in the face of uncertainty. This is impossible to avoid when hard uncertainty is present, as the effects could only be disaggregated if the objective probabilities of climate change damage were known, which is a rather circular chain of reasoning!

The 'precautionary principle' enshrined in Article 3.3 of the FCCC can be broadly read as a simple statement advocating risk-adverse decision making due to the existence of hard uncertainty about irreversible and potentially catastrophic outcomes. Interpreting this provision in terms of formal quantifiable criteria has proved unsurprisingly fraught with difficulty, because of the scale of unknowns in the problem. Strategies to deal with hard uncertainty include augmenting optimal decision making techniques for risk aversion, looking at the extent of realistic insurance against catastrophic outcomes and using portfolios of redundant strategies to offset different potential outcomes (Collard 1988, Manne and Olsen 1994). All these methodologies tend to use subjective probabilities in order to produce tractable quantitative answers, however they emphasise the unreliability of this data, the likelihood of surprises and non-marginal impacts.

The logic of irreversible climate change damage and hard uncertainty implies that governments should reduce emissions substantially now, while they wait for a better understanding of global warming science to come about. Analogous to the option value approach, this position is often termed the 'no regrets' policy because cuts in greenhouse gases are expected to be virtually costless due to the diffusion of cheap, efficient technologies, the removal of energy market distortions and the secondary benefits of reducing local pollution, congestion and so on. Unsurprisingly, the emissions targets consistent with such an easy choice are very contentious, but obviously this hedging strategy will fail if the 'no-regrets' options do not reduce CO_2 emissions to a point where concentrations begin to stabilise. In this case irreversible future damage is still increasing, as is the possibility of a catastrophic outcome, and so in avoiding hard decisions a false sense of security can be fostered. A portfolio strategy approach would augment the no-regrets abatement options with an incremental programme of research and investment designed to mitigate the potential impacts of climate change; these could include raising sea defences, developing drought resistant crops and increasing water supply capacity standards as the existing infrastructure needs to be replaced. Again the aim is to gain maximum amelioration of future risks at little or no cost.

Unfortunately, analysis of these options has generally been non-rigorous, and has avoided non-marginal changes in economic factors. When looking for insurance against a risk the basket of goods being used as a hedge must be weighted for its links to the source of uncertainty; for example, if the bulk of a country's income comes from export crops it has no viable internal insurance against extreme climate change. This is analogous to contingent claims analysis in finance theory, where risk of a stock is measured by comparison to a basket of other stocks with different relationships to the economic cycle, commodity prices and so on. The idea of a transcendent riskless rate of return does not exist. Therefore, under hard uncertainty the value of extreme events is better calculated as a scenario, rather than as a marginal monetary value, but this involves much more modelling effort than the simple case.

SUMMARY OF OPTIMISATION ISSUES

Anthropogenic climate change can be characterised as a problem which is highly uncertain, potentially catastrophic and operationally irreversible and the impact of which is globally heterogeneous. All of these features combine to undermine the validity of policy prescriptions derived from the type of simple cost/benefit analyses attempted to date. The results of these studies have been largely determined by assumptions about discount rates, environmental transition paths and extrapolation of very uncertain, partial damage estimates. Of these assumptions, the first two form a sub-set of the determinants of a sustainable growth path, and will depend on what particular form of intra- and inter-generational equity in resource consumption is assumed. As this is essentially an ethical and political issue, the definition of these parts of the optimal framework is a matter of political economy, and especially the need to ensure global agreement, not pure economics. Measurement of damage costs and their use in optimisation, on the other hand, is a more technical issue.

Given an appropriate political and ethical framework, 'optimal' strategies will only be useful if far more work is done to assess the very long run impacts of rising temperatures, and whether they compromise the sustainability of the majority of current ecosystems, including agricultural crops and animals. The operational irreversibility of these damages, and our ethical attitude towards future generations, will greatly influence the value placed on leaving environmental options open to the future. Therefore, this information will greatly change the prices associated with environmental damage, and thus our current policy prescriptions. Additionally, given the dependence of optimal strategies on the timing of abatement, a much more accurate description of the dynamics of the emissions/temperature linkage, using objective data, is needed from study of previous geologic events, so that reliance on subjective probabilities may be reduced. While there is a reasonable probability of better *scientific* information in the future, translating this into economic impacts on human and social systems will remain highly speculative, and is indeed the weak link in the chain of analysis. Because of this in the short run it will be more productive to concentrate on quantifying the secondary benefits of reducing GHG emissions, as these are more easily measured, and will provide more certain figures to offset the costs of abatement strategies. However, in the medium to long term it is likely that easy, no-regrets choices will disappear and real trade-offs between material welfare and the direct costs of climate change will have to be made.

While the most neglected area in many studies has been consideration of damage costs and sustainability, the role of uncertainty, though greatly studied, has still not been satisfactorily addressed. The methods of stochastic optimisation tell us that present uncertainty, combined with non-linear behaviour in the climate system, can increase the value of present abatement compared to either a 'best guess' or an expected value calculation of future damages. However, under several sets of plausible conditions, such as an inability to learn more in the future about damage unless concentrations of GHGs rise, we may want to postpone abatement. As these policy prescriptions depend mostly on second and third order effects the sophistication of dynamic optimisation techniques will be of little or even detrimental use, while the price dynamics and transition costs of climate change remain as uncertain as they are now. In fact, differences in damage assessments much smaller than currently exist could switch the optimal abatement policy from being environmentally sustainable to one that implies constantly increasing atmospheric concentrations of GHGs in the future. In this environment of uncertainty it is more rational for policy makers to take a risk-averse attitude, which will increase the attractiveness of near-term abatement measures to prevent climate change.

The FCCC emphasises that irreversibility and uncertainty should influence current policy, and it is right that economics should contribute to the policy debate on their implications, even though quantitative modelling is still in its infancy. Current knowledge is not good enough for the legitimacy of decisions to be determined by use of the correct *process* for optimisation, and there should be more emphasis on the type of future world we want to see occur. It must be recognised that there is no answer to the question of how much abatement is optimal which does not depend on our attitudes towards what type of environment is right and fair to bequeath to future generations. Therefore, assessment of climate policy is teleological, in that our psychological wishes for a certain type of future will determine our assessment of the value of physical changes. If we value future generations having options over the use of environmental goods, then we will value irreversible damage more, and so abate more in the present than a selfish assessment of damages would predict. Technically this may just result in a change in how environmental assets are priced, but essentially optimisation is being driven by our visions of the future, not the other way round.

Given this context the kinds of economic questions that must be answered in the short run are: should emissions automatically be lowered now as a precautionary measure? what are the benefits and potential for international co-operation given the likely range of uncertainty around the distribution of impacts? and how will learning affect current strategies? In the next chapter we attempt to model some of these questions inside EGEM using a simple model of the relationship between emissions and climate change, and a first (or zero!) order estimate of the range of possible damage costs. These are based on some of the studies discussed above and amended to reflect some more obvious short-comings. The aim of the modelling is to give a feel for the numerical size of the uncertainties involved and their potential influence, and not to calculate any optimal policy regimes for the future!

APPENDIX 7.1: DISCOUNTING AND GDP GROWTH

In theoretical models growth paths are often optimised by integrating welfare over an infinite time period, but discounting from the point of view of current consumers. This is a tautological exercise because the integral will only be finite, and therefore tractable, if the discount rate is larger than the growth rate. In turn this implies that current consumers do not value the long distant future at all, because any non-zero valuation would lead to a non-finite integral up to infinity. If this is the case then it is likely that current generations will exhaust environmental stocks because they will not value their future use.

Formally the present value of one tonne of carbon can be approximated by:

$$D_c = \sum_{t=0}^{t=\infty} D_0 \frac{(1 + y_t - a_t + r_t)^t}{(1 + \rho + y_t - P_t)^t}$$

where D_c is the total present value of damage costs for one tonne carbon; D_0 is the cost of damage costs in one year (once equilibrium temperature is reached); y_t is the economic growth rate; a_t is the rate at which damage decreases due to absorption of CO_2 into sinks in the carbon cycle; r_t is the rate of increase of the damage per tC as concentrations increase; ρ is the pure rate of time preference; and p_t is population increase.

It is evident that this sum will only converge if:

$$1 + y_t - a_t + r_t < 1 + \rho + y_t - p_t$$

$$\text{i.e. } r_t - a_t < \rho - p_t$$

The parameter r_t depends on the shape of the damage/concentration curve (as concentration is proportional to emissions) over time. A roughly linear curve would make $r_t = 0$ in the longer term, but if damages increase faster than concentrations then $r_t > 0$. The value of a_t has been estimated by Nordhaus (1993) as 0.8 per cent, from data on carbon emissions and concentrations over the past 125 years. If $\rho = 0$, as some suggest, then for any population growth rate greater than 0.8 per cent and assuming $r_t = 0$, the sum would not converge and the cost of a tonne of carbon would be infinite. More reasonably, if $\rho = 2$ per cent and $p = 1.73$ per cent (the rate for 1985–90, WRI 1990), emissions could continue at a rate making $r_t = 1.07$ per cent p.a. before this occurred.

It is clear that the terms will not rapidly vanish despite discounting, and so very long term impacts do matter. There are also situations with apparently finite probability which could incur infinite costs. These must be excluded from the calculation if it is to converge.

APPENDIX 7.2: CLIMATE CHANGE AND SUSTAINABLE GROWTH

The growth model derived here is inspired by a more complex treatment in Bovenberg and Smulders (1994), which has a more detailed discussion of the use of policy control instruments and the dynamics of these types of models.

Defining a model of instantaneous economic *welfare* U over time, when welfare depends on the quantity of environmental assets N and composite polluting man-made assets (human capital, material capital and extracted natural resources) K_U which exhibit constant returns-to-scale:

$$U = n(N).f(K_U)$$

where: $dn(N)/dN > 0$, $df(K_U)/dK_U > 0$ and where the changes in environmental productivity with the environmental stock are not necessarily monotonic, but can be inflected. The environmental stock changes in each period according to:

$$\partial N/\partial t = N' = g(K_U, K_T) + \nu(N),$$

$$\partial g/\partial K_U < 0, \ \partial g/\partial K_T > 0$$

$$\nu(N) > 0 \ \forall \ N.$$

where K_T is the stock of abatement technology and $\nu(N)$ is the ability of the environment to assimilate pollution.

Steady state conditions

When the economy grows at a steady state the value of the polluting and abating sectors must remain in the same proportion, while the environmental stock is constant:

$$K = K_U + K_T \Rightarrow K_U = \alpha.K, \; K_T = (1-\alpha).K$$

$$\therefore K'_U/K_U = K'_T/K_T,$$

$$N' = 0.$$

Consumption and saving decisions are made by the representative consumer based on the maximisation of expected utility and so the steady state growth rate η is given by the Ramsey Rule:

$$\eta = \sigma_I.(r - \nu)$$

where σ_I is the intertemporal elasticity of utility, r is the marginal return on the composite man-made asset stock K_U and υ is the pure rate of time preference. From the production function:

$$r = n(N).\frac{df(K_U)/dK_U}{K} = \alpha.n(N).\frac{df(K_U)/dK_U}{K_U} = \alpha.n(N).\varsigma$$

where ς is the constant rate-of-return on each unit of capital. Therefore, given that σ_I and υ are constant and exogenous to the model the rate of utility growth will be determined by the sustainable environmental stock n(N) and the proportion of man-made assets diverted away from consumption and into the abatement sector.

Given an initial environmental stock N_1 the choice of α is made by maximising r:

$$\text{Objective} = \max[\alpha.n(N).\varsigma]$$

If $n(N) > 0$ at $N = 0$, that is, the carbon sensitive environmental stock is not essential to welfare, then the following *equilibrium* solutions are possible.

Single equilibrium

If $d\nu(N)/dN = 0$, that is, CO_2 assimilation is constant with concentration, then:

- If $n(N_1) - n(0) > 1 - \alpha^*$, where α^* is the solution of $g(\alpha^*) = -\nu(N) =$ constant, then environmental productivity is higher than the drop in man-made productivity needed to stabilise concentrations. The environmental stock will tend towards N_{max} over the very long time as an incremental increase in abatement will eventually reduce concentrations to pre-industrial levels (Environmental Trade-Off scenario with maximum environmental stock).
- Otherwise $\alpha = 1$, and $N' < 0$ and $N = 0$ at equilibrium; that is, the cost

of maintaining stable concentrations is higher than environmental productivity (Environmental Destruction).

Multiple equilibria

If CO_2 assimilation *increases* at higher concentrations:

$d\nu(N)/dN < 0$, and sustainable equilibria exist for all $g(\alpha^*) = -\nu(N^*)$ between:

$$[g(\alpha_0) = -\nu(0)] > \alpha^* \geq [g(\alpha_{max}) = -\nu(N_{max})]$$

- Equilibrium at $\max[\alpha^*.n(N^*)]$ unless (Environmental Trade-Off, Technological Utopia).
- $n(0) > \max[\alpha^*.n(N^*)]$ when $a = 1$, then $N' < 0$ and $N = 0$ at equilibrium.

If CO_2 assimilation *decreases* at higher concentrations:
$d\nu(N)/dN > 0$, therefore $\partial\alpha^*/\partial N > 0$ two equilibria exist;

- If $n(0) < \max[\alpha^*.n(N_{max})]$ when $\alpha = 1$, then $N' > 0$ and $N = N_{max}$ at equilibrium.
- If $n(0) > \max[\alpha^*.n(N^*)]$ when $\alpha = 1$, then $N' < 0$ and $N = 0$ at equilibrium.

Dynamics of transition to equilibrium

If optimal natural stock N^* is less than starting stock N_1 then the optimal growth path can always be feasibly reached by incremental optimal decisions because:

$$\alpha^* > \alpha_{trans}, \text{ for } N_1 \geq N \geq N^*$$

$$\therefore \max[dU/dt] \text{ is on transition path}$$

If optimal natural stock N^* is greater than starting stock N_1 then achievement of the optimal growth path will not be the result of market price signals if:

$$\frac{\partial U}{\partial \alpha} \cdot \frac{\partial \alpha}{\partial g} > \frac{\partial U}{\partial N} \text{ at } N = N_1$$

In this case investment in polluting commodities will be more profitable than investment in abatement technologies, based on instantaneous shadow values, therefore $N' < 0$ and $N = 0$ at equilibrium.

APPENDIX 7.3: ILLUSTRATIVE WELFARE LOSSES FROM CLIMATE CHANGE

In the face of falling productivity due to climate change, actual welfare losses will depend on the shape of the demand and supply curves. Consumers are unambiguously worse off if they pay a higher price for less goods; producers may face either an increased or decreased surplus depending on the nature of the shift in the supply curve.

For a simplified case, the economic welfare loss can be calculated for the case where:

$$\text{Demand } D = aP^{\alpha}$$

$$\text{Supply } S = bP^{\beta}$$

where α and β are demand and supply elasticities for global food production and P is price. The equilibrium price P^* is given by:

$$S = D = aP{*}^{\alpha} = bP{*}^{\beta}$$

$$P* = (b/a)^{1/(\alpha-\beta)}$$

If the yield is reduced due to climate change, the shifted supply curve is:

$$S = \lambda bP^{\beta}$$

where λ is the reduced yield under changed climate conditions, given a certain set of other inputs at the same price (e.g. for a yield loss of 10 per cent, $\lambda = 0.9$). Quantity supplied will be:

$$Q' = aP'^{\alpha}$$

So the new equilibrium price and quantity will be given by:

$$P' = (\lambda b/a)^{1/(\alpha-\beta)} = P{*}\lambda^{1/(\alpha-\beta)}$$

$$Q' = a[(\lambda b/a)^{1/(\alpha-\beta)}]^{\alpha} = Q{*}\lambda^{\alpha/(\alpha-\beta)}$$

The changes in consumer and producer surplus can now be estimated as proportions of initial market volume, P*Q*. A linear approximation to the loss in consumer surplus ΔC is given by:

$$\Delta C = (P'-P*)(Q*-(Q*-Q')/2)$$

$$= P{*}Q{*}(\lambda^{1/(\alpha-\beta)}-1)(1+\lambda^{\alpha/(\alpha-\beta)})/2$$

This will be large when $\lambda^{1/(\alpha-\beta)}$ is large, i.e. α small and negative and β small and positive, or inelastic demand and supply. The change in producer surplus can similarly be approximated as:

$$\Delta S = (P{*}Q{*} - P'Q')/2$$

$$= P{*}Q{*}\,(1 - \lambda^{1/(\alpha-\beta)+\alpha/(\alpha-\beta)})/2$$

If demand is inelastic ($\alpha < 1$), this expression will result in a net gain to

Table 7.1 Changes in producer and consumer surplus as agricultural yields decrease

Market scenario	*Yield loss (%)*	*Price increase (%)*	*Loss in consumer surplus*[a]	*Gain in producer surplus*[a]	*Net welfare loss*[a]
$\alpha = -0.2$	−7	13	−13	5.1	−7.6
$\beta = 0.4$	−15	31	−30	7.5	−18.0
	−37	116	−108	43.0	−65.0
$\alpha = -0.1$	−7	27	−27	12.0	−15.0
$\beta = 0.2$	−15	72	−70	31.0	−39.0
	−37	370	−340	150.0	−190.0

a Expressed as a percentage of the total original market value.

producers, because the increased price more than makes up for the loss in yields. However, the difference will always be smaller than the loss of consumer surplus.

The value of the total welfare loss may thus be calculated for various values of α, β and λ. The results depend upon the assumptions on the shape of demand and supply curves, but we can give an illustrative example of the possible magnitude of these effects by guessing possible parameters. α would be expected to be negative and small, as overall demand for food is inelastic, and β will be positive but also low. Table 7.1 illustrates sample calculations for two possible sets of market conditions. The yield losses correspond to declines of 7–15 per cent in agricultural production under the IPCC's 'best guess' $2 \times CO_2$, 2.5°C warming, and a minimum of 37 per cent under Cline's very long term 10°C scenario.

It is clear that the yield loss alone is not an adequate indicator of total loss, as it fails to take price changes into account. Under the first scenario, overall welfare loss is slightly higher than the yield losses, and much higher than the long term case of 37 per cent yield losses. Under more inelastic conditions, losses are much greater even for the low yield losses considered by the IPCC, while Cline's long term warming would bring devastating losses.

The increase in overall loss increases non-linearly with the loss in yield. A regression of $\ln(\Delta L)$ against $\ln(\Delta Y)$ where ΔL is the total welfare loss and ΔY the loss in yield shows a fairly consistent coefficient on ΔL of ≈ 1.25 for several elasticities. This indicates much greater non-linearity in losses with temperature change than have been usually considered in extrapolations of damage results. Damage is usually modelled as proportional to ΔT^{γ}; where ΔT is the change in temperature, and γ is the temperature-damage exponent, normally taken as between 1.2 and 1.5 in past studies, and represents both physical and economic non-linearities. This implies that values of $\gamma \approx 1.25$ only account for damage increases due to the effects of changing prices on consumer surplus. Therefore, including the non-linearity of crop responses to temperature, that are also conflated inside γ, could make a figure of $\gamma = 1.5$ to 1.9 more applicable, raising the projected costs of climate change considerably.

The supply curve used here passes through the origin (unlike a classical Ricardian curve), which may be justified by observing that globally the lowest production costs faced by farmers of basic staples are very small compared with the current equilibrium price. This assumption increases the gain in producer surplus over the more general case with a minimum supply price greater than zero. If the 'reservation price', below which no production occurs, is significantly greater than zero and increases with climate damage then producer surpluses will be more prone to decline.

In reality there would be considerable complications caused by market distortions such as the CAP, lack of free trade, and subsistence farming. There is of course no global equilibrium price; prices vary by country and foodstuff. The elasticities are likely to change at different volumes and with incomes, as would supply parameters, and the very simple functional form used here is only illustrative. However, this analysis does reflect the consequences of non-marginal changes in provision of basic resources essential for life. The impact of these extra costs would be particularly harsh in developing economies. If a developing nation spends 50 per cent of its income on food and water (including imputed labour costs from subsistence farming) and the overall increase in food and water costs from climate change is 10 per cent, this will reduce 'surplus' non-food GDP by 10 per cent (i.e. $10 \times 50/(100 - 50)$). For the same productivity reduction in a rich nation which spends 10 per cent of income on food and water, non-food GDP would only reduce by 1.1 per cent.

APPENDIX 7.4: UNCERTAINTY, IRREVERSIBILITY AND ABATEMENT DYNAMICS

In this appendix we formally prove the results linking uncertainty, irreversibility and abatement dynamics using a simple model of decision making under uncertainty. This model is similar in structure to those derived in Ulph and Ulph (1994a) and Dixit and Pindyck (1993), however our interpretation of the conditions and results is rather different. When considering two period models both of these studies consider the second period as being potentially knowable, whereas we define it as generic future uncertainty, thus it will always move ahead of the current period and will never be experienced. This may seem a rather subtle distinction, but on reflection the difference affects both the form of optimisation undertaken and the interpretation of results.

Full and partial stochastic optimisation

Considering a two-period world where we make decisions on abatement now (period 0) which will effect uncertain damage costs in the future (period 1), our aim is to define the optimal current strategy so as to maximise expected future utility.

Therefore, the objective function to be optimised is:

$$max\{E[B_0(e_0) + \rho.(B_1(e_1) - D(\lambda e_0 + e_1))]\} \qquad (7.4:1)$$

where E is the expectations operator, $B_{0,1}(e_{0,1})$ are the benefits from emitting GHGs, $e_{0,1}$ are the amount of GHGs emitted in each period, $D(\Sigma_n e_n)$ is the stochastic damage function (e.g. $D(\Sigma e) \sim LgN[\mu,\sigma^2]$) dependent on the stock of GHGs in period n, ρ is the single period discount factor ($\rho < 1$) and λ is the decay constant of GHGs in the atmosphere ($\lambda < 1$). Details of the functions being:

$$dB_{0,1}/de > 0;\ d^2B_{0,1}/de^2 < 0;\ B_0(e) \neq B_1(e);$$

$$dD/d(\Sigma e) > 0;\ d^2D/d(\Sigma e)^2 < \text{or} > 0.$$

Therefore, emissions in the first period, having decayed by a factor λ, cause an uncertain level of damages in the future which are then discounted by a factor ρ. Equation (7.4:1) represents the *full stochastic optimisation* of the climate change system, which should be compared to equation (7.4:2) which is the *partial* optimisation used in most cost/benefit studies:

$$max\{B_0(e_0) + \rho.(B_1(e_1) - E[D(\lambda e_0 + e_1))]\} \qquad (7.4:2)$$

In the partial optimisation the expected value operator is only associated with the damage costs, not the whole system; therefore, non-linear interactions between the components of the optimised system will not be considered. This becomes clearer if we simplify the distribution of uncertainty surrounding D by assuming two discrete future states, D_H and D_L, with probabilities α and $(1 - \alpha)$ respectively. That is:

$$D_H(\Sigma e) > D_L(\Sigma e),\ \forall\ \Sigma e.\ P[D = D_H, \alpha],\ P[D = D_L, (1-\alpha)]$$

The multiplicative probability weighting is a very simple form because the probability of damage does not depend on the past level, or rate, of GHG emissions/taxes; thus the probabilities are exogenous to the system which is unlikely in the case of climate change. In this simplified case the expected value and optimality conditions for (7.4:1) are (note: ` denotes a first derivative):

$$EV = B_0(e_0) + \alpha.\rho.(B_1(e_{1H}) - D_H(\lambda e_0 + e_{1H})) + (1-\alpha).\rho.(B_1(e_{1L}) - D_L(\lambda e_0 + e_{1L})) \qquad (7.4:3a)$$

$$\therefore \partial EV/\partial e_0 = B`_0(e_0) - \rho.(\alpha.D_H`(\lambda e_0 + e_{1H}) + (1-\alpha).D_L`(\lambda e_0 + e_{1L})) = 0 \qquad (7.4:3b)$$

$$\partial EV/\partial e_{1H} = B`_1(e_{1H}) - \alpha.D_H`(\lambda e_0 + e_{1H}) = 0 \qquad (7.4:3c)$$

$$\partial EV/\partial e_{1L} = B`_1(e_{1L}) - (1-\alpha).D_L`(\lambda e_0 + e_{1L}) = 0 \qquad (7.4:3d)$$

The optimality conditions for (7.4:2) are given by:

$$EV = B_0(e_0) + \rho.(B_1(e_1) - \alpha.D_H(\lambda e_0 + e_1) - (1-\alpha).D_L(\lambda e_0 + e_1)) \quad (7.4{:}4a)$$

$$\therefore \partial EV/\partial e_0 = B`_0(e_0) - \rho.(\alpha.D_H`(\lambda e_0 + e_1)) + (1-\alpha).D_L`(\lambda e_0 + e_1) = 0 \quad (7.4{:}4b)$$

$$\partial EV/\partial e_1 = B`_1(e_1) - \alpha.D_H`(\lambda e_0 + e_1) - (1-\alpha).D_L`(\lambda e_0 + e_1) = 0 \quad (7.4{:}4c)$$

Comparing the optimality conditions (7.4:3b–d) and (7.4:4b,c) shows that, while in the partial optimisation the expected value of the damage function determines the optimal strategy, in the full optimisation the expected value of the *system*, including the optimal response of emissions in period 1, determines the control strategy in the first period. A mean preserving spread in damage uncertainty implies D_H increasing and D_L decreasing proportionately. By definition this will not change the amount of first period abatement in the partial optimisation as the expected damage value is unchanged. For full optimisation the effect of this on period 0 emissions is ambiguous, and will depend on how the optimal amount of emissions changes with increasing damage, and whether this increases or decreases the expected marginal damages in equation (7.4:4a). If the cost of abatement in period 1 rises sharply (i.e. $\partial B^3{}_1/\partial e^3{}_1 > 0$) then increasing uncertainty will tend to raise period 0 emissions as the cost of extra abatement in the future rises faster than the level of extreme damages.

If the damage function is virtually linear in the stock of emissions, that is, the marginal cost of an emission does not change much with GHG concentration, then the result of the two methods will be identical as $D_H`(\lambda e_0 + e_{1H}) \approx D_H`(\lambda e_0 + e_1)$ etc. The importance of using full optimisation therefore depends on second and third order effects, and in the face of large uncertainties about even first order magnitudes of the damage function these subtleties may be considered irrelevant for policy making purposes. If this is the case, partial optimisation will suffice as a way of crudely estimating the quantitative level of optimal emissions, though no intuition about more complex reactions to uncertainty and irreversibility can be inferred from it.

Optimisation with abatement dynamics

If the cost of abatement in period 1 depends on the amount carried out in period 0, because of investment dynamics, learning-by-doing or other factors, this will alter the optimality conditions given in (7.4:3). Defining the benefits from emissions (Benefits = 1/Abatement Costs) in this case as:

$$\text{Benefits in period 1} = B_1(e_0,e_1);$$

$$\partial B_1/\partial e_0 < 0;\ \partial B_1/\partial e_1 > 0;\ \partial^2 B_1/\partial e^2{}_0 < \text{or} > 0;\ \partial^2 B_1/\partial e^2{}_1 > 0.$$

The optimality conditions become:

$$\partial EV/\partial e_0 = B`_0(e_0) + \rho.(\alpha.(B`_1(e_{1H}) - D_H`(\lambda e_0 + e_{1H})) + (1-\alpha).(B`_1(\partial e_{1L}) - D_L`(\lambda e_0 + e_{1L}))) = 0 \quad (7.4{:}5a)$$

$$\partial EV/\partial e_{1H} = B`_1(e_{1H}) - \alpha.D_H`(\lambda e_0 + e_{1H}) = 0 \quad (7.4{:}5b)$$

$$\partial EV/\partial e_{1L} = B`_1(e_{1L}) - (1-\alpha).D_L`(\lambda e_0 + e_{1L}) = 0 \quad (7.4{:}5c)$$

Simplifying the intuition of these conditions by assuming that the magnitude of the different functions implies the same equilibrium marginal damage costs and benefits in period 1 as in the solution of (7.4:3), we can look at the effect of the extra terms in $\partial B_1/\partial e_0$ in (7.4:5a). By definition $\partial B_1/\partial e_0 < 0$, therefore the expression inside the brackets in (7.4:5a) $\rho.(.\ .)$ will be more negative than in (7.4:3a) and for the equality to hold $B`_0(e_0)$ must be higher implying more abatement in period 0.

Alternatively the equations could be rewritten in a dynamic form, so as to make abatement costs in period 1 depend on the difference in abatement required in the two scenarios, by defining:

$$\text{Benefits in period } 1 = B_1(e_1,(e_0-e_1));$$

$$\partial B_1/\partial(e_0-e_1) < 0,\ (e_0-e_1) > 0;\ \partial B_1/\partial(e_0-e_1) > 0,\ (e_0-e_1) < 0$$

$$\partial B_1/\partial e_1 > 0;\ \partial^2 B_1/\partial(e_0-e_1)^2 < \text{or} > 0;\ \partial^2 B_1/\partial e^2_1 > 0.$$

In this formulation both installing and removing abatement devices causes penalties. In this case abatement in period 0 will depend on the amount of adjustment needed to reach the equilibria at D_H and D_L from the original (undynamic) optimal strategy, and the shape of $\partial B_1/\partial(e_0-e_1)$. Emissions will be higher in period 0 if it is harder to dismantle abatement measures which are not needed ($\partial^2 B_1/\partial(e_0-e_1)^2 < 0$), and lower in period 0 if the converse is true.

Optimisation and damage irreversibility

In the solution of (7.4:3) it is quite possible for optimal second period emissions e_{1H} or e_{1L} to be less than zero; however, as no feasible form of CO_2 scrubbing exists, this is rather unbelievable, and for realism emissions should be considered as being irreversible. Irreversibility adds a set of constraints to the optimality conditions in (7.4:3b–d) of e_0, e_{1H} and $e_{1L} > 0$, to give:

$$\partial EV/\partial e_0 = B`_0(e_0) - \rho.(\alpha.D_H`(\lambda e_0 + e_{1H}) + (1-\alpha).D_L`(\lambda e_0 + e_{1L})) + \zeta = 0 \quad (7.4{:}6a)$$

$$\partial EV/\partial e_{1H} = B`_1(e_{1H}) - \alpha.D_H`(\lambda e_0 + e_{1H}) + \theta_H = 0 \quad (7.4{:}6b)$$

$$\partial EV/\partial e_{1L} = B`_1(e_{1L}) - (1-\alpha).D_L`(\lambda e_0 + e_{1L}) + \theta_L = 0 \quad (7.4{:}6c)$$

$$\partial EV/\partial \zeta = e_0 = 0,\ \partial EV/\partial \theta_H = e_{1H} = 0,\ \partial EV/\partial \theta_L = e_{1L} = 0. \quad (7.4{:}6d)$$

Obviously, $\zeta = \theta_H = \theta_L = 0$ if the optimal level of emissions never drops to zero; the irreversibility constraint is not binding and the first period optimal strategy e_0 is the same as for the reversible case (7.4:3). In the terminology of Ulph and Ulph irreversibility is not *effective* in this case.

If for reversible emissions the optimal level of emissions in the high damage case (e^*_{1H}) is negative, in the irreversible case $\theta_H > 0$, as it accounts for the difference between marginal costs and damages as we move away from the optimum in (7.4:6b). This will have an ambiguous effect on the expected marginal damage cost of emissions in period 0; there are three possible cases:

- If $d^2D_H/d(\Sigma e)^2 > 0$, dD_H/de increases with e and e_0 decreases.
- If $d^2D_H/d(\Sigma e)^2 < 0$; $dU_1/de \neq dD_H/de$, $\forall\ 0 > e_1 > e^*_{1H}$; and $dU_1/de < dD_H/de$, $\forall\ e_1 < e^*_{1H}$; then (7.4:6b) defines a benefit minimum at $e_{1H} < 0$, the maximum lies at the next positive value of e where (7.4:6b) holds, or when $dU_1/de = 0$; the irreversibility constraint does not bite.
- If $d^2D_H/d(\Sigma e)^2 < 0$; and $dU_1/de < dD_H/de$, $\forall\ e_1 < e^*_{1H}$; dD_H/de decreases with e and e_0 increases.

The intuition of these cases is not that straightforward, as again they depend on second and third order effects. In the first case marginal damage increases with concentrations, so the inability to abate optimally in period 1 increases concentrations and thus the expected marginal damage costs in period 0, hence abatement in period 0 is higher. The non-intuitive converse of this argument holds in the third case where marginal damages decrease with concentrations, and optimal period 0 emissions will rise; this is because raising period 0 emissions reduces the amount of optimal abatement needed in period 1 by reducing marginal damages, thus the constraint is relieved.

For the partial optimisation method we augment equations (7.4:4a,b) with the relevant constraints and can see that abatement in period 0 will only be affected by irreversibility if this constraint bites for the *expected* value of the damage function. If this is true the same set of three cases holds as described above, but just replacing $D`_H$ with $\alpha D`_H(\ldots) + (1-\alpha)D`_L(\ldots)$. Thus, first period emission could rise or fall in response to irreversibility, but the requirements for the constraint to be effective are harder, leading to too much pollution in many cases.

The effects of increasing the variance of the damage function (mean preserving) are similar to before. For full optimisation increasing the variance will strengthen the reaction of period 0 emissions as it increases the constraint; e_0 will therefore rise or fall more compared to the reversible case. As before, increasing the variance of the damage distribution does not affect the result of a partial optimisation.

Summary of results

Partial optimisation fails to capture many of the effects of future uncertainty and irreversibility on current optimal strategies, and in many situations will recommend too little current abatement. The two methods are only equivalent if the marginal damage caused by emissions is unchanged with GHG concentrations, that is, the damage function is linear.

Using full system optimisation the effects of increasing uncertainty on optimal strategies are ambiguous, but generally if marginal damage increases with concentrations of GHGs then higher uncertainty implies higher current abatement, and vice versa. If abatement is subject to dynamics, that is, future abatement is cheaper if more is done now, then again a more aggressive abatement strategy will be preferred; if future costs of abatement depend on the *rate* of abatement needed then the calculation of an optimal strategy is more complex and will depend on second and third order changes in cost and damages. If the irreversibility of emissions is important, that is, under some possible future states of the world we would want to have negative emissions, then current abatement will be higher if marginal damages increase with GHG concentrations, and visa-versa; higher uncertainty over future costs will strengthen these reactions in their appropriate directions. In general, if irreversibility is important a partial optimisation will understate the amount of extra current abatement needed to react to this possibility.

8

QUANTITATIVE MODELLING OF OPTIMAL INTERNATIONAL ABATEMENT POLICIES

INTRODUCTION

This chapter investigates some of the issues surrounding the optimisation of climate change policy, by running illustrative scenarios on the EGEM macroeconometric model. The aim of performing these numerical simulations is to calculate a first-order assessment of how the different factors outlined in Chapter 7 affect climate change policy, not to produce policy prescriptions for optimal abatement levels. The areas covered include: the scope of international co-operation, methods of welfare aggregation, irreversibility of GHG emissions, the investment dynamics of abatement, and uncertainty over future damage costs. Implicitly this means we are examining issues not only of intra-generational equity, by considering how climate damages should be aggregated across countries with different incomes and preferences, but also inter-generational equity, by asking how knowledge about the future damages should affect our current decisions to spend money on GHG abatement.

The fact that EGEM only has details of abatement costs in the developed countries means that none of the optimisations carried out in this chapter can stabilise global emissions of CO_2, let alone stabilise GHG concentrations. Therefore, we are unable to investigate issues of sustainability in any detail, and especially how different damage costs assumptions will affect convergence to environmentally sustainable, or destructive, equilibria. These important issues will need a more global approach, integrating models of climate, ecosystems and economy, and as such are more suited to the integrated assessment models mentioned in Chapter 3, than a macroeconometric model such as EGEM. Despite this limitation, the optimising behaviour of the developed countries is interesting as they will be the first to accept truly binding abatement commitments, and their actions now will be vital in securing future abatement obligations in developing countries. Furthermore, the questions of how the dynamics of abatement and damages interact in decision making are similar in all countries; so modelling the OECD countries gives an indication of the importance of these effects in the rest of the world, while keeping results based in firm empirical data. The nature of EGEM as a model based in empirically measured data also means we cannot investigate the use of adaptive strategies towards climate change, as opposed to GHG abatement. This is a major omission, but unavoidable given the paucity of data on adaptation and how this could affect future damage costs.

Modelling optimal policy making requires that the equations describing the macroeconomy in EGEM be supplemented by a climate change damage function, relating emissions of GHGs to economic and welfare costs in the future. We do not consider the secondary benefits from energy taxation, in terms of local pollution reductions and so on, as the model does not produce forecasts of pollutants such as SO_x and NO_x. However, this omission is not vital given that we are more interested in the differential effects of various factors, not absolute levels of abatement. The methodology used to project climate damages demonstrates the immaturity of research in this area, and we claim no superior realism for our numbers; dependent as they are on a priori assumptions about discount rates, growth rates and price dynamics. However, putting some numbers on the problem does allow illustration of complex effects which are often confusing when tackled in an abstract manner. Additionally, the results of the empirically based parts of the model – especially the energy sector – will suggest the appropriate policy response towards different levels of forecasted climate change damage, however these projections have been derived.

After surveying the limited literature on damage cost estimates, we construct a model which quantifies the marginal cost of climate change damage, based on current estimates from a variety of sources. The properties of this model are examined in detail, and a stochastic distribution of damage costs is calculated, which covers a wide spectrum of initial assumptions. Combined with forecasts of energy-related CO_2 emissions from EGEM, this model allows the total future environmental costs of various levels of CO_2 emissions to be calculated. Numerical optimal control algorithms are then used to find the carbon tax levels that each country grouping would impose under different informational and co-ordination regimes. In the terminology of Chapter 7 all the calculations carried out here are 'partial stochastic optimisations'; that is, they find optimal policies by deterministic modelling using the expected value of greenhouse damages, but do not optimise the expected benefits of greenhouse policy (full stochastic optimisation). To highlight this limitation we also investigate some of the properties of the damage and abatement functions which would give a large divergence between the results of full and partial optimisation. These simulations indicate where the problems with partial optimisation studies may lie, and give a first guess as to which are the critical aspects of the climate change problem to study quantitatively in the future.

The first simulations look at how optimal abatement in the developed countries changes depending on the scope of co-operation, and how damages to developing countries are included in their objective functions. The effect of abatement dynamics and irreversibility on current hedging abatement are then explored, by running extreme damage scenarios under different informational structures to see if these constraints will bite in the near future. Following Ulph and Ulph (1994b) we then consider how learning and damage cost uncertainty affect abatement under different co-ordination regimes.

MEASURING AND PROJECTING THE DAMAGE COSTS FROM CLIMATE CHANGE

The basic methodology for measuring the costs of climate change taken by existing studies is to consider the scenario of CO_2 concentration doubling, described under various definitions, and then calculate the effects of this on a range of sectors considered vulnerable to increased temperature; usually, agricultural and forestry productivity, water supply, sea level rise, ecosystem damage, and mortality/morbidity effects. The reason for this approach is purely historical, as CO_2 doubling was the basic test scenario considered by climate modellers. When used in computer modelling exercises the costs for CO_2 doubling are extrapolated, and interpolated, using a variety of simple models linking pollution, temperature rise and damage. The components of these models are discussed below, but first we discuss the estimates of damage for the basic scenario.

The costs of CO_2 doubling

Most studies to date have been enumerative, that is, the consequences of a certain average temperature rise or climatic shift on a certain sector has been estimated, and then the results for all sectors considered have been added up. This ignores interactions between related sectors, and each sector and the rest of the economy, as these accounting exercises are basically static. Some work is now being done on integrated assessment to account for feedbacks and feed-forwards, especially in adaptation and changing consumption patterns; however, this work is still at an early stage.

Of the enumerative studies most have only quantified the different sources of damage for the USA, and have all come to fairly similar totals of 1–1.5 per cent of GNP for a warming of 2.5 °C, which is the central best guess for the temperature rise associated with CO_2 doubling (Fankhauser 1994b). However, there is a wide variation in the estimates for the different categories that make up these totals. For instance Titus (1992) sees forestry as the greatest source of loss at \$38bn, followed fairly closely by air and water pollution at \$23.7bn and \$28.4bn respectively, with agriculture only suffering \$1bn losses (all for 4 °C warming). By contrast Cline (1992) believes agriculture to be the most important at \$15bn, with forestry a mere \$2.9bn, air pollution \$3bn and water pollution not worth mentioning. Fankhauser (1993) has human life/morbidity largest followed by water supply, while Nordhaus (1991b) believes sea level rise to be most important. As all these estimates are derived independently, and each justified with equally valid reasoning, it would be quite possible for a different researcher to come up with a total equal to the sum of all the highest, or all the lowest, estimates for each category; giving a range of 0.4–2.7 per cent of US GDP when these figures are adjusted to 2.5 °C warming.

As the error margin for each component figure is also rather wide, often up to an order of magnitude, a reasonable range for damage in the USA might in fact be around 0.2–5 per cent of GNP for 2.5 °C warming. As the relationship between temperature and emissions is also imperfectly understood, the impact of CO_2 doubling may in fact be anything from 1.5–4.5 °C of warming, thus increasing the error range even further. Even these widely different figures do not include the influence of longer term impacts past CO_2 doubling, or 'greenhouse catastrophes' such as melting of ice-caps, changes in deep sea currents or major ecological collapse. It is to cope with this sort of uncertainty that so many analysts resort to 'best guess' figures, and few have even attempted Fankhauser's (1995) simple expected value approach shown in Figures 7.4(a) and 7.4(b).

Outside the USA there are very few comprehensive studies, apart from extrapolating directly from USA figures according to percentage of GNP. Fankhauser (1993) however does give estimates for six world regions, and generally shows damage to be most severe in developing countries, because they are more reliant on agriculture and other climate dependent sectors. One of the major problems in these evaluations is that non-market goods represent about two-thirds of the impacts measured, and these are usually valued based on a proportion of the country's GDP (or GDP per capita) where they occur, as they are not tradable products. As mentioned in Chapter 7, when dealing with mortality and morbidity effects this leads to debates over defining the 'value of a life' for people with different incomes, which depends on ethical and political criteria, not economic ones. The political debate over these issues has been relatively fruitless to date, because commentators have not distinguished between normative statements about what *should* be the value of saving a life and descriptive statements about what people in a country are actually prepared to sacrifice to gain environmental and health benefits. Until there is some clarity as to the attitude of rich countries to the lives of people in poor ones, this issue will be unresolved and a matter of conjecture in damage cost estimates. We will consider below how much different approaches to cost derivation and aggregation can affect optimal abatement.

Modelling climate change damage

Having estimated a range of damage figures for CO_2 doubling, these must then be extrapolated into the future if the impacts of a unit of carbon dioxide emitted now are going to be calculated over its ≈ 100–200 year life in the atmosphere. This involves constructing a simple relationship between emissions, atmospheric concentrations, temperature rise and damage costs. Temperature rise as a function of emissions is calculated from the following equations (from Cline 1992, pp. 50–54 and Table 2.1; Nordhaus 1993). Increasing atmospheric CO_2 stocks and concentration due to energy-related emissions are taken as:

$$S_t = S_{t-1} + (1 - 0.64).E_t - \nu(S - S_0)$$

$$C_t = 0.4707.S_t$$

where S_t is atmospheric stock, with a baseline $S_0 = 750$GtC, E_t are greenhouse gas emissions, 0.64 is the proportion of emissions that are rapidly absorbed by ecosystems, and ν is the proportion of 'excess' carbon above the pre-industrial level that is taken up by carbon sinks each year, estimated as 0.008. Concentration is then calculated from a physical constant estimated at 0.4707 ppm/GtC. Radiative forcing, the amount by which the transparency of the atmospheric to outgoing infra-red radiation changes with increased carbon concentrations, is given by:

$$RF = 6.3 \ln(C_t/C_0)$$

where RF is radiative forcing in Wm^{-2}, C_t is the current CO_2 concentration, and C_0 is the pre-industrial concentration level of 279ppm. This formula only considers the effects of fossil carbon dioxide emissions however, so an adjustment must be used to account for other greenhouse gases and for CO_2 from other sources, principally deforestation. The 'true' radiative forcing is calculated by projecting emissions of these other gases up until 2050, and including them as a CO_2 equivalent concentration. From 2050 onwards, an approximation derived by the Intergovernmental Panel on Climate Change (IPCC) is used:

$$RF = 0.1058\, C - 0.506 \times 10^5.\, C^2 - 23.995 \qquad \text{until 2050;}$$

$$RF = 1.447 \times 6.3 \ln(C_t/C_0) \qquad \text{henceforth.}$$

with C_t representing energy-only CO_2 emission-related concentration, the second formula implicitly assumes the emission paths for other greenhouse gases are in some way proportionate to CO_2 emissions from energy use. This is a highly suspect conjecture because the other major gases, especially agricultural methane, have natural constraints to their growth which will bind well before fossil fuels are exhausted.

The equation for radiative forcing is then converted into one for temperature by calibrating it to the equilibrium temperature rise calculated inside global climate models (GCMs) for CO_2 doubling. Radiative forcing is 4.4 when CO_2 doubling is reached, therefore the warming commitment, the amount by which equilibrium temperature will rise as a result of a certain level of radiative forcing, is given by:

$$\Delta T_c = \tau.RF/4.4$$

where τ is the climate sensitivity, estimated by IPCC GCM modelling studies at between 1.5 and 4.5 °C/Wm^{-2}, with a central value of 2.5 °C/Wm^{-2}. The actual amount that measured temperatures will rise will depend on the rate of heat transfer between atmosphere, land, sea surface and deep oceans, and the equilibrium temperature will be slowly approached over a number of decades. This dynamic can be approximated by assuming that the actual temperature change gradually approaches the warming commitment at 2 per cent per year:

$$\Delta T = \Delta T_{t-1} + 0.02.(\Delta T_c - \Delta T_{t-1})$$

Similarly, sea level rise will not reach equilibrium until after a very long delay, as it can take thousands of years for higher temperatures to affect deep oceans fully. For example, the IPCC (as quoted by Cline 1992) assume that by 2100 mean sea-level will be 66–100 cm higher for their base case, by which time actual warming will be 4.2 °C, which corresponds to a higher warming commitment.

Given this connection between emissions and observed temperatures the costs of CO_2 doubling can now be extrapolated into the future. Expressing damages as a percentage of world GDP the extrapolating relationship can be written (Fankhauser 1995) as:

$$D_t = D_2.(\Delta T_t/\tau)^{\gamma}.1.006^{(t^*-t2)} \qquad (1)$$

where: D_2 = percentage GDP loss under $2 \times CO_2$; τ = climate sensitivity; t^* = time for doubling to occur; t2 = time assumed for doubling in cost estimates for D_2. Damage costs are therefore assumed to be proportional to ΔT^{γ}; where ΔT is the warming commitment, and the damage-temperature exponent γ is usually set to 1.2–1.5 depending on sector, although values of 2–3 have been used. The parameter γ can be interpreted both as reflecting the relationship between temperature and physical damage, and as a proxy for the changes in the price of climate sensitive environmental assets as their stock decreases. Therefore, as was discussed in Chapter 7, its value will be critical in determining the ability for any model to converge to a sustainable solution in the future. The final part of equation (1) accounts for the fact that costs may also increase with the rate of change of climate, as economic and natural systems have less time to adapt to changed conditions. Fankhauser (1994a) assumes costs to be proportional to $1.006^{(t^*-t)}$, where t^* is the time assumed in costing studies for CO_2 doubling to occur. In other words costs will be 0.6 per cent lower for each year that doubling can be delayed past this date.

Little is known about the critical parameter linking damage and temperature, because most studies have focused on CO_2 doubling and so have no other points to plot on the damage/temperature curve. As demonstrated in Figure 8.1, its precise value is evidently critical, particularly in the very long term (model calibrated using the very long term warming figures of Cline 1992).

As warming commitment depends on radiative forcing, which is physically proportional to the log of atmospheric CO_2 (equivalent) concentration, each additional unit of CO_2 has less effect on radiative forcing and thus temperature; rising concentrations make the atmosphere more opaque

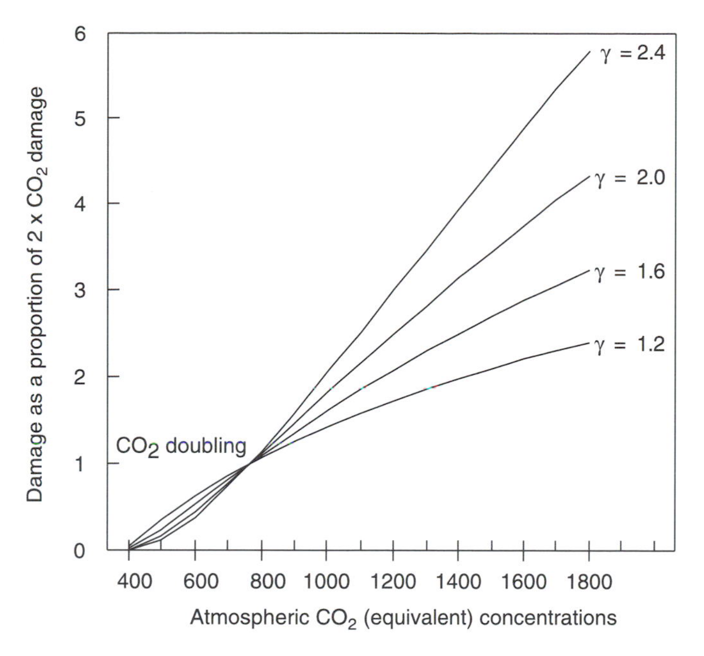

Figure 8.1 Increase in damage costs with CO_2 (equiv.) concentration and γ

to the infra-red wavelengths absorbed by CO_2. Therefore, the relationship between climate *damage* and emissions is determined by the balance between the log relationship of concentration and temperature, and γ – the relationship between temperature and damage. At lower values of γ (e.g. $\gamma = 1.2$) damage will increase more slowly than concentration, and so the marginal damage caused by a unit of emissions will decrease. At higher values of γ the relationship between damage and concentration becomes roughly linear, at least over the range from $2 \times CO_2$ onwards, as illustrated in Figure 8.1. Thus, while the marginal cost per tonne of carbon will depend on future emissions, which affect concentration, it is not clear whether higher emissions will increase or decrease marginal costs.

A significant problem with using this polynomial extrapolation is that while a high value of γ will tend to make marginal costs increase if future emissions increase, the normalisation of costs at CO_2 doubling means that at concentrations below $2 \times CO_2$ a high value of γ reduces the marginal damage cost, and with discounting this can significantly reduce the present value of damage costs. An example of this effect is shown in Figure 8.2 which plots the discounted marginal damage cost of a unit of CO_2 emitted in 1990, 2000, 2025 and 2025 (for example, the future costs caused by a unit of CO_2 emitted in 2000 are projected for 200 years into the future, and then discounted back to 2000), for two values of γ (1.0 and 2.0) and two different discount rates – 1.0 per cent p.a. and 4.0 per cent p.a. In this emissions scenario CO_2 concentration doubling occurs in 2033, and it is clear that

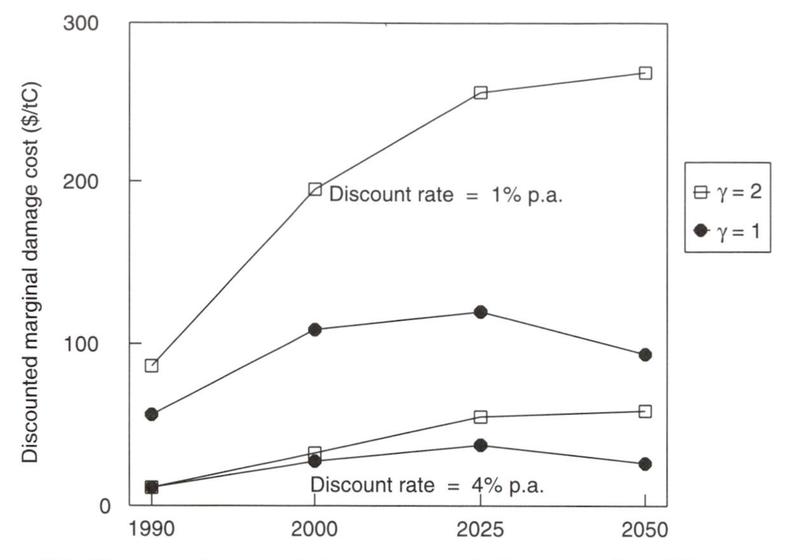

Figure 8.2 Discounted marginal damage cost of CO_2 emitted at different times

discounting reduces the damage costs of emissions before this point proportionately more than it does for emissions further into the future.

To minimise this effect, and disaggregate the role of increasing prices and the damage/temperature relationship, equation (1) is expanded here to include explicitly the proportionate linkage (η) between damage and GDP growth:

$$\text{Damage Costs} = (\text{GDP})^{\eta}.D_t$$

This form allows future damage values to increase strongly, without overly depressing the contribution of damage before $2 \times CO_2$ is reached. Of course, the link with GDP will vary with the future emissions path, but as we are not considering global stabilisation scenarios here this is not of vital importance given the other approximations involved.

Climate change damage is extrapolated here as a percentage of GDP, and so like any exponential process it is very sensitive to small changes in the parameters governing the evolution of underlying trends. Parameters which are positively correlated with marginal costs are the global growth rate, η the linkage between GDP and damages, and γ the damage/temperature relationship; the size of marginal damage costs is reduced by the discount rate. The initial damage value at $2 \times CO_2$ is also critical, especially at high discount rates where early damage values are proportionately more important. The influence of these various effects are shown in Table 8.1 which presents the discounted cost of emissions[1] in a certain year under a number of different parameter values; the discount rate is set to the GDP growth rate + 2 per cent, which at current rates of per capita GDP growth implies a pure rate of social time preference of ≈ 3.73.

In all these cases the damage/temperature exponent (γ) was set to 1.4,

Table 8.1 Influence of parameter assumptions on marginal damages (1990 $/tC)

	2 % global GDP growth rate			*3 % global GDP growth rate*		
Parameter assumptions	*2000*	*2025*	*2050*	*2000*	*2025*	*2050*
1 % GDP at 2 × CO_2						
η = 1.00	16.1	32.1	47.2	25.9	62.1	118.6
η = 1.05	29.1	58.6	88.6	50.2	122.7	244.1
4 % GDP at 2 × CO_2						
η = 1.00	64.4	128.5	188.7	103.7	248.2	474.2
η = 1.05	116.3	234.6	354.3	200.8	490.7	976.2

the climate sensitivity (τ) to 2.5 °C/Wm^{-2}, and emissions are such as to make CO_2 doubling occur in 2033. In can be seen that even with small variations in assumptions, which are well within the current range of uncertainty, the projected marginal costs of a unit of CO_2 emitted in the same year can differ by up to an order of magnitude. The optimal policy regimes calculated from such models are therefore purely a function of the damage cost input parameters, and how any expected values were formulated, because the uncertainties surrounding the costs of controlling emissions are very much smaller than those for damage costs.

Table 8.2 shows how marginal costs vary with future emissions, and with the temperature/damage exponent γ. The figures give the percentage change in marginal cost between the base case, or 'business-as-usual' scenario, where emissions continue to increase, and a stabilisation scenario where global emissions are constant at ≈ 1995 levels (5.5GtC per year) from 2000 onwards. The form of the damage equation means that these per centage changes will be virtually constant whatever the damage/GDP relationship, or rate of economic growth. Given projected increases in developing countries' global emissions stabilisation appears very unlikely in the near future, but the comparison bounds the potential magnitude of changes in marginal costs.

Even though base-case emissions are 700 per cent higher than the stabilisation level in the last relevant period, this has a proportionately small effect on marginal cost, with the magnitude depending on the value of γ. Cost is least sensitive to emissions when the damage/concentration curve is close to being linear, which happens for values of ≈ 1.9, as shown

Table 8.2 Change in marginal damage costs with stabilisation of emissions and γ

Temperature/damage exponent	*%age change in marginal costs with stabilisation* 1990	2000	2025	2050
γ = 2.4	−32.7	−31.1	−32.2	−33.2
γ = 1.8	−3.0	−1.0	0.4	3.4
γ = 1.4	10.3	12.3	15.2	20.8
γ = 1.2	15.3	17.4	20.9	27.8

in Figure 8.1. With lower values of γ, emissions stabilisation increases the marginal cost of GHG emissions, although total costs decrease. This results from the log relationship between concentration and radiative forcing, which makes an incremental tonne of carbon cause more damage at low GHG concentrations. With values of γ higher than 1.9 stabilisation reduces marginal costs, because damages rise faster with temperature than the change in temperature per unit of GHGs falls as concentrations increase.

Therefore, the logged relationship between atmospheric temperature and the concentration of CO_2 removes much of the non-linearity we might expect to find when perturbing a complex system such as the global climate. It also implies that the critical measure determining 'optimal' policy – the marginal damage cost of emissions – is relatively robust to things under human control, for example the absolute level of emissions, but very sensitive to parameters outside direct control, such as economic growth rates and the damage/temperature linkage γ.

In order to derive a stochastic model of climate damage we decided to vary only two major parameters, γ and η, as these seem to be the main factors which determine damage costs. For ranges of these parameters marginal damage costs were calculated assuming constant 3 per cent growth in world GDP, a 5 per cent discount rate, and that damage equals 2.75 per cent of world GDP at CO_2 doubling with constant prices (i.e. η = 1.0). The distribution of parameter values is log normal, with γ having a range of 1.0 to 2.4, and η a range of 0.99 to 1.06.

A summary of the results is given in Table 8.3, for both a business-as-usual emissions path (BAU) and stabilisation at 1995 emissions. The expected value is the sum of the expected marginal damage costs associated with a unit of CO_2 emitted in each relevant year. The average value is the marginal damage cost calculated using the expected value of the parameters (γ = 1.66, η = 1.023), and the 'best guess' estimates are found using the mode of the distribution of the parameters (γ = 1.4, η = 1.01). The values at the edges of the stochastic distribution are also given at the 5th percentile and 95th percentile.

The most obvious feature of these results is that the values are very high compared to Fankhauser's. This is because the discount rate is lower, the global growth rate higher and the link between prices and growth stronger.

Table 8.3 Stochastic model of expected discounted marginal damage costs

Year	*Expected discounted marginal damage costs (1990 $/tC)*							
	Expected value				*Average value*		*Best guess*	
Emissions path	*BAU*	*1995 levels*	*5th percentile*	*95th percentile*	*BAU*	*1995 levels*	*BAU*	*1995 levels*
1990	71	70	36	130	66	67	47	52
2000	129	127	60	261	117	122	81	93
2025	324	326	133	602	295	313	195	231
2050	701	714	227	1266	613	679	376	475

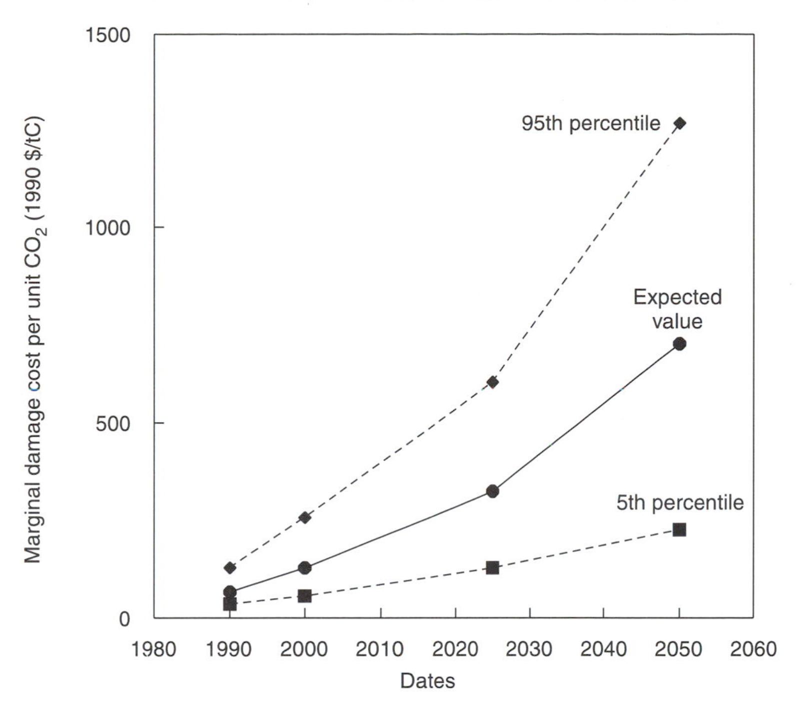

Figure 8.3 Stochastic distribution of marginal damages from climate change

To put these figures in perspective, in the year 2050 discounted climate change damage is equal to ≈ 3 per cent of world GDP at the 5th percentile of the distribution, and 15 per cent of GDP at the 95th percentile, which could be considered quite a conservative range of results for long run climate impacts. Figure 8.3 shows graphically how marginal costs evolve over time, by plotting the 5 per cent and 95 per cent confidence intervals around the expected value; the increasing uncertainty over the spread of damages is readily apparent.

The other important feature of the results is that the expected and average costs are very close, showing that the type of damage model constructed above is virtually linear in emissions and concentrations. This is backed up by the very small changes in marginal costs for emissions stabilisation. The percentage difference between the results widens in later years due to the non-linear influence of GDP growth and price effects, but as they are damped by discounting this is also not that marked. In contrast the best guess estimate is much lower than the properly calculated expected value result, due to the skewness of the log normal parameter distributions. The most important feature of the best guess model is that marginal damages decrease strongly with GHG concentrations, due to the low value γ in this case. Under the logic explained in Chapter 7, if these values were used to determine optimal emissions, current emissions could be driven *upwards* under increasing uncertainty and irreversibility; a policy prescription would not be upheld if the expected value damages were used, as they include counter-balancing cases where γ is high. This result illustrates how, when using partial optimisation, the likelihood of calculating an optimal path which is ecologically sustainable will be highly sensitive to how the central values of the damage distribution are derived, and what this implies about their second differentials.

Linearity in the damage function, and the subsequent insensitivity of marginal costs to concentration levels, may undermine the usefulness of complex integrated assessment models of climatic, ecological and economic systems. The added value of full system optimisation over partial models depends on there being strong non-linear connections between the parts, but the feedbacks and forwards calculated here seem likely to have only small effects on the final optimum solution. More detailed partial equilibrium studies may therefore be as likely to give useful information as more highly aggregate 'mega-modelling' approaches. This result will be robust as long as radiative forcing is considered to evolve logarithmically, and the transition path for prices is linked to a small exponent of GDP growth. On the other hand, it may be that this type of damage cost extrapolation misses large non-linear aspects of damage and climate reactions, and/or vital dynamics in probability; such as whether the likelihood of future damage depends on the *rate* of emissions production, and not just its level.

OPTIMAL POLICY MODELLING IN EGEM

The model developed above was incorporated into the full EGEM model, described in Chapters 4 and 5, by defining piece-wise linear functions for the discounted marginal damage cost of emissions which depend on time and emissions. This approximation to the damage function allows us to use EGEM's dynamic optimisation algorithms to investigate behaviour under different conditions. The basic objective function is to minimise the net cost of climate change – that is, climate damages minus the macroeconomic and welfare costs of emissions control – by setting carbon tax rates in the different regions of the OECD over the simulation period 1995–2030.

The only economies modelled in detail are North America (Canada and the USA), Japan, and the Core European Union (Germany, France, Italy and the UK); the remaining OECD countries' percentage emissions abatement and marginal control costs are assumed to mirror Europe's in the objective function. Only OECD abatement is modelled, as only these countries are committed to controls under the FCCC. As non-OECD emissions are projected to overtake OECD emissions in around 2003, and are then expected to continue growing strongly, there is no potential for these optimisations to achieve a sustainable level of *global* carbon emissions. For the BAU scenario considered here, the OECD share of global CO_2 emissions drops from 43 per cent to 16 per cent between 1995 and 2030. This more accurately reflects the OECD's share of global population, but it is still expected to produce over 60 per cent of global product, underlining its greater energy efficiency.

Despite these limitations of the model's scope, there are several issues that can be usefully investigated in this framework:

- The effect of co-operation between the OECD regions and the world.
- Incentives for joining a co-operative treaty.
- The effects of uncertainty, abatement dynamics and irreversibility on current abatement.
- The effects of learning more about the distribution of future damage costs on co-operation.

Modelling of macroeconomic costs

The costs of controlling emissions using carbon taxes consist of macroeconomic distortions caused by changed input prices, and welfare costs caused by reductions in the direct consumption of energy by consumers. There may also be terms of trade effects and competitiveness externalities, and these are discussed more fully in Chapter 9. EGEM has no explicit utility function, so welfare losses from decreased energy consumption are given by assuming a linear welfare function, and positing that a fixed proportion of the gross consumer surplus lost by reduced energy use (calculated from the energy demand equations estimated in Chapter 4) is actually lost in substitution to other goods. The proportions vary between regions based on purely *ad hoc* estimates of the relative dependence of consumers on fossil fuel use; therefore, North America loses 30 per cent of gross consumer surplus, Europe 20 per cent and Japan 10 per cent.

EGEM has two different supply side structures which can be used to measure the macroeconomic distortions of imposing carbon taxes: a conventional three-factor production function, and one that incorporates endogenous technical change in labour and energy productivity. In these simulations the endogenous technical change model is always used; detailed simulation properties are given in Chapter 5. The main feature of this model which will influence the results of an optimal control exercise is that it assumes that costless, price induced technical improvements occur in the energy sector in the long run. Therefore, imposing a carbon tax leads to high unit abatement costs in the short term, but these rapidly decrease to a much lower level towards the end of the simulation. The short run costs are caused by an immediate diversion of investment from labour-saving to energy-saving technology when the taxes are imposed, but price induced technological change means that these investments do not need to be repeated to ensure continuing emissions reductions. Macroeconomic costs are further reduced by recycling carbon tax revenues through employers' labour taxes, reducing unemployment and raising output relative to recycling through income taxes.

Therefore, there are strong dynamics in abatement which link present and future decisions. The influence of these interactions on optimal policy has not been considered by other optimal control exercises, as they have usually taken a static optimisation approach. It will be seen that these dynamics will have a non-trivial effect on optimal abatement regimes in the OECD.

When optimising, EGEM minimises the undiscounted sum of climate damages and abatement costs across the simulation period. Abatement costs are calculated as the instantaneous difference in macroeconomic output and welfare from the base-case, and damage costs are the discounted expected value of damages caused by emissions in that year. Tax levels are then determined for nine-year intervals of the simulation period, in a way that minimises the sum of costs and benefits over the whole period. The dynamics of the model mean that setting taxation periods any shorter leads to anomalous effects in the final periods. Macroeconomic costs are left undiscounted because cost and damages both grow roughly in proportion with GDP, and the opportunity cost of investment in energy efficiency has already been taken into account inside the supply side of the model (see Chapter 5). The optimisation can be considered closed-loop, in the sense of Ulph and Ulph (1994a), as the interactions over the whole simulation period are taken into account, including the value of investing now in order to achieve future reductions in emissions. The limited length of the simulations means that there will be end-effects, and the total cost of abatement will be over-estimated, because the benefits of price-induced technical innovation in energy efficiency are not taken into account beyond the last period.

CO-OPERATIVE AND NON-COOPERATIVE ABATEMENT LEVELS

When balancing the costs and benefits of climate change countries can consider many different sets of damage costs, depending on how they decide to co-operate. Perfect co-operation implies internalisation of the global external costs of emitting another unit of CO_2; that is, the sum of the damage costs incurred in all countries from that extra pollution. Under perfect co-operation, therefore, countries will abate until the marginal cost of controlling emissions equals this marginal global damage cost.

Perfect co-operation maximises global welfare, but not necessarily each country's welfare. A country with high emissions, but low *national* climate change damages, may well find itself worse off by co-operating, while low emitters which suffer high damages (e.g. low-lying developing countries) unambiguously gain from perfect co-operation. Theoretically, the winners can merely redistribute their welfare gains to the losers, because total welfare always increases enough to compensate high abating countries (that is, co-operation is a win–win game). However, as explained in Chapter 7, in the real world things are more complicated. If climate change damages are mostly welfare, or non-cash, costs in the developing world, there will probably not be enough financial resources to compensate abators in the developed countries. In this circumstance, the objective function of abators will focus on the *financial* value of climate change damage, which can be reallocated between parties to any agreement, and abatement will be consequently much smaller than if global welfare/utility was being optimised.

The difficulty in providing actual cash flows to abators is re-enforced by the timescales of the problem, as much of the expected damage costs seen now actually occur far into the future, and represent the willingness-to-pay of future consumers to avoid climate change. These consumers will have far higher incomes than current generations, especially (with luck?) in the developing world; therefore, compensation should be funded through inter-generational financial debt which will be paid off in the future. In a world of IMF conditions on government debt stocks this seems unlikely, even though it is the optimal strategy for developing countries if climate change is seen as being as important as other problems such as primary health care and education.

Here we are calculating the hypothetical optimum levels of abatement in the developed world, even though they cannot unilaterally stabilise global emissions in the medium to long term. The policy justification for modelling these scenarios is that OECD abatement is a political pre-requisite to wider control of emissions, so it is important to assess the likelihood of this occurring. If the OECD countries aim to promote co-operation in the rest of the world they will have to act as if this is in place already, and thus set their marginal abatement costs to the global benefits of abatement, based on either financial or welfare measures. If they just decided to abate unilaterally, without taking into account damage in the developing world, this would not encourage future co-operative behaviour as the OECD is not bearing any extra costs beyond what it would do just considering its own narrow interests. Of course, the worst case scenario is that even the regions inside the OECD cannot agree to co-operate, and thus each region calculates abatement without taking into account any of the damage outside its own territories.

Therefore, there are four potential ways of including the damages from climate change into the objective function of the developed countries:

- Global damages based on the welfare value of climate change (GWV).
- Global damages based on the financial value of climate change (GFV).
- OECD damages based on the financial value of climate change (OFV).
- OECD *regional* damages based on the financial value of climate change (RFV).

In the terminology of Chapter 7 using the GWV of climate change damage assumes that global utility is aggregated to give a form of co-operation where every citizen of the world has an equal weight put on their life and preferences, regardless of actual financial income. This gives the perfect 'democratic' outcome, and ensures intra-generational equity; however, it is likely that rich countries which abate a lot will be worse off financially. In political terms, this case implies that the developed countries highly value the co-operation of the South, and are prepared to bear large upfront costs in order to demonstrate their good faith and willingness to co-operate.

Using the GFV condition assures that actual financial compensation can be paid to abating countries, subject to the ability to borrow freely against future generations. Using the GFV of climate change, abatement will be lower than with the GWV if per capita utility loss is higher in developing countries than in developed ones, because this utility is valued at its monetary level which depends on per capita incomes. If utility loss is equal in both North and South, the GFV will be equal to the GWV when utility functions are linear in income. If the climate sensitive environment is a superior good, and utility losses are equal, using the GFV will result in higher abatement than using the GWV of damages. GFV assumes a more cautious policy by the OECD, accounting for some of the global damages of climate change, but not giving the developing world equal weight in its deliberations. This policy is less likely to stimulate global co-operation than using GWV, unless the environment is a strongly superior good, or utility damages in the developing world are small.

OFV occurs when the OECD is unilaterally abating, and not considering damage outside its own region. As all developed countries have approximately equal income levels, using financial valuations of climate damage should maximise OECD welfare, including non-market utility. Again, depending on the distribution of utility losses, and superiority of environment goods, abatement may be higher or lower than in the GWV case, but will always be lower than the GFV case. OFV implies that the developed world is not interested in stimulating global co-operation on climate abatement, and so will only work to prevent damage to its own economies. Alternatively, it could result from the developed world's aiming to free-ride on abatement efforts in developing countries, if expectations of climate damage become unexpectedly severe there in the short to medium term. Using the RFV of climate damage assumes there is no co-operation even inside the OECD, which will happen in the absence of compensating side payments, or without the use of policy instruments such as tradable permits that can serve this type of function. Each region therefore considers its own costs and damages, and abatement will be lower than in the co-operative cases.

Table 8.4 the gives cumulative drop in CO_2 emissions from the baseline case over the simulation period (1995–2030), expressed as a percentage of the baseline, for the four different scenarios. Damage costs were modelled using the expected value of the damage distribution derived above, which was calculated assuming equal percentage GDP impacts from climate change in all parts of the world. Full damages are used in the GFV case, and damages are weighed for the OECD's share of global output in the OFV case. RFV damages are proportional to regional output (North America, Japan and Europe) and only regional CO_2 emissions are considered, with the simulation being solved as a pure uncooperative game

Table 8.4 Optimal abatement levels for different objective functions

Region	*Percentage change in CO_2 emissions*			
	GWV	*GFV*	*OFV*	*RFV*
OECD CO_2	−36.7	−21.9	−16.6	−0.11
World CO_2	−9.56	−4.52	−3.51	−0.02
North America CO_2	−44.8	−26.2	−16.8	−0.16
Japan CO_2	−8.7	−8.74	−6.83	−0.04
Europe CO_2	−25.7	−17.8	−12.7	−0.01

between the different regions. In all simulations marginal global damages are a function of cumulative global emissions and time.

To approximate the GWV case we assume that the developing world suffers twice as much per capita utility loss as the OECD, and the environment is not a superior good (see Fankhauser 1995 for examples of developing country damage costs). Using the same measure of global per capita damages as is implied in the GFV case, the GWV costs can be calculated as a simple multiple of the financial costs, which here is an average of 1.87 × GFV over the simulation period.[2]

Figure 8.4 puts these results into perspective by showing annual OECD carbon emissions over the simulation period for each case. Stabilisation of OECD emissions at 1990 levels implies a reduction of ≈ 23 per cent from

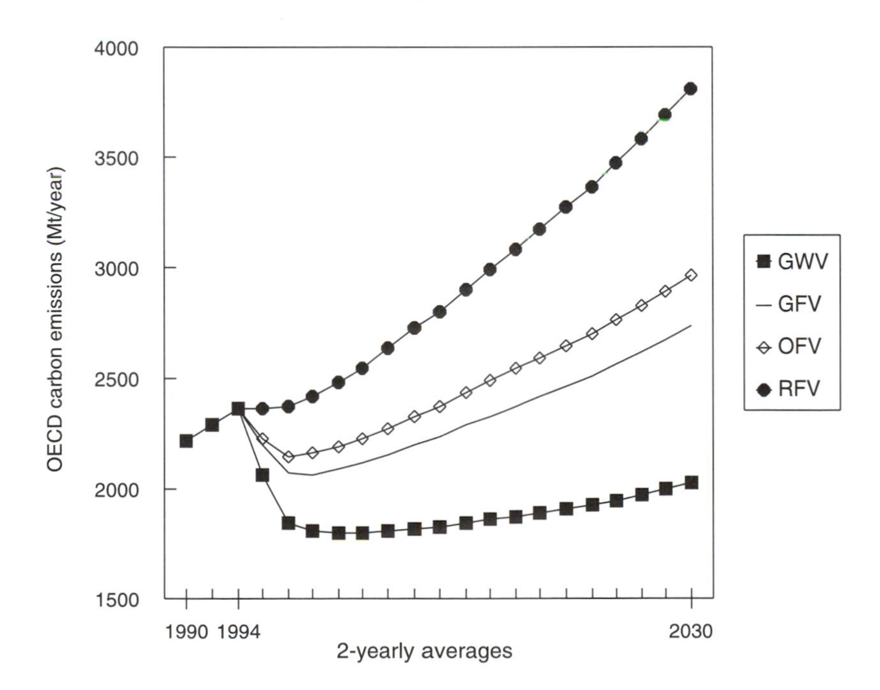

Figure 8.4 OECD carbon emissions for different optimisation regimes

the baseline case, and so with GFV damages OECD abatement approximates to the current FCCC obligations extended until 2030. However, demand is still trending up at the end of the period (baseline emissions in 2030 are 41 per cent above 1990 levels), and so stabilisation beyond 2030 implies higher tax levels than imposed here. In the GWV case OECD emissions are virtually stabilised across the whole simulation period, with price driven increases in the growth of energy efficiency being large enough to cancel out demand growth. A small tax increase from these levels would stabilise emissions into the future, without the need for increasing tax levels over time. It must be remembered that in traditional econometric and GE models an equilibrium condition of stable emissions and non-growing prices only occurs if an arbitrary backstop technology has been included. If energy use only responds to price elasticities then after-tax fossil fuel prices must rise in line with other input prices, such as labour, to ensure stabilisation (see Chapter 4 for analytic equilibrium conditions).

From these results the importance of the co-operative stance taken by the OECD can be clearly seen. If the utility costs of the developing world are taken into account this produces a significant increase in abatement over the GFV case where only financial costs are considered; nearly stabilising OECD emissions into the future. If only the damage costs to the OECD are considered, abatement drops slightly, but the dynamics of abatement mean that much irreversible abatement is committed to at the beginning of the simulation period when the OECD bears a proportionately larger share of the financial value of climate damages. As the South grows in population and wealth (relative per capita incomes in the South grow by 20 per cent over the simulation period) the difference between global and OECD-only co-operation grows, but the resulting change in abatement commitment is partially masked by the previous irreversible efficiency improvements.

The most dramatic change comes when the OECD countries themselves fail to co-operate, with abatement dropping to virtually zero. This is because, in the OECD-only case, countries are merely optimising to a smaller benefit function whereas, in the RFV case, the model is solved as a multi-player game, so they are also taking into account the abatement efforts of the other OECD regions as well. Therefore, this case represents the Nash-equilibrium solution to GHG abatement in the OECD, and emissions fall dramatically as countries try to free-ride on the abatement efforts of others.

It is interesting to compare the range of OECD abatement under different degrees of co-operation, with the range of abatement defined by the stochastic damage function derived above. Table 8.5 gives the optimal abatement strategies (percentage summed reduction in CO_2 over the simulation period) in the OECD for the expected value of the distribution (EV), the best guess value (BG), and the high and low values defined by the 95th and 5th percentiles of the distribution. These are all

Table 8.5 Optimal abatement levels for different damage functions

Region	Percentage change in CO_2 emissions			
	EV	*BG*	*95%*	*5%*
OECD CO_2	−22.70	−13.80	−34.60	−11.15
World CO_2	−4.52	−2.40	−9.07	−2.74
North America CO_2	−26.20	−19.00	−43.80	−14.02
Japan CO_2	−8.74	−0.10	−8.51	−3.40
Europe CO_2	−18.10	−6.89	−25.10	−9.98

global damage cost scenarios (GFV), assuming that the OECD internalises all the financial costs of its polluting activity.

Comparing Tables 8.4 and 8.5 shows that the economic uncertainty surrounding damage cost estimates leads to a similar range of optimal abatement as the political uncertainty surrounding co-operation between countries and regions. In particular, the difference in abatement caused by using the 95th percentile of the damage distribution, instead of the expected value, is smaller than that caused by changing the aggregation method used to combine financial and welfare damages in the North and South. This highlights the importance of the political process in defining and agreeing objectives, and the fact that climate change policy will not be solely determined by which scientific and economic information is used.

Table 8.6 gives the total macroeconomic and welfare costs (undiscounted) per tonne of CO_2 abated for the scenarios in Table 8.4, and the average tax levels (weighed by CO_2 emissions) imposed in each region. It can be seen that tax levels rise very high in the GWC case, though average abatement costs rise more slowly. Tax levels are steady or declining across the simulation period, which results from not discounting the model's objective function back to 1995. Running sensitivity simulations with different discount rates showed that the tax levels only became skewed towards the later periods at rates of 4 per cent and above; that is, rates above the aggregate growth rate of OECD output (3.15 per cent over the simulation period).

Abatement in Japan rises more slowly than in the other regions because there are strong trade effects between Japan and North America. There-

Table 8.6 Abatement costs and tax levels for different objective functions

Region	Costs in 1990 US$/tC			
	GWV	*GFV*	*OFV*	*RFV*
World total cost	456	296	274	238
North American tax	625	293	169	26
Japanese tax	181	165	139	6
European tax	332	217	157	34

fore, the high American tax levels in the GWV case depress domestic demand and Japanese exports; this increases the perceived cost of abatement in Japan and lowers the equilibrium tax level. This effect could be significant in the 'real world' if trade linkages lowered the willingness of Japan to agree to strict controls, even if all abatement was carried out elsewhere. At high damage cost levels North America imposes by far the highest taxation levels, because its marginal cost of abatement rises far more slowly than in Europe. However, this situation is reversed in the more non-cooperative cases, with lower damage costs, and Europe's tax rates become proportionately higher. This stems from Europe's lower abatement costs at low tax levels, a result of recycling carbon taxes through employers' labour taxes; the subsequent increase in employment is more important in Europe because of its higher aggregate labour productivity and unemployment levels. At high tax levels these recycling benefits become swamped by Europe's larger costs of substituting away from energy, caused by higher pre-carbon tax energy prices than North America.

Though these optimisation results are reasonably robust when significant abatement is called for, comparing the non-cooperative results in Table 8.4 with the low damage cost scenario in Table 8.5 probably overvalues the consequences of non-cooperation. The results in Table 8.6 show that EGEM has high non-zero costs of abating the first unit of CO_2. These arise from the high productivity of energy estimated in the supply sides, and the demand side effects of energy taxation as detailed in Chapter 5. Therefore, there is a certain level of greenhouse damages that will stimulate no abatement measures at all, giving a threshold effect which is plainly shown in the RFV case in Table 8.4. Microeconomic studies show that many CO_2 saving measures will be costless, or cost saving, to individual economic actors and this undermines the relevance of EGEM's results at low damage cost levels, unless we assume that the model is capturing unseen externalities and/or transaction costs. Despite this there is some plausibility in the idea of a threshold effect for abatement measures, but this is more likely to be a function of government inertia, and the inability to agree on international co-operative measures which would produce relatively little gain (3 per cent of world GDP in 2050, for the low damage case), than the non-existence of cheap abatement options.

UNCERTAINTY, ABATEMENT DYNAMICS AND IRREVERSIBILITY OF EMISSIONS

Chapter 7 explained how uncertainty, abatement dynamics and the irreversibility of emissions interact to influence current levels of abatement, even if decision makers are risk-neutral. The usual method of determining optimal policy under uncertainty, previously termed partial optimisation, involves controlling emissions until the cost of abatement equals the expected value of the distribution of future damages. Therefore, for risk-neutral decision makers, changes in the variance of the damage function only affect current abatement by the amount they shift the *mean* value of damage costs. This usually implies that high damage/low probability events have little influence on current policies.

However, if damages turn out to be at the extreme end of their probable distribution it is likely we would want very high levels of abatement, or even negative emissions. Negative emissions are impossible and so form a constraint, and massive abatement requires qualitatively different types of investment and R & D than more moderate targets. To be truly optimal the cost implications of these possible constraints and control strategies should be reflected in current period abatement. A full stochastic optimisation will take these interactions into account by modelling the optimal abatement regimes associated with each probable damage cost scenario, and weighting their importance by the probability of that scenario. To approximate this method we model optimal responses to the extremes of the damage cost distribution, in order to see if the results are sufficiently different from the average case to merit consideration in deciding current abatement.

Dynamics and uncertainty

In EGEM abatement costs depend not only on the level but also on the rate of emissions reduction, because increased fuel prices permanently raise the rate of productivity growth in the energy sector. Therefore, it is likely that rational decision makers will commit to pre-emptive abatement so as to lower costs in the future. That is, present emissions will be set in a way that equalises the marginal costs of present abatement, and its effects on future costs, with current and future damage.

To test the importance of this effect in EGEM we used two different informational scenarios: firstly, optimisation was done over the whole simulation period so all present and future costs were considered simultaneously; the second scenario optimised the model in nine-year periods, with only the damage costs in that period, and abatement in the ones before, affecting current abatement levels. Therefore, in the second scenario decision makers are myopic, and do not look forward to see how quickly damage costs grow over time. If abatement costs in each period are completely independent, as generally assumed in putty–putty general equilibrium models, then both scenarios will be identical. Otherwise, the penalty for this lack of information, or foresight, will depend on the speed by which damage costs increase, and the temporal interdependence of abatement costs. The scenarios were run using the expected value of the damage function, and the damage costs at the 5th (low damage) and 95th percentile (high damage) of the damage function. This range of costs should give a feeling for the size of error incurred by using partial optimisation based around the expected value figures.

In the simulations above costs and damages were undiscounted over the simulation because both are rising approximately in line with GDP. However, as we are specifically looking at how abatement changes over time, in

Table 8.7 OECD abatement with myopic and pre-emptive decisions

Time period	*Percentage change in OECD CO_2 emissions from base* 1995–2030	1995–2003	2004–2012	2013–2021	2022–2030
Expected value					
Pre-emptive	−17.9	−11.1	−15.7	−19.8	−22.7
Myopic	−14.0	−8.4	−16.6	−15.8	−13.5
High damage					
Pre-emptive	−32.7	−20.8	−28.9	−35.9	−40.8
Myopic	−16.2	−7.5	−8.9	−16.7	−26.2
Low damage					
Pre-emptive	−10.2	−6.6	−9.3	−11.2	−12.4
Myopic	−8.1	−9.1	−8.7	−6.2	−8.1

these simulations all costs are discounted back to 1995 at 3 per cent per annum. In the base case this reduces the optimum level of abatement slightly from the figure given in Table 8.5.

Table 8.7 gives the results of the optimisations, showing the percentage decrease in OECD emissions in each of the periods, and the summed percentage drop over the whole simulation. In all cases total abatement is clearly lower in the myopic scenario, showing that pre-emptive abatement does save costs in the long run, and as would be expected these savings are much larger in the high damage case. Savings are quite small in the low damage and expected value cases, with the largest difference showing in pre-emptive abatement in the third period, as this is when damage costs start to rise quickly.

Figure 8.5 shows the process of decision making which produces these results in the high damage case. Abatement is far smoother in the pre-emptive case and costly increases in the rate of abatement are avoided. Contrastingly, the myopic decision maker is constantly trying to catch up after setting taxes which are too low in the early periods. This means the marginal cost of abatement is higher at any particular abatement level, and it is never optimal to abate as much as in the pre-emptive case.

The importance of pre-emptive abatement in the high damage case means that, if optimal policy is calculated using just the expected damage value, our options will be dangerously limited in the future. If damage does turn out to be very bad, then trying to increase abatement quickly to respond to this knowledge will be very costly, and so we will have to settle for greater damage than is optimal in the forward-looking case. On the other hand, there seems little risk of over-abating even in the low cost scenario, because damage costs increase over time at both ends of the damage distribution. These results imply that increasing uncertainty over the variance of damage costs should lead to greater current abatement than that predicted from a partial optimisation. However, the precise influence of these non-linear effects on current abatement strategies will depend on how likely such extreme scenarios are.

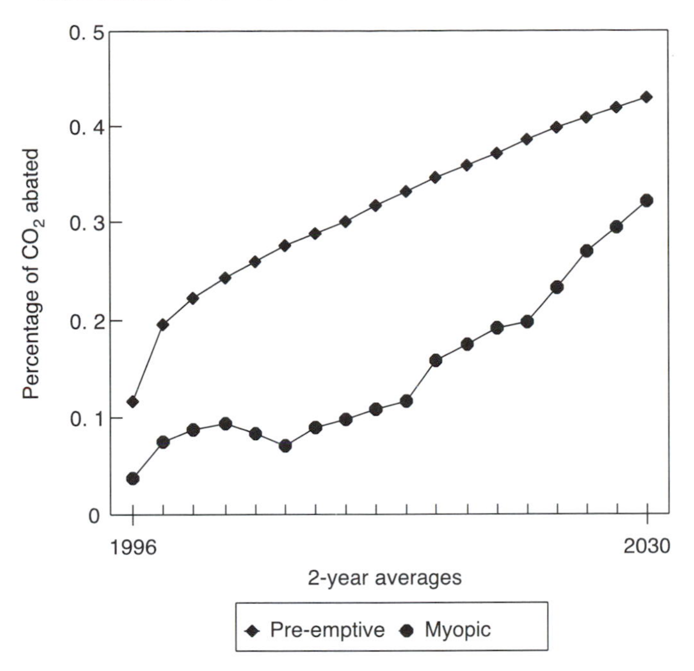

Figure 8.5 OECD emissions reductions: high damage case

Irreversibility and uncertainty

Conversely, the issue of irreversibility of emissions does not seem to be important over this time span. At the high end of the damage distribution maximum abatement in the last period is 41 per cent of current period emissions, and even if the OECD used global welfare values in their optimisation it is unlikely that they would want negative emissions given these figures. The fact that emissions are irreversible does not therefore produce an *effective* constraint inside EGEM, and changes in the variance of future damage costs will not alter optimal current period emissions due to irreversibility effects. However, these results may not be that reliable, because by 2030 the share of fossil energy in OECD production has fallen by approximately 75 per cent from 1995 levels. Therefore, the macroeconomic costs of abatement are probably being overstated, as the model is far from the values it was estimated around.

Even if these cost estimates are reliable this result only concerns irreversibility of emissions, not the damage caused by emissions. Irreversible damage has to be accounted for by projecting prices which reflect the true cost to future generations of not having certain irreplaceable environmental assets; that is, future generations' willingness-to-accept the complete loss of ecosystems and species. The damage function derived here implies these price increases parametrically through the GDP multiplier (η) on damage costs, and there been little empirical research into these issues.

More systematic analysis of these effects may greatly increase the range of damage costs, and so make the irreversibility constraint on emissions effective.

As was mentioned in Chapter 7, the influence of damage and emissions irreversibility on current abatement strategies will also depend on how reversible abatement policies are. If pre-emptive abatement would be hard to undo if damages are discovered to be low, then the greater the uncertainty over the variance of damage costs, the lower current abatement should be. The persistence of abatement costs once energy taxes, or other policy instruments, have been removed will depend on the approach to modelling macroeconomic costs, and as Chapter 5 explained this is mainly decided by the a priori assumptions of researchers.

In a traditional production function, permanent reductions in energy intensity in the economy, such as are produced by the price-driven energy efficiency model used in EGEM (Chapter 4), lead to continual decreases in productivity and rising costs. Conversely, if modellers use an energy demand model with direct elasticities these costs will disappear when the carbon tax is reduced, and energy demand eventually returns to its baseline level. Therefore, different modelling approaches can lead to differently biased results. If a traditional production function is used, which assumes no price driven technical progress in energy use, then the long run costs of abatement will be overstated, but the dynamics of abatement will only have a minor effect on optimal abatement strategies. If technical progress is price-driven, or backstop technologies exist, but the production function includes direct use of energy (not energy services), then the cost of committing to pre-emptive abatement will be high as costs will carry on occurring once the tax has been removed. Again this will suggest lower current abatement than if price driven costless technical progress is included in both the production function and the energy demand functions.

The relative costs, and reversibility, of abatement and damages are therefore ill-defined in most models of climate change economics. The future costs of irreversible damage are not taken into account by damage functions which depend solely on the concentration of CO_2, and not past concentrations. The dynamics of abatement costs can only be assessed if the relationship between productivity growth, factor prices, and investment is modelled in each sector of the economy, and we have attempted to do this in EGEM. Current models often have an inherent bias towards conservative abatement strategies because they underestimate irreversible damages, and overestimate irreversible abatement costs. If a proper assessment of optimal strategies under uncertainty, as opposed to partial optimisations, is to be undertaken, these faults will have to be remedied for the results to be of use to policy makers.

LEARNING, COOPERATION AND OPTIMAL ABATEMENT

Future damage costs from climate change are uncertain, but we can potentially learn more about them in the future. The effect of expecting improved knowledge on current behaviour is ambiguous, but current international policy under the FCCC implies a wait-and-see approach, where abatement efforts are low, but research efforts high.

One danger with a wait-and-see policy is that knowledge will only evolve slowly. When reasonable certainty is reached we may already be committed to high irreversible damage, so the delay in implementing abatement efforts may be expensive in the long run. The results from EGEM identify the dynamics of abatement as probably the most important influence on these decisions, but this could be because of our ignorance of the irreversible damages from low levels of climatic change.

Therefore, optimal policy needs to balance the funds spent on learning about the effects of climate change, and investment in pre-emptive abatement, adaptation and R & D. The mere possibility of learning itself will not affect current abatement levels if countries act co-operatively, and are risk-neutral, because the probability of each level of damage costs occurring is already fully described by the damage cost distribution. Given that these probabilities are accurate – which is impossible to determine – full stochastic optimisation will adequately define optimal policy, *without* extra consideration of learning. If policy makers are risk-averse, either psychologically or as a reflection of their uncertainty over the accuracy of the damage cost distribution, then the possibility of learning will decrease current abatement levels. This is because learning, by definition, decreases future uncertainty and so the future costs of climate change to risk-averse policy makers.

The story becomes more complex when several countries are involved, and co-operation may not exist. Learning may allow identification of winners and losers from climate change, or more probably differentiate between countries with high and low damage costs. If high emitting countries find they have low damage costs they have an incentive to renege on abatement obligations, which were undertaken when damage was assumed to be spread equally. By reducing their emissions they will force high damage countries to abate more, raising the cost of control and increasing greenhouse damages (Ulph and Ulph 1994b). Therefore, the ability to learn about future damages increases the gains from co-operation, and the importance of constructing institutional mechanisms to encourage compliance; for example, side payments to low damage/high polluting countries.

The effect of the potential for future learning on current emissions policy will depend on each country's perceptions of how international damages are linked. For example, North America may assume that its large climatic range and size gives more flexibility in responding to climate change than more geographically restricted states. This implies that if it suffers high damages so will other OECD nations, but it may have low damages while other countries have high ones, and not vice versa. Put another way, North America may assume there are only three possible distributions of damages between itself and the rest of the OECD: high/high, low/high and low/low. With these expectations America has an incentive to free-ride on GHG abatement in other OECD countries, because it expects them always to want a greater, or the same, level of global abatement as it does. In a co-operative agreement this free-riding could result in America's demanding increased side payments. With non-cooperative behaviour, if America finds it has low damages then it will actually reduce its abatement. On the other hand, if country damages are correlated quite closely there are no extra free-riding incentives produced by the possibility of learning, and the difference between non-cooperative and co-operative cases reduces to how climate damage functions are defined, which we have considered above.

Therefore, learning about climate change has only positive effects if co-operation can be guaranteed, but may have negative effects if countries act independently. To look at the size of potential negative effects we model the case where learning causes the breakdown of an existing co-operative agreement. Simulations are split into two periods: from 1995 to 2013 all regions of the OECD optimise based on damages to the whole OECD, because precise impacts cannot be identified with each country; from 2014 to 2030 countries act based on their own damages, which have now been allocated definitively between regions. In both periods the total marginal damage cost of CO_2 emissions in the OECD is the same, the only difference is the knowledge of its distribution.

To simplify the permutations of breakdown we consider the case where damages turn out to be low either in North America or Europe, with Japan always experiencing high levels of damage. Therefore, three scenarios were modelled: co-operation over the whole simulation period; America learns it has low damages in period two; Europe learns it has low damages in period two. In both periods the model solves for the non-cooperative Nash Equilibrium between the regions, the only difference is that the damage cost function seen by each player changes from including all OECD costs to just that region's damage costs.

If there is a strong free-riding effect we would expect the net gains from abatement (that is, saved damages minus abatement costs) to be much lower when countries learn the unequal distribution of damage. For each scenario, Table 8.8 gives the abatement levels in the whole OECD, America and Europe over the whole simulation and the percentage difference in damage costs, abatement costs and net benefits compared to the no-learning case.

The surprising feature of the results in Table 8.8 are that, though abatement falls with non-cooperative learning – especially when America has low damages – the net benefits of abatement do not change markedly compared to the co-operative case. The main reason for this is that

Table 8.8 Abatement levels and benefits with learning

Region	*No learning*	*America low damage*	*Europe low damage*
		Percentage change from base	
OECD CO_2	−42.3	−29.9	−40.2
North America CO_2	−49.4	−28.2	−50.5
Europe CO_2	−38.4	−41.2	−30.7
OECD costs		*Percentage change from no learning*	
Change in damage costs	0	30.2	5.8
Change in abatement costs	0	−35.8	−5.9
Change in net benefits	0	−0.54	−0.54

marginal damage costs are virtually linear in emissions, and so the drop in total abatement in the second period due to learning does not produce large changes in the abatement choices of the regions facing high damages. The differences between the scenarios come more from the first-order effects of non-cooperation, that is, the redefinition of regional objective functions, than from the free-riding interactions of the regions. By definition, these first-order effects will not be that large because the total marginal damage cost in the OECD is the same in both scenarios, and only the country with the lowest damage costs has defected. Therefore, the co-operating countries still face the majority of damage costs, and most of the fall in abatement is caused by the increase in average abatement costs brought about by the defection of one region which was previously abating.

The changes in abatement are also much smaller than the changes in tax levels in the regions; in North America tax levels drop by 98 per cent with learning about low damage, and in Europe they drop by 94 per cent. Though there is some increase in abatement in the high damage cost countries, this is relatively small. The changes are masked by the abatement inertia from the first period when co-operation gave high taxes, and so if the simulation period was extended the differences between learning and no-learning would become wider.

However, even given the importance of abatement inertia in these simulations it seems that the linear nature of the damage cost function removes most of the secondary effects of free-riding caused by learning. The major problem facing policy makers is how to ensure efficient abatement if one country faces low damages but can abate at a low cost. This is a question of institutional design and is considered in detail in Chapter 11, where detailed modelling of different policy instruments is carried out.

CONCLUSIONS ON THE NUMERICAL OPTIMISATION OF CLIMATE CHANGE POLICY

The problems surrounding the numerical optimisation of climate change policy were qualitatively outlined in Chapter 7, and some particular aspects investigated here using the EGEM model. As mentioned before, the validity of any optimising approach depends not only on the reliability of its inputs but also on using the correct methodology to process these figures; current research has been markedly deficient in both areas.

Economic assessments of optimal policy can only be defined once the ethical and political framework surrounding the valuation of climate change impacts has been agreed; particularly attitudes towards future generations and damage in developing countries. This is not an economic problem, where a 'correct' solution can be found, though economic techniques may be used to assess the implications of different valuation schemes. Once the valuation framework has been settled, the remaining uncertainties influencing policy can be split into first and second order effects: first-order effects depend on the absolute level of climate change damages, and the marginal damage cost caused by each unit of GHG emissions; second-order effects depend on how marginal damage costs change with greenhouse gas concentrations in the atmosphere, and marginal abatement costs with the rate of emission reduction.

The main first order question is: given a best-guess estimate of future damage costs, how much should we abate? The modelling above showed that defining a damage cost figure is fraught with factual and political uncertainty, and the choice of welfare aggregation method is as important as the derivation of damage cost estimates. EGEM calculates the range of possible optimal abatement in the OECD, caused by uncertain damage cost estimates, as 11–34 per cent of total CO_2 emissions in 1995–2030. This assumes that countries internalise the *global* cost of GHG emissions, and so co-operate. Using the average damage cost estimates, if the OECD regions (North America, Japan and Europe) do not co-operate, that is, they only consider their own damage costs when abating, then emissions fall by only 0.1 per cent. If the OECD considers that its collective damage costs abatement rises to 21 per cent, and if welfare costs to the South are included, the optimal policy is a 37 per cent drop in emissions, and virtual stabilisation into the future.

The FCCC target of maintaining OECD emissions at 1990 levels implies a 23 per cent reduction in CO_2 emissions in 1995–2030. This seems compatible with the central estimates of damages and costs given here, but is too weak if developing country damages are given an equal financial weight in damage cost calculations. If developing countries are to be persuaded to agree to abatement obligations, before they reach OECD wealth levels, then these figures would imply that the FCCC targets need to be set below 1990 stabilisation levels. With higher abatement the OECD would be signalling that they were taking into account developing country damages – both financial and welfare – when setting abatement levels, and are therefore making a first move towards global co-operative behaviour.

First-order effects are defined by gross estimates of costs, and have little to say about how to incorporate uncertainty over damages into current abatement policy. The simplest approach to including uncertainty is to enumerate all the possible effects of climate change, and then weight them by subjective probabilities to give an expected value figure for climate damage. However, the discussion in Chapter 7 showed that this methodology will not give optimal abatement if there are significant non-linear relationships between climate damages, emissions and abatement costs – and these can be grouped together as second-order effects.

If second-order effects are important this implies that the marginal cost of abatement and damage in the future will depend on the policy options undertaken now. For example, if marginal damage costs increase with GHG concentrations, then if one country abates first this lowers the future marginal cost of damages to other countries, who in turn will abate less. Given that damage costs in the future are uncertain, for each potential damage cost there is an optimal policy reaction, and the results of these policies will affect current abatement. Therefore, if second-order effects are large the whole span of the damage distribution must be modelled in order to define an optimal policy, not just its expected value.

The logarithmic relationship between GHG concentrations and radiative forcing, and hence temperature, implies that the marginal effect of each unit of emissions declines as concentrations rise. Working against this effect is the plausible assumption (which has not been tested empirically) that the marginal damage caused by a rise in temperature increases as temperatures move away from current levels. The estimates used here result in a damage cost distribution in which marginal damages rise slowly with GHG concentrations at its high value end, but fall with concentrations at the low value end. Expected marginal damage values are virtually linear in concentrations, but change over time with GDP growth.

A linear damage function means that the subtleties of optimisation under uncertainty disappear, and using a simple expected value approach will produce a reasonable approximation to a fully optimal policy. The results from EGEM looking into these type of effects reflect this, even when considering complex gaming behaviour between different regions of the OECD. However, if current models are understating the non-linear nature of irreversible climate damage, then second-order damage effects could be very important – as current research into damage costs has not addressed these issues properly, no assessment can be made of which case is most likely.

The only second-order effect which is important in EGEM is the dependence of abatement costs on the rate of abatement, and past energy prices. This results from the empirically estimated energy demand functions (Chapter 4), which allow price induced increases in the growth rate of energy efficiency, a proxy for stimulating technical progress. Modelling showed that if damage costs are at the extreme end of the distribution, the cost of abatement can be markedly reduced by pre-emptive abatement; that is, reducing emissions below the point where marginal costs equal marginal damages at the beginning of the simulation period. A similar simulation using the expected value of the damage distribution showed few advantages from pre-emptive abatement, as costs are not rising that quickly in this scenario. Abatement policy defined by the expected damage value will therefore be too slack because, if damages turn out to be high in the future, abatement would have been cheaper if previous emissions were lower. The influence of this second-order effect will depend both on the probability of such high damage costs occurring and on when they occur, but they need to be included inside optimal policy calculations.

Rightly, most international debate over climate change has focused on the enumeration of first-order damage effects and on the potential for abatement and adaptation. The economics of this are fairly straightforward and will be influenced far more by the political economy of international decision making, and co-operation, than by subtle economic effects.

Debates over second-order policy, for example whether uncertainty over climate change implies pre-emptive abatement or a wait-and-see approach, are more dependent on rigorous economic reasoning, and a multitude of hard to measure variables. The most important factor influencing decisions over uncertainty is how the financial and welfare costs of climate change evolve over time, and in response to irreversible and catastrophic damage. The second most important variable is an assessment of the dynamics of climate change abatement, especially in the context of massive ongoing urbanisation in developing countries, which will be hard to wean off fossil fuel use unless energy efficient options are planned for now.

Unfortunately, we need to know the answer to the second-order problems in order to deal correctly with limitations in our knowledge of the first-order magnitudes. Mathematically, the second derivative of the damage cost function contains less information than the full function, but in the real world it is hard to measure unless the full extent of damage has been derived. Luckily, we can answer the questions about abatement dynamics with more certainty, as these depend on things nominally inside human control, such as technology and economics. However, projecting the size of different effects in the future is still not straightforward.

In conclusion, the techniques of stochastic optimisation are relevant to a world where we can model and understand uncertainty; for example, predicting the likely outcome of rolling a dice, or spinning a roulette wheel. In this type of problem accurate distributions of outcomes can be calculated, which are related to causal activity in a reliable sense. Climate change uncertainty is not like this. The uncertainty surrounding damages is systemic and not well defined; it is the probability distribution which is uncertain, not just the actual outcome. By definition, the 'true' probability distribution of climate damages contains all the uncertainty we know about, but cannot contain what we do not know. Simply using stochastic optimisation therefore leaves out an important part of the problem, and will tend to result in too little abatement; even if the optimisation is done correctly, and not in a partial manner.

In response to the amount of uncertainty surrounding the damage function, rational policy makers should be risk-averse when setting abatement targets; the amount of risk aversion being based on qualitative assessments of the paucity of knowledge surrounding climate impacts. When combined with full stochastic optimisation techniques this gives a framework in which to consider all potential problems, including irreversibility in damages and abatement, catastrophic events and other non-linear behaviour. Such an approach will probably *not* result in 'no-regrets' strategies, because effective risk-aversion a priori implies a marked slowing of the build-up of GHGs in the atmosphere. This will involve real costs, and the optimum level of abatement will not be determined by the available range of low or no-cost abatement, but by our best assessment of climate change risks and the current state of knowledge.

9

CARBON ABATEMENT IN INCOMPLETE INTERNATIONAL AGREEMENTS

INTRODUCTION

The Framework Convention on Climate Change proposes stabilisation of emissions in Annex I countries (the OECD plus CIS and Eastern Europe), while allowing developing countries to continue increasing their emissions. This split in responsibility recognises that restrictions on developing countries would be unfairly detrimental to their economic development, and this is certainly justifiable on equity grounds, especially as the majority of past emissions have come from the developed world. However, if the incompleteness of the treaty causes carbon emissions to increase in developing countries both the effectiveness and the stability of any agreement will be greatly reduced. This lowering of effectiveness will in turn adversely impact on developing countries, as their reliance on the agricultural sector makes them highly vulnerable to climatic change.

Increases in carbon emissions by non-signatories, brought about by measures to reduce emissions in one group of countries, are commonly termed 'carbon leakage', as emissions leak from the agreement via non-participating countries. The possibility of leakage has been a major factor in reducing the acceptability of agreements, together with related fears about international competitiveness.

There are two channels for carbon leakage: via downwards pressure on world fuel prices (in particular oil), and via migration of energy intensive industries away from the OECD. Barrett (1994a) suggests a third route, by which abatement might reduce the marginal environmental damage, and so reduce the incentive to abate. However, in the case of climate change the uncertainties surrounding damage costs make the impact of this effect marginal in practice.

The fuel price effect arises as reduced fossil fuel use in the controlled areas depresses the price of fuels traded in international markets, and so increases consumption elsewhere. The magnitude of these effects will depend on the scale and efficiency of the markets in each fuel, in particular those outside the OECD. For gas and coal these markets represent only a small proportion of current fuel use, as most supplies are both produced and used domestically. The main developed gas markets are within the EU, and between Canada and USA, and so would not affect leakage in an OECD-wide agreement. However, global markets in oil are relatively well developed, if far from being free and complete, so a significant reductions in oil use by the OECD would probably produce downwards pressure on world oil prices (Verleger 1993). The extent of fuel price leakage is thus essentially limited to oil, and as it is dependent on decreasing world demand it could never exceed the reduction in oil use by OECD. Lower oil prices may in fact reduce global carbon emissions by promoting substitution away from coal (particularly in China and India). The magnitude of the effect will depend on oil demand and supply elasticities, the behaviour of OPEC and interfuel price and substitution effects.

Leakage from changes in industrial mix and trade patterns may potentially be the larger of the two effects, and the more difficult to quantify. Within the OECD imposition of instruments such as carbon taxes will produce an overall increase in the price of manufactured goods, which could lead to reduced exports; however, the competitive advantage of the advanced countries may be enough to absorb this price increase with little loss of trade. If tax revenues are recycled through employers' labour taxes this will cancel out some or all of the price rise, though this effect will mainly benefit less energy intensive industries. The most visible impact of the tax will be a shift away from energy intensive processes as demand for energy intensive products declines and industries relocate to lower-cost areas. The main unknown determinant of the process is the relationship between energy prices and the decision of a company to relocate to another country. Certainly, historical data and the determinants of energy intensity reductions found in Chapter 4, show that the movement of energy intensive processes to less developed countries has been a main factor in the *apparent* energy efficiency increases in the developed world. However, detailed research into these processes has pointed to market access, and the particular demands of different stages of development (see Chapter 3), as the main motivation for these shifts, not differences in relative non-labour factor prices.

There will also be additional effects due to changes in income, especially reduced income in energy-exporting countries and reduced demand for imports to developed country markets. This will tend to reduce energy consumption world-wide, counteracting leakage, albeit by the undesirable means of reducing incomes.

In analysing these shifts in prices and trade patterns, results are generally expressed as a leakage rate, defined as the percentage by which emission abatement within the OECD is neutralised by increases elsewhere. Attempts to quantify carbon leakage have produced a wide range of results, with estimates of leakage rates ranging from highly negative to 100 per cent. However, work to date has been based on a variety of theoretical assumptions, and there is a notable shortage of empirical evidence. In the next section we review previous work on modelling leakage rates for various forms of OECD co-operation, highlighting the differences in assumptions and parameters which have driven the dispersion of results. We then discuss how recent research into the determinants of oil market behaviour and industrial location might impact on the modelling of leakage, before deriving a model based on these insights for use inside EGEM.

This model is then examined in detail and the relative magnitude of each effect compared. Finally the interaction between treaty stability and leakage is explored to see if the magnitude of industrial relocation is likely to undermine, or enforce, international agreements.

PREVIOUS ESTIMATES OF CARBON LEAKAGE RATES

Estimates of carbon leakage in the literature have used models with different assumptions and levels of disaggregation, and while some concentrate purely on fuel price effects others consider the trade effect more important. The main features of each study are summarised in Table 9.1.

Pezzey (1992) uses the Whalley–Wigle general equilibrium (GE) model to produce some of the highest estimates of leakage rates, at around 60 per cent for a unilateral cut of 20 per cent in annual CO_2 emissions over the base case up to 2100 by OECD, or 70 per cent for similar EU unilateral action. EU production of energy intensive goods falls by 5.4 per cent or 8.1 per cent in the two scenarios respectively. However, the model does not distinguish between the three fossil fuels, and assumes free trade and perfect competition in world energy markets. That this is far from the truth does not need restating.

Using the OECD model GREEN with a more sophisticated treatment of world trade, Oliveira-Martins *et al.* (1993) find much lower leakage rates, as does the earlier work by Burniaux *et al.* (1992). Stabilisation of CO_2 emissions in OECD leads to leakage rates of 1.4 per cent by 2050, or 2.2 per cent for the EU alone. The rate varies over time, generally decreasing, and may be negative. Energy intensive industries show losses of 1.6 per cent in the OECD scenario, and 2.4 per cent in the EU. For the EU alone, the greatest leakage is via other OECD countries, reflecting their stronger trade links and freer markets. The main factors reducing emissions in non-OECD countries appear to be the reduction in demand by oil-exporting countries, and a shift away from coal in China and India. However, the model does not allow for international capital mobility, which will lead to an under-estimate of location effects, although the authors stress that this may be to some extent allowed for in changes in trade due to competitive advantage.

Perroni and Rutherford (1993) find a typical leakage rate to be around 10 per cent for a unilateral OECD reduction of 5 per cent from 1990 levels, using the Carbon-Related Trade Model (CRTM), a static general equilibrium model based on Global 2100 (Manne and Richels 1990). Production of basic materials within OECD drops by 7 per cent, and by 10 per cent in the USA. An updated version of CRTM uses a recursively dynamic approach by Felder and Rutherford (1993), who stress the marginal leakage

Table 9.1 Summary of carbon leakage studies

Study	*Model and solution type*	*Industrial disaggregation*	*Fuel market model*	*Policies simulated*	*Leakage rate*
Pezzey (1992)	Whalley–Wigle GE model	Energy intensive/ non-energy intensive	Free trade; no distinction between fuels	20% cut in OECD 20% cut in EU	60% (2100) 70% (2100)
Oliveira-Martins *et al.* (1993)	GREEN	Energy intensive/ non-energy intensive	International market for oil only; inter-fuel substitution considered	OECD stabilisation EU stabilisation	1.4% (2050) 2.2% (2050)
Perroni and Rutherford (1993)	CRTM, based on Global 2100; static GE	'Basic materials' + other industry	International for oil; regional markets for oil, gas and electricity	OECD stabilisation	5–15% (2020)
Felder & Rutherford (1993)	CRTM, based on Global 2100; recursive dynamics	'Basic materials' + other industry	International for oil; regional markets for oil, gas and electricity	OECD cut of 2% p.a. OECD cut of 4% p.a.	0–40% (1990–2100) −30 to +35% (1990–2100)
Manne and Rutherford (1994)	Global 2100; full forward-looking intertemporal dynamics	None	Competitive international oil market; gas trade between ex-USSR and OECD; some limits on fuel substitution	20% cut in OECD	Increasing to 30% (2050)

rate, that is, the leakage from a further 1 per cent reduction by OECD, which may vary considerably from the average rate. Marginal rates vary over time, reaching a maximum of 45 per cent depending on the level of OECD abatement, and falling after 2040, becoming negative with a minimum of −180 per cent in one case. Average leakage rates are from 0 to 40 per cent, of which a fairly constant 10 per cent leakage rate comes from the trade effect with the remaining much less predictable element from oil prices and fuel substitution effects. These results depend on the use of a backstop coal-derived synthetic fuel with a very high carbon content, use of which depends on resource depletion rates in oil and gas.

An alternative adaptation of Global 2100 is presented by Manne and Rutherford (1994), using a solution method of intertemporal equilibrium, determining prices and quantities simultaneously over a sequence of time periods up to 2050. Again it is assumed that a carbon-free backstop is more expensive that one that is carbon emitting, and hypothetically free international markets in oil and gas are modelled. They produce leakage rates increasing over time up to 35 per cent for an OECD policy of 20 per cent cutbacks, though this arises only from fuel market effects as they do not look specifically at trade in energy intensive goods.

Horton *et al.* (1992) model stabilisation of emissions by 2000 for the OECD and estimate leakage to be around 10 per cent. They then study specific highly affected sectors in more detail, modelling imperfect competition, and they find much higher leakage rates, up to 100 per cent. However, as this is just for a single energy-intensive sector, this is exactly what one would expect and it reveals little of the equivalent full economy rate.

There are a number of single country studies. For the UK, for example, Barker *et al.* (1993) find that UK iron and steel and chemical industries would lose only around 1.5 per cent of their export markets under an OECD carbon/energy tax of $10/bbl, with other non-energy intensive industries showing gains. This result is due to general GDP gains from energy taxation (a function of Barker's more detailed treatment of tax revenues) and the modest size of the tax, which is not sufficient to stabilise emissions.

As with the divergent results on the macroeconomic costs of stabilising emissions reviewed in Chapter 3, there appears to be little consensus on the magnitude of leakage. These differences comes from both alternative model structures and different a priori parameter assumptions which alter the base case emissions levels and abatement scenarios. All authors show higher leakage rates when fewer countries are included in the agreement, and leakage rates vary over time as well as with the level of reductions within the abating group of countries. However, little attention is paid to the effects of different tax recycling measures which could have important implications for industrial competitiveness.

All numerical results are naturally highly dependent upon input assumptions, and sensitivity analyses have been performed for both GREEN and

CRTM. It appears that the low estimates of leakage rates by Oliveira-Martins *et al.* are due in part to the use of high supply elasticities for coal; when a low value is used leakage reaches around 25 per cent. Higher Armington elasticities (for substitution between domestically produced and imported goods) also increase leakage. Perroni and Rutherford (1993) show leakage is higher with increases in basic materials supply elasticities or growth rates. Winters (1992) in a review of some of this work believes that GREEN and CRTM under-estimate leakage, due to strong product-differentiation between OECD and non-OECD in both models. Treatment of energy demand, interfuel substitution outside OECD, and technological options for backstop fuels are also critical areas of difference, which past 2040 or so really become a matter for speculation rather than forecasting.

Policies to reduce leakage, which were studied using the above models, include protection of energy intensive industries by exemption from the tax, as in the European Union proposal (CEC 1992), which Oliveira-Martins *et al.* find has no long term impact on leakage rates as a higher tax is then required throughout the rest of the economy. Felder and Rutherford consider export subsidies on OECD basic materials as a means to reduce leakage, but find them relatively ineffective. Perroni and Rutherford (1993) use the CRTM to analyse trade in carbon emission rights, and find that this can reduce leakage by reducing changes in trade of basic materials.

Obviously, the different model structures and dispersion of results rest on disputes about the basic economics of the leakage routes through world oil price reductions and industrial relocation. Before we try to construct a model for use in EGEM we review the wider literature relevant to this problem, especially that on historical patterns of industrial relocation and long run oil price forecasting.

LEAKAGE THROUGH WORLD OIL MARKET RESPONSES

Modelling the world oil price is an endeavour that many have attempted and few have achieved with any great accuracy. Huntington (1994) notes that a number of price forecasts in 1980 generally estimated the 1990 price to be around three times its actual value. He attributes some of the blame to the unforeseen increase in non-OPEC supply and low exogenous GDP forecasts pre-1986, but more of the blame to mis-specified demand responses after 1986. Others (e.g. Austwick 1992) believe that oil markets are essentially too complex to model quantitatively and one should confine oneself to inter-disciplinary estimation of lower and upper bounds and discussion of factors affecting movements.

Gateley in his 1984 review assesses various views on the forces behind the 1973–4 and 1979–80 price shocks, including such issues as the development of OPEC as a cartel and the stimulus to long term research and development from high prices. He favours modelling behaviour using reaction functions, because he believes OPEC to be 'groping towards an unknowable "optimal" price path', and he also considers OPEC's power to be short term, with more fundamental supply and demand forces driving the market. Other theoretical backgrounds to past forecasts have included Hotelling-style resource scarcity rents, or some form of OPEC target capacity utilisation rule. Gateley and other authors (Baldwin and Prosser 1988; Adelman 1989; Greene 1991) consider the Hotelling rule quite inapplicable in practice due mainly to uncertainty and/or sub-optimality in OPEC's decision making, and long lags in supply and demand responses. Models using some form of reaction function to a target capacity utilisation, with OPEC adjusting price depending on whether the market is tight or slack, have been generally more successful. However, this type of model is only useful in the relatively short term, as long run capacity depends on investment and R & D, which are an endogenous function of prices.

More recent work tends to place importance on the growth of non-OPEC supply and the objectives and market power of OPEC (see review by Bacon 1991). From 1977 to 1985 non-OPEC production grew steadily at 5 per cent p.a. (Parra 1994), causing a reduction in OPEC's market share from 50 per cent in 1970 to 31 per cent in 1986 (Al-Sahlawi 1989), as demand growth was far less than expected. In fact in the OECD demand declined from 1973 to 1985, due to both active energy conservation policies and substitution of other fuels (chiefly natural gas). In December 1985 OPEC decided on a change of policy described by Parra (1994), and tried to recoup some of its markets, influenced strongly by Saudi Arabia which was suffering the effects of reduced demand, being the swing producer. Netback pricing was introduced, production increased, prices dropped through the floor, and have remained low ever since. This may be effective as a long term strategy, because non-OPEC operating costs are fairly low but exploration and development costs are high, and so the effect of low prices since 1986 has been to prevent new fields from being opened up rather than forcing existing ones to close down. OPEC's market share has crept back up, to 41.1 per cent in 1993 (BP 1994), as non-OPEC sources have come to the end of their natural lifetimes, and low prices have inhibited the growth of new fields. Morgan (1987) stresses that some of the effects of the 1970s' price shocks are irreversible, including technological innovation and the 'realisation of strategic vulnerability' on the part of oil importers. Certainly, though the real price of oil is near pre-1973 levels, it is no longer the universal energy source it once was in industry and electricity generation, although it still dominates transportation.

Though current low prices could be a deliberate strategy by OPEC, another possibility is that they reflect a lack of co-operation between OPEC members in controlling output and maintaining high prices. The strength of OPEC and its chosen policy has certainly varied over the years. With its decision making being rather less than transparent, it is difficult to say whether objectives have altered or just its means of achieving them. It has even been suggested that it is in OPEC's interest to promote volatility in oil markets (Morgan 1987), to shatter confidence and prevent non-OPEC from making any secure long term investments.

Several authors use models to attempt to analyse issues of OPEC's market power. Alsmiller *et al.* (1985) developed the World Oil Market model, treating OPEC as a von Stackelberg cartel, based on a dynamic Hotelling model; that is, suppliers attempt to maximise the net present value of all future revenues. They compare scenarios with OPEC acting as a cartel and as competitive producers, and show that co-operation within OPEC could increase the price by 25 per cent. Although the oil price forecasts made of $35–70/bbl (1983 prices) now appear unrealistic, they were very much in line with other forecasts at the time – the discrepancy is due mainly to the Hotelling framework and unforeseen developments in world markets. An alternative approach is to use a recursive simulation framework, in which decisions are made based on past and present rather than future prices, as exemplified by Baldwin and Prosser (1988). In their model OPEC attempts to maintain market share while maximising revenues. They look at the trade-off between market share and revenue, as well as the more traditional price/volume trade-off, and evaluate OPEC's behaviour in terms of these objectives. An analogous approach is taken by Greene (1991), who uses a static model to calculate short and long run price curves. These depend upon OPEC's market share as this determines its monopoly power, and the rest of the world's supply and demand responses. As a cartel OPEC may then choose a price that lies anywhere between the two curves, which accounts for some of the volatility since there is a large difference between short and long term market responses. The analysis is repeated assuming that a core membership of Arab OPEC controls the price, a model which actually fits more closely to observed data from 1974 to 1990.

These studies appear to verify the observation that OPEC acts as some sort of cartel, but falls short of acting to maximise its long term revenues to the theoretical co-operative optimum. OPEC's market power is limited by its lack of perfect information, in particular on non-OPEC supply costs and technological developments, and by political difficulties within OPEC leading to its failure to impose output quotas or prices upon its members. Since 1986 OPEC's target prices have generally not been achieved. A fairly modest target price for a basket of oil products was set at $18 in 1987, a price intended to allow recovery of market share. It increased to $21 in 1990, but was only exceeded for six months during the supply disruption of the Gulf war in 1990 and never again since (Parra 1994), indicating the weakness of market power OPEC now exhibits.

Empirical work on non-OPEC supply is fairly rare, but an econometric estimation by Al-Sahlawi (1989) estimates long term supply elasticity to be 0.6. This figure may be an under-estimate, as it is from pooled data including the USSR and China which were highly inelastic, although they may not remain so as their economies are now more market-led. Single country figures are as high as 2.6 for the UK (although this figure was not significant), 2.1 for Egypt and 0.85 for Norway. Short run elasticities are much lower, often below 0.1, as would be expected given the long time-delays in changes in production. High long run elasticities reflect the ability for falls in prices to choke off the development of new fields by non-OPEC producers. Once investment has been made in productive capacity elasticities will be lower, especially because the public revenue needs of some producer countries means that sales volume is more important than maximising long term value.

To summarise it is clear that accurate forecasting of either short or long run price reactions is fraught with uncertainty. However, in explaining the observed data non-OPEC supply response and the level of OPEC's market power – in terms of both its market share and its internal cohesiveness – seem to be the most important determinants. If a carbon tax were adopted by the OECD it seems unlikely that in the long term OPEC would be able to maintain prices much higher than those dictated by the market.

Any effective carbon-dioxide abatement policy has to ensure that a substantial amount of oil stays in the ground. Reliance on physical resource scarcity increasing the price cannot be effective until too late. Oil (and other fossil fuel) producers must be forced into seeing lower prices to reduce supply, while at the same time the tax wedge between supply and consumer prices must increase to decrease demand. An important corollary of this is that, unless OPEC can restrain production and increase prices, there will be a transfer of wealth from oil producing to consuming countries, as the economic rent from consuming less than the free-market equilibrium level of oil is taken more in the form of consumer taxes rather than above-cost producer prices. This process is already occurring as high excise duties on petrol have caused complaints from the Saudi Arabians that the developed countries are removing their 'legitimate' rents (Verleger 1993). It is to be expected, and has already been observed at the 1995 Berlin meeting of the FCCC, that OPEC will do anything in its power to prevent a carbon tax from being successful. This institutional resistance perhaps reflects the inability of OPEC to raise prices as a way of capturing these rents, and analysis based round this possibility seems to be based on a memory of past power in the 1970s that no longer exists (for example, Wirl 1994).

RELOCATION OF INDUSTRY

General environmental regulation is thought to have a fairly small impact on competitiveness, according to a review for the USA by Jaffe *et al.* (1994). Some studies even find positive impacts on trade, brought about by regulation stimulating innovation and eliminating the oldest, least efficient plant, as well as benefiting those firms that supply environmental goods and services (Ekins and Jacobs 1994). However, some industries are harder hit than others, and these unsurprisingly tend to be the sectors which consume and convert raw natural resources: electricity generation, metals and chemicals.

If a significant carbon tax is imposed the most energy intensive industries will have an added incentive to relocate to areas outside the agreement, where already energy prices are generally lower than within the OECD. Manufacturing accounts for around 30 per cent of energy use in OECD countries, and typically 10–15 per cent of total primary energy consumption is used in the 'top three' energy intensive industries: iron and steel, chemicals and other metals (UN 1992). Other industries such as paper/pulp mills, glass and cement are also important. Energy costs in these industries may be around 15 per cent of input costs, compared with 3–5 per cent for the economy as a whole. It must be remembered, however, that only manufacturing and primary materials production are prone to migration, while domestic, transport and service sector consumption cannot for physical reasons. Therefore, the maximum extent of this relocation leakage is limited to 10–15 per cent of energy use, which calls into question some of the high leakage estimates obtained by previous studies.

Motivations for relocation in the energy intense industries

Given this upper bound for the effects of industrial relocation the proportion actually prompted to leave is still very uncertain. Competitiveness and relocation depend on many more important factors than energy prices; including labour costs, product quality, market proximity, the extent of capital and technological mobility, existing trade links and trading conditions, such as import restrictions or tariffs.

Keeling (1992) provides an interesting discussion of these trends in the world steel industry. Steel consumption in the OECD has declined since 1973 and developing country demand continues to grow. Production has proportionately shifted towards the developing countries, and this increase has mainly been in the newly industrialised countries of South-East Asia. Steel intensity, like the intensity of energy and other raw materials, tends to increase during a certain phase of development connected with industrialisation, city expansion and infrastructure development, and level off as growth in the economy comes more from services and less steel-intensive sectors. Therefore, the development of steel production outside the OECD appears to be linked strongly to the location of demand for steel, much of it being for heavy structural uses.

OECD producers have also lost some of their markets because of the much lower wage costs in competing countries and the diffusion of new and state-of-the art technology, especially the advent of electric 'mini-mills' which makes small competitors more viable. The OECD industry has survived by a combination of productivity improvements, protectionist policies, and product differentiation into high quality and specialised steels for use in car manufacture and high grade engineering.

It is not clear to what extent the situation will be exacerbated by energy taxes. As energy is still a relatively small part of costs, it will be less important than labour costs or technological expertise. World steel prices tend to be fairly volatile, as demand varies cyclically with capital investment throughout the economy. For this reason, as well as others, levels of trade are relatively low at 26 per cent of world output with most industries relying greatly on their domestic markets. Transport costs are often high and markets restricted – Keeling (1992) considers steel to be amongst the most protected industries in the world, with most OECD countries trying to keep their often monopolistic and nationalised industries going. Even if barriers were removed, the extent to which developing countries can penetrate Western markets will depend on their producing the appropriate quality steel products at competitive prices, and suffering the fluctuations in demand and world prices.

The chemical industry has faced similar problems, with stagnating Western markets and increasing competition from low-wage countries in addition to rising environmental costs. Increasingly OECD chemical production is moving into higher added-value fine chemicals, rather than energy intensive bulk products. However, in the remaining bulk industry the concern over rising environmental costs is demonstrated by the vocal opposition of American chemical manufacturers to the Clinton 'Btu tax' by claiming it would cost 10,000 jobs in the USA (Storck 1993).

By the very fact that energy costs are high in these industries, energy efficiency is already of primary importance and tends not to suffer from the types of market failures found elsewhere. Howarth *et al.* (1993) show that for five OECD countries energy use per unit output in manufacturing decreased from 1973 to 1988 by 14–30 per cent, while in other sectors (transport, domestic, other industry, services) the reductions were smaller and in some countries there were increases. In many cases the largest reductions were in countries/sectors where there had been successful policies to reduce fuel consumption by regulation, such as the US car fuel economy standards and Danish home insulation. There is also some evidence that in manufacturing the reduction in energy use is driven more by the introduction of new process technologies than by energy prices. For instance, Boyd and Karlson (1993) use an econometric model to look at the diffusion of new technologies in the US steel industries and find that, although energy prices are a significant determinant of the date of adoption of the new furnaces, the coefficient is very small compared to other factors. This implies that for these industries improvements in general productivity and product range are more important than improvements in the use of a single input factor, and that this will tend to reduce the propensity of companies to relocate in response to a carbon tax.

Policy responses to relocation

Given that some, but seemingly not a vast amount, of carbon leakage will occur through industrial relocation the question arises of how policy makers should take this into account. The now moribund EU carbon tax proposal allowed for protection of the most sensitive industries by making them exempt from the tax. This has the obvious disadvantages of reducing tax revenues and the direct effectiveness of the tax, but would have alleviated the possible economic loss to the EU from migration of these industries, and by reducing carbon leakage would have improved the environmental effectiveness of the proposal. This is because these industries, while exempt from general taxation increases, would have had to submit to energy auditing and management to ensure that they improved their efficiency to a similar extent as they would have under the tax regime, although the problems with enforcing this are palpable.

Using exemptions to prevent leakage has one potential advantage that, as industrial energy intensity is currently considerably higher in low-cost developing countries, the migration of these industries could result in higher overall energy use as the more efficient OECD plant is replaced. However, this may be a feature of past development patterns as new plant in developing countries tends to have efficiencies which are comparable with the OECD. In fact there are examples of entire plants exported to low-cost countries – for instance British Steel's Ravenscraig steelworks which has been sold to Malaysia (Clifford 1993), or the mothballed DSM melamine works at Geleen in the Netherlands, only built in 1992, which is considering a move to Indonesia (Chemical Marketing Reporter 1994). In the context of carbon leakage we are considering the early retirement of the least efficient (and hence higher energy cost) OECD plant, and its replacement by new growth elsewhere. Hence it appears a reasonable assumption that efficiencies are at least similar between the two, despite the difference in energy prices.

One problem with this exclusionary approach is that a carbon tax may produce its most important effect by reflecting the full external costs of energy throughout the economy, rather than just affecting the price of directly consumed energy. Since a carbon tax is intended to internalise an external cost, exemption prevents the full costs of energy being passed on into energy-rich materials, masking the signal to reduce consumption of these energy-intensive products. This type of material efficiency has proved to be an important component in the reduction of energy intensity in the past. For instance, the quantity of steel used in car manufacturing has decreased significantly since the 1970s as steel quality and vehicle design allow lighter bodywork, though this trend has been reversed recently with the standard fitting of increased safety and comfort features (Keeling 1992).

Higher prices increase the incentives for recycling of paper and glass, and promote the substitution of less energy-intensive materials, such as plastics, for steel. Balancing this concern is the fear that if industry relocates as a response to increased energy prices the price rises will also not be fully passed on. The only apparently effective solution would be to tax imports of energy embodied in raw materials, and presumably also the raw materials embodied in products; for example, the energy used in making the steel that made the car exported from Korea to Europe. Despite its theoretical attractions this kind of policy is, for all practical purposes, impossible due to the problems of measuring energy inputs, and would be likely to be in conflict with the GATT, or at least be challenged under the GATT rules.

EFFECTS OF REVENUE RECYCLING AND INTERNATIONAL AGREEMENTS

Assuming carbon taxes are recycled through reductions in existing input taxes to industry, and not income taxes, an increase in costs for energy intensive industries will be balanced by a decrease for much of the remainder of the economy (see Pezzey 1991 for an example). As it is a small proportion of the economy that is relatively hard hit, the rest is likely to see a much smaller gain; for example, if the 'Big Three' face price increases of 5–10 per cent, most of the economy may see real decreases of 1 per cent or less, depending on the means of revenue recycling. This could mean that most of economy would benefit from increased competitiveness, balancing the loss seen by the energy-intensive sectors. This type of zero-sum shift depends upon the response of trade to prices and critically whether trade elasticities are constant with respect to price changes. The concern over competitiveness appears to assume that a 10 per cent increase in prices may lead to significant loss, while a 1 per cent improvement would be negligible, leading to a net loss in competitiveness. An alternative analysis may accept that overall *equilibrium* losses would be small, but contend that transitional costs in making the change in industrial/economic structure could outweigh the long term impacts.

A more worrying problem is if taxes were only imposed in one region, for example the EU, producing the possibility of leakage within the OECD. This is potentially more serious, as it would allow countries such as the USA that are similar in terms of development of labour force, industrial infrastructure and markets to compete with European industry under the added advantage given by a tax. It would appear likely that leakage would be higher, and the few studies that have taken this into account confirm this (Pezzey 1991; Oliveira-Martins *et al.* 1993). There would thus be rather more justification for a tax-exemption of basic industries in this case. However, energy prices already differ markedly between the OECD regions and there is no evidence that the high cost countries, such as Japan and Germany, suffer a loss of competitiveness in manufacturing relative to low cost countries such as the USA.

Of course, any type of international agreement that aims to share the abatement burden efficiently among the developed countries will also face this problem. The countries abating most will face not only higher direct abatement costs but also reductions in competitiveness compared to other countries in the agreement. Mabey (1995a) shows that this has ambiguous effects on treaty stability; if trade costs are low or similar to the benefits from co-operation they tend to encourage defection from any agreement and non-cooperative solutions. However, if trade costs are very large they may support an agreement because defection of one party could cause the rest to leave, as the loss in competitiveness overrides the environmental benefits of remaining in the treaty. In this case the only stable international agreement would be one that imposed similar trade costs in each country, and this may not be compatible with efficiently allocated abatement.

MODELLING CARBON LEAKAGE WITHIN EGEM

Estimation of emissions and responses to carbon taxes in EGEM concentrates on the nine major OECD countries,[1] that together make up 88 per cent of OECD emissions. With regard to current policies this emphasis is appropriate as it is the developed countries that will be expected to bear the brunt of CO_2 reductions in the short to medium term. However, in order to assess carbon leakage, and to consider the full global perspective, some estimates must be made of emissions from the rest of the world, the behaviour of world oil markets and the shift in trade patterns.

For the nine OECD countries, an aggregate model of energy demand has been econometrically estimated, together with fuel share models for gas, coal and oil that represent substitution effects between the three as functions of their relative prices. The aggregate fossil fuel demand depends on price and GDP, with a term for technical innovation induced by price changes. The model is described in more detail in Chapter 4. The developed countries are assumed to operate in three country coalitions (Europe, North America and Japan) which may be designated as being 'in' or 'out' of an agreement to allow analysis of carbon taxes in different areas and their impact on competitiveness.

EGEM divides the rest of the world into a number of regions: OPEC; the remainder of the OECD not modelled explicitly; South and South-East Asia; Latin America; Africa; and the remaining less developed countries including China. Each of these regions is modelled as a trade block, with equations representing import and export volumes, value and prices (with slightly more detail for the OECD). The latter four are grouped together as the less developed countries (LDCs).

Carbon dioxide emissions in the RoW

Carbon dioxide emissions are modelled for LDCs on a basis of export goods volume, on the assumption that this is an indicator of the commercial energy use in the economy; this is an accurate reflection of global warming influence because traditional energy sources, such as wood and dung, are carbon neutral over their whole use cycle. An elasticity of 0.8 is assumed, as trade has grown more rapidly than GDP, and energy use per unit of GDP has remained fairly constant in most of these countries. Current emissions are taken from WRI (1992) data for 1987. Table 9.2 shows the respective elasticities of emissions with respect to world oil price, which are weighted according to the share of oil in energy use.

Table 9.2 Elasticity of CO_2 production with respect to real price of oil

Region	*Elasticity*
Africa	−0.24
Latin America	−0.22
South and South-East Asia	−0.13
China and rest of LDCs	−0.06
Former Communist countries	−0.09
OPEC	−0.17

OPEC countries are modelled similarly but emissions increase with respect to export value, rather than volume, with an elasticity of 0.6; this is because exports are predominantly oil and hence change due to variations in the oil price.

The resultant response equation for each region is written in error correction form, to allow a lag in responses to a change in prices:

$$\Delta\ln(CO_{2t,r}) = 0.8\Delta\ln(CO_{2t-1,r}) - 0.1(\ln(CO_{2t-1,r}) - 0.8\ln(XGI_{t-1,r}) + s_r\Delta\ln(P_t) + k) \qquad (1)$$

where: $CO_{2t,r}$ = CO_2 production for region r, $XGI_{t,r}$ = exports goods index, s_r = oil price elasticity, P_t is the real price of oil, and k is a constant.

For other OECD consumption is assumed to be proportional to consumption in Europe, as these countries are similarly developed being mainly either current or future EU members, with the exceptions of Switzerland, Norway, Australia, New Zealand, and Iceland.

Oil price leakage effects

The model of the world oil market used here is based on a supply/demand balance, where OPEC as the price setter aims to reach a target market share. Oil demand within the nine OECD countries is modelled explicitly, again as detailed in Chapter 4, whereas elsewhere it is assumed to be proportional to the CO_2 emissions as modelled above.

Non-OPEC producers act as price takers, adjusting their output with an elasticity of 0.8 (based on the estimate from Al-Sahlawi 1989, adjusted upwards for reasons given above), while OPEC adjusts the price to keep their long term market share stable at 50 per cent. No estimates of resource depletion are made, as over the time horizon of the forecast (1995–2030) it is not thought that resources will reach critically low levels; similarly there is no explicit backstop energy source. However, the structure of the model forces up the real price if and when demand increases, and an implicit backstop will occur at a sufficiently high price. The advantages of this model are its relatively simple form and its ability to replicate observed long term market trends; however, it cannot forecast instability in prices or sudden shocks (for example, due to dynamic ‘pinches’ in supply and demand investment) that may be of great importance in the short to medium term.

Oil supply is given by:

$$\Delta \ln(S_{ROW,t}) = 0.8.\Delta \ln(P_t) + 0.5.\Delta \ln(S_{ROW,t-1}) \quad (2)$$

$$S_{OPEC} = D_{TOT} - S_{ROW} \quad (3)$$

where: S_{OPEC} = supply from OPEC; S_{ROW} = supply from the rest of the world; D_{TOT} = total world demand (including stock changes); P_t = world oil price, given by:

$$\Delta \ln(P_t) = 0.6\Delta \ln(P_{t-1}) - 0.2.(0.5 - S_{OPEC}/(S_{OPEC}+S_{ROW})) \quad (4)$$

Therefore, this model will estimate the oil price effect on carbon leakage, under the assumption that international coal and gas markets are not developed enough to have a significant impact. The key parameters are the non-OPEC supply elasticity and oil demand elasticity for LDCs. These can be tested by comparison with other forecasts of oil price and LDC energy demand. Interfuel substitution in LDCs is not taken into account, as it is implicitly assumed that fuel mix remains the same as at present.

Modelling industrial relocation and competitiveness effects

GEM is not ideally suited to measuring these types of trade effects as industrial production is not disaggregated by sector. The overall impact on trade flows will be quite adequately assessed, but the shift away from energy intensive industries is less straightforward. However, for each of the G9 countries, the proportion of energy used in manufacturing is known, as is the proportion in the 'Big Three' industries of iron and steel, other metals and chemicals. From this data the reduction in these sectors in the G9 can be estimated, and the CO_2 emission reductions then attributed to other countries.

Trade leakage is divided into that caused by any overall changes in trade volumes and shares, including income effects, and that from the structural shift away from energy-intensive processes within the OECD economies. The overall trade changes are made at an aggregate level, dependent on overall producer prices, incomes and markets. It is implicit that if the overall impact is small, some sectors will benefit from the reductions in existing taxes, while others pay through the energy tax. For the countries modelled, the proportion of the manufacturing industry and the proportion of the 'Big Three' to relocate is estimated from the percentage increase in energy costs.

For each of the nine countries modelled, the proportion of CO_2 arising from heavy industry, allowing for relocation, is:

$$\Delta \ln(CO_2H_t) = 0.7.\Delta \ln(CO_2H_{t-1}) - 0.05.(\ln(CO_2H_{t-1}) - \alpha.C_e.\Delta \ln(RP) + k) \quad (5)$$

where: CO_2H = proportion of CO_2 from heavy industry; $\Delta \ln(RP)$ = change in real price of energy from the base-case; α is the 'relocation elasticity' (2–4 depending upon the number of countries participating in the treaty); C_e is the share of industrial costs going to energy (≈ 0.15) and k is a constant derived from the current proportion of energy used in the 'Big Three'.

This change in CO_2 emissions becomes an increase in emissions from non-signatory countries which is added to the total CO_2 for the rest of the world, under the assumption that emissions per unit of goods produced is the same as in the OECD. Unless specified this does not affect overall trade volumes or prices in the model, as it alters only the net energy intensity of production, as represented by the proportion of energy consumed in energy intense industries.

Once the total leakage from an OECD policy has been estimated, the results of leakage within the OECD (e.g. an EU only emissions reduction scenario) are assessed. The 'size' of the treaty is taken into account by adjusting the trade elasticity, allowing the stronger intra-OECD leakage effects to be accounted for. This accounts for the greater possibilities for relocation, and freer markets in fuels and basic materials between OECD countries.

LEAKAGE RESULTS FROM EGEM

In order to assess the magnitude of the different leakage effects, several simulations were run over the period 1995–2030. Initially, income effects from changes in OECD GDP are excluded, allowing identification of the oil price and industrial relocation effects caused by a carbon/energy tax within the OECD, or its constituent regions. Later simulations endogenise the macroeconomic impact of the tax, incorporating changes in GDP within the OECD and their knock-on effects on world trade and hence carbon emissions elsewhere.

Leakage rates with OECD and regional taxes

Figure 9.1 shows the leakage rate brought about by the European Union proposal of a $10/bbl tax split 50:50 between carbon and energy, phased in over 1995–2002. The figure shows four scenarios for a tax applied throughout the OECD, and in three constituent regions alone: the European Union, North America and Japan. The rate is expressed as a proportion of the total carbon abatement in the region(s) to which the tax applies. This fairly modest tax level is sufficient to reduce carbon emissions within OECD by 14 per cent from their base case level in 2030, which does not achieve stabilisation as in the base-case OECD emissions rise by 48 per cent from 1990 to 2030. Meanwhile non-OECD emissions increase by a factor of 7, due to their stronger economic growth and higher energy intensity – developing country emissions overtake the OECD in 2002.

With a tax throughout the OECD, the leakage rate reaches about 16 per

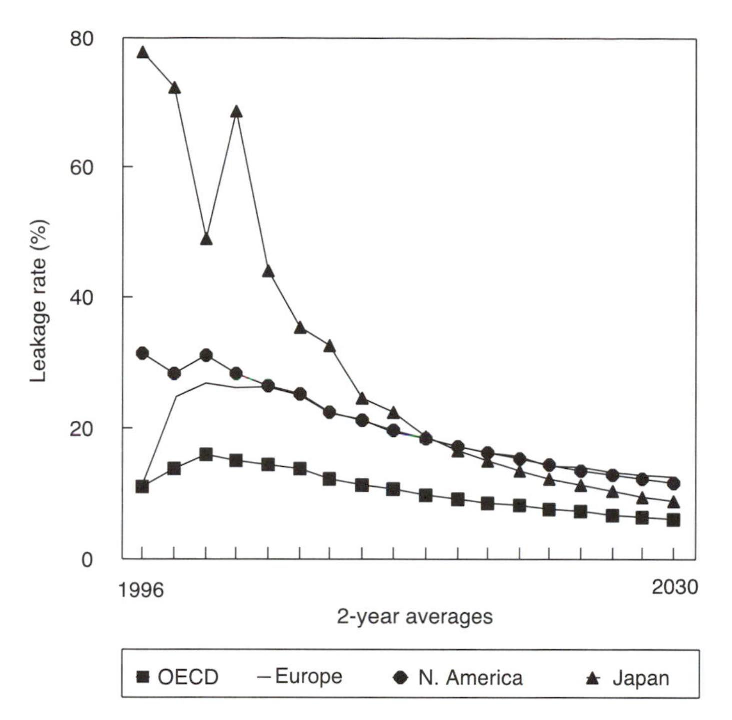

Figure 9.1 Leakage rates for OECD-wide and regional carbon taxes

cent at its height and then falls to around 6 per cent. The rate of increase declines as the level of abatement continues to rise, while the leakage losses stabilise after around ten years. For a tax in any one of the three regions the rate is generally significantly higher, as they will face competition from other OECD countries without the tax, increasing the industrial relocation element of leakage. Oil price effects will depend upon the characteristics of the country(ies) with the tax; that is, the share of oil in energy consumption and the price elasticities of oil and total energy. The figure shows fairly similar rates for Europe and North America, with rates up to 30 per cent but declining to around 10 per cent. For Japan the initial rate is much higher, at up to almost 80 per cent in the first year, but again declines to under 10 per cent in the long term. These high rates are due to the high share of oil in Japan. Initial carbon abatement is almost entirely in reduced oil use, which then reduces world price and other countries' consumption. After some fluctuation, the market then adjusts to a lower total oil consumption level so the leakage rate declines.

Disaggregating leakage effects

Figure 9.2 shows for the case of the OECD-wide tax, detailing how the leakage is broken down into the three effects; namely the oil price effect,

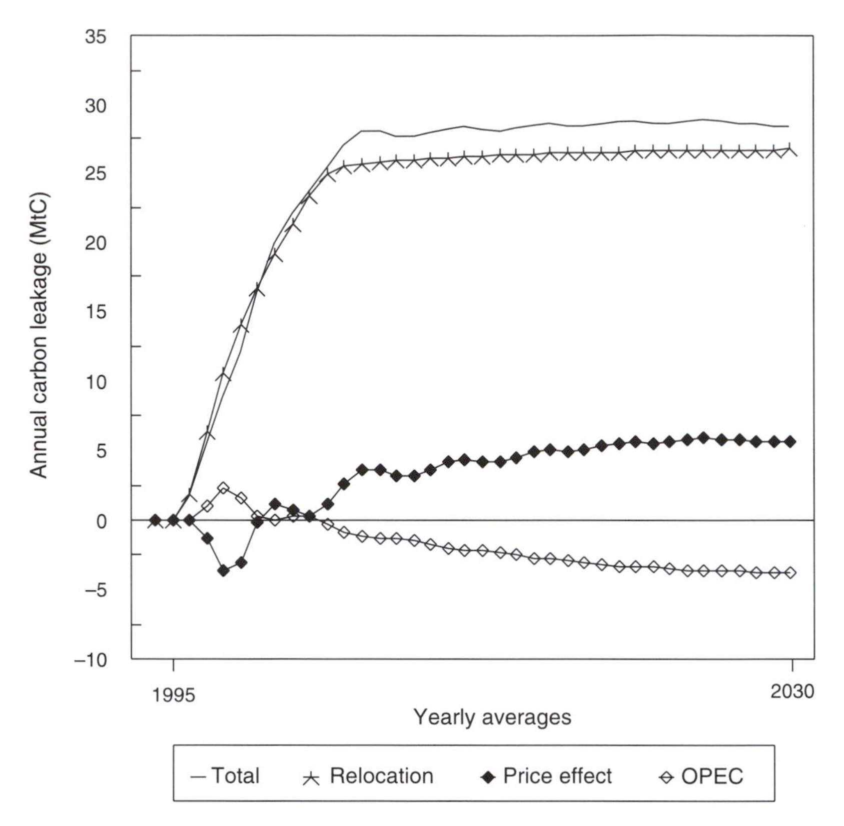

Figure 9.2 Leakage through different channels

trade relocation and OPEC-income effect. The latter is brought about as reduced oil demand and lower world oil prices significantly reduce oil revenues to oil exporting countries, and thus reduce their domestic energy consumption. This effectively contributes a negative leakage, as the reduction in oil demand from OECD is reinforced by reduced OPEC demand.

The graph shows differences from the base case in MtC. To put the figures in perspective abatement by the OECD increases to 484 MtC by the year 2030, so the maximum amount of leakage through each avenue is under 5 per cent of total emissions. It can be seen that industrial relocation is the greatest contributor to the total, but after adapting to the new price the amount levels out. This comes about as the tax causes OECD countries to lose a proportion of their heavy industries, but once they have adapted to the altered price this proportion then remains constant. The oil price effect is significant but smaller, and fluctuates over time as the market comes to a new equilibrium. The lost OPEC revenues and concomitant reduction in emissions from OPEC's own consumption reduce the overall leakage significantly.

Looking at the oil price effect in more detail, Figure 9.3 shows how the

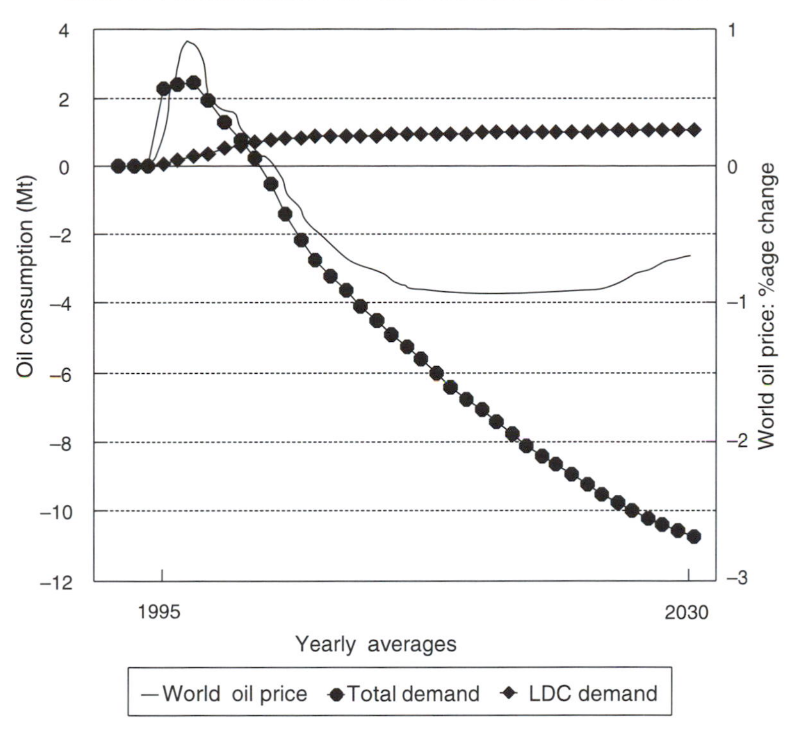

Figure 9.3 World oil price and demand: changes from base-case, 1995–2030

world oil market is modelled. The initial impact of the tax is to cause a slight *rise* in oil demand, as it is substituted for coal in some countries, causing a rise in world oil price. After around 2000 OECD oil demand starts to fall below that in the base-case, and the price follows, causing the smaller increase in oil demand by the LDCs. The oil price is maintained at about 1 per cent less than the base-case, sufficient to depress production to a lower level as oil demand overall continues to decrease from its base-case level.

Leakage rates and taxation levels

As the tax applied so far has been the same $10/bbl, constant after 2002, a comparison was made to assess the effect of the size of the tax on leakage. Figure 9.4 shows the forecast leakage rate in 2030 for four different carbon tax rates throughout the OECD. The smallest is the tax of $10/bbl in 2002 already simulated; the three further scenarios represent a tax of $10/bbl in 2002 increasing at $1 p.a. thereafter, and taxes of $20/bbl and $30/bbl increasing at $2 and $3 p.a. respectively. The first of these is sufficient to stabilise OECD emissions until 2007, resulting in a 10 per cent increase by 2030 over 1990 (compared with 48 per cent for no tax), while the two

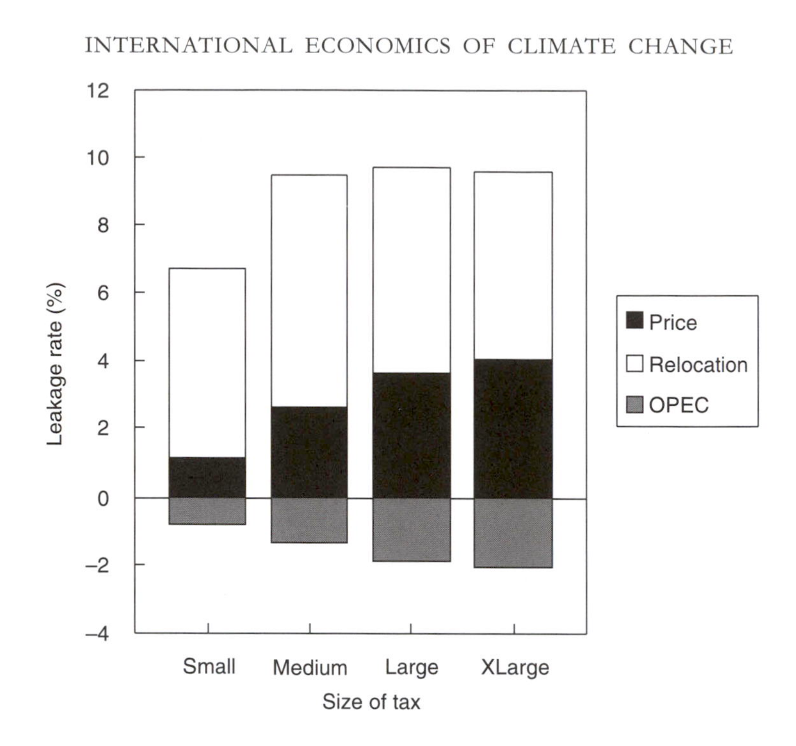

Figure 9.4 Variation in leakage rate and composition with size of tax

higher rates reduce emissions by 6 per cent and 15 per cent from their 1990 levels by 2030.

It can be seen that the overall leakage rate increases only slightly with the larger taxes; that is, the amount of leakage in tonnes of carbon is increasing at a similar rate to carbon abatement. The component of leakage from industrial relocation does not vary greatly, remaining at 5–7 per cent. The oil price effect however does increase with the larger taxes, although this is mitigated by the reduction in emissions from OPEC countries brought about by the depressed oil market.

Income effects and carbon leakage

For simplicity the above simulations have excluded the possibility of income effects from the OECD, but imposing a carbon tax will have macroeconomic impacts in the countries concerned, which will cause changes in trade and GDP in non-OECD countries, thus affecting their CO_2 emissions. The magnitude of these trade effects within the OECD will depend upon policies in terms of recycling of tax revenues as well as interest rate and exchange rate reactions.

In order to assess income effects, the full EGEM macroeconomic model was used; as EGEM only models the G7 economies in detail the effect of taxes on the rest of the OECD was scaled from the cost of abatement in the European Union. As described in Chapter 5, there are two different

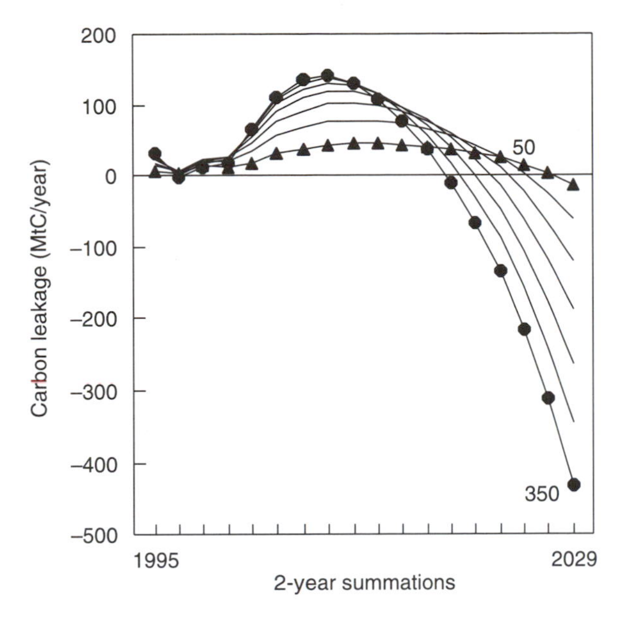

Figure 9.5 Carbon leakage from OECD (1995–2030) with income effects

supply side models available in EGEM. The simulations performed here use the endogenous technical progress model, which produces smaller output changes than a more conventional production function approach. The macroeconomic effects of carbon taxes are further reduced by recycling tax revenues through employers' labour taxes. The income effects calculated here are therefore on the lower end of the possible range of impacts calculated from a macroeconometric approach. However, as econometric estimates of price and substitution elasticities tend to be lower than the very long run values assumed in general equilibrium (GE) models, these costs will probably be higher than those calculated from a GE model over the same period.

A range of taxes, US$50–350/tC, were imposed in the OECD, and these were constant over the period 1995–2030; this corresponds to a pure carbon tax of $6.5–45.5/bbl of oil. Emissions in the OECD are stabilised at 1990 levels over this period by a tax of $275/tC, so this range includes any tax that is likely to be proposed in the near future.

The results of these simulations are shown in Figure 9.5, where a positive value indicates increased emissions in the non-OECD countries. It can be seen that income effects greatly reduce leakage, and totally outweigh the positive leakage effects from trade and oil prices by the end of the simulation period. Because imports decline more than proportionately with consumption, the income losses in the OECD significantly depress demand for products from the rest of the world. As the fastest growing developing countries are dependent on Western export markets for most of their growth, this reduces their export revenues and also has a multiplier effect

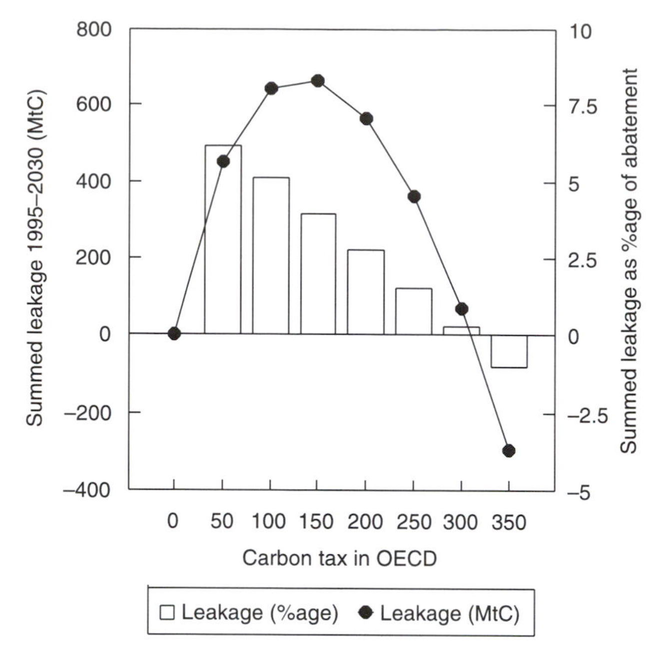

Figure 9.6 Leakage summed over simulation period, 1995–2030

as imports are reduced from other developing countries. For instance, 24 per cent of exports from South and South-East Asia go to other countries in the same group, but 58 per cent go to OECD countries. Reducing OECD imports thus reduces exports, which then reduces imports as they are constrained by export revenues, reducing exports further. The net effect is a reduction in CO_2 emissions, or negative leakage.

Figure 9.6 shows leakage summed over the simulation period for the range of taxes described above. Leakage is still positive, if greatly reduced, at most taxation levels. However, as a proportion of OECD abatement, leakage declines very quickly as taxes rise, and with a stabilisation tax it only amounts to ≈ 1 per cent of OECD abatement.

The uncertainties underlying these results are generally applicable to any model which uses an estimated trade sector. EGEM models world trade as flows between the G9 countries and the seven regional groupings mentioned above. Trade is divided into visible and invisible sectors, with invisibles being further subdivided into non-factor services, returns on overseas assets and unrequited transfers. Volumes of trade are based on 1980 trade patterns and are affected by costs and market growth. Essentially visible exports from each trading block depend on demand in traditional import markets and relative labour and 'export' prices in the trading countries. 'Export' prices reflect the mix of manufactures and commodities in each country's trade; manufacturing prices are determined by imports, domestic energy and labour costs; while traded commodities are priced in world markets. Visible imports into a country are determined by domestic real incomes, export prices in traditional trading partners and relative labour costs. Invisible imports and exports are determined by growth in world income and relative consumer prices.

In response to a rise in energy prices exports will decrease from countries imposing a tax, and their imports from low price countries will increase, as long as final demand does not drop too sharply. This leads to a trade imbalance which would usually be equilibrated by a currency devaluation in the countries imposing the tax. Devaluation restores a country's competitiveness but it also decreases welfare because its population's ability to buy imports has reduced. EGEM has no direct welfare measurement so modelling this exchange rate response would 'lose' a cost of carbon taxation inside the model. Therefore, we fix the real value of exchange rates, the relevant deflator being the factory gate price of manufactured goods, and so GDP falls due to trade effects.

The reliance of this type of model on past trade patterns means that it probably overstates the future effects of reductions in OECD income on non-OECD countries. It is likely that, especially in Asia, growth in domestic and regional markets will be faster than that in exports to the OECD, because domestic consumption is increasing and saving rates are falling. In this case trade leakage will be higher than shown in Figures 9.5 and 9.6. The true figures seem likely to lie between those unadjusted for income which reached $\approx$ 9.5 per cent, and those of Figure 9.6 which peak at $\approx$ 6 per cent. For a stabilisation tax of $275/tC the appropriate range of leakage would be 9.5−1.0 per cent, which is quite small and unlikely to undermine seriously any agreement given the existing inaccuracies present in judging the impact of carbon taxes, and hitting long term emissions targets.

TRADE LEAKAGE AND THE STABILITY OF INTERNATIONAL AGREEMENTS

Up to this point we have mainly considered how carbon leakage affects the ability of the OECD to hit an emissions target successfully. If substantial leakage occurs then the unit cost of abatement will rise and countries will either have to pay more to meet specified targets, or conversely they will commit to lower targets. These issues are vital for the effectiveness of any international agreement, but to date the issues of industrial relocation and competitiveness effects have probably been the largest impediment to implementation of an OECD-wide treaty. This debate has largely focused on the potential loss of income from relocation, rather than on changes in the aggregate global amount of CO_2 emitted.

The effect of income losses from industrial relocation, or changes in competitiveness, on the stability of any non-coercive agreement is ambiguous, and has been analysed theoretically by Barrett (1994b) and Mabey (1995a). Mabey's analysis is more pertinent for the case of global warming because it does not focus on tariff implementation as Barrett's does, but looks at how changing competitiveness affects the propensity of countries to leave an existing agreement.

Firstly, considering the case where there are *no* competitiveness effects from unilateral abatement: if several countries commit to abate emissions then there will always be an incentive for each one to free-ride, save its abatement costs, but gain from the emission reductions of the remaining co-operators. Standard analysis, such as Barrett (1994a,c), has concluded that this effect will lead to the breakdown of any treaty containing a large number of players because they will all sequentially defect from, or conversely fail to join, the treaty. The incentive to free-ride depends on the remaining countries continuing to co-operate, otherwise there are no extra benefits over the un-cooperative case and full co-operation is the utility maximising strategy. With no competitiveness effects, it is reasonable for each incremental free-rider to assume that the other countries will remain in the agreement, because they can *always* renegotiate new, lower, abatement levels that still leave them with increased utility compared to the non-cooperative case. It is therefore an empty and non-credible threat for the co-operating countries to say they will leave the agreement *en masse* in response to free-riding.[2]

However, given an existing agreement, if energy intensive industries relocate to the countries that initially free-ride, the subsequent GDP losses may outweigh the environmental benefits of co-operation, causing the other countries to leave the treaty. The treaty will therefore not break down beyond the point where this 'trade leakage' exceeds environmental benefits, because there is no incentive to free-ride if the act of doing so destroys the remaining agreement. In this case the threat of breakdown is credible, because there is no feasible re-negotiation which will leave the co-operating countries better off than abandoning their abatement commitments. Therefore, it is in the co-operating countries' own interest to leave the treaty. In this way high levels of industrial relocation can serve to *stabilise* an abatement treaty, as long as it initially contains more than the critical number of countries.

Of course, the practical realisation of this result depends on the defecting country understanding the consequences of its actions, the other actors being able to observe its behaviour and the reaction functions of the co-operating countries being commonly known. However, the institutional structure of the FCCC should mean this information is clear and should allow adequate communication of every party's intentions. Institutionally, these could be codified into a self-enforcing ratification level at the equilibrium number of free-riders, thus leaving the consequences of a free-rider's action in no doubt.

The critical defection point could occur after any number of countries have left the agreement, but if changes in competitiveness are small this effect will never be significant enough to exceed abatement benefits. Following Barrett's (1994a) analysis, in this case abatement will gradually fall to zero as countries successively free-ride. Figure 9.7 shows a numerical

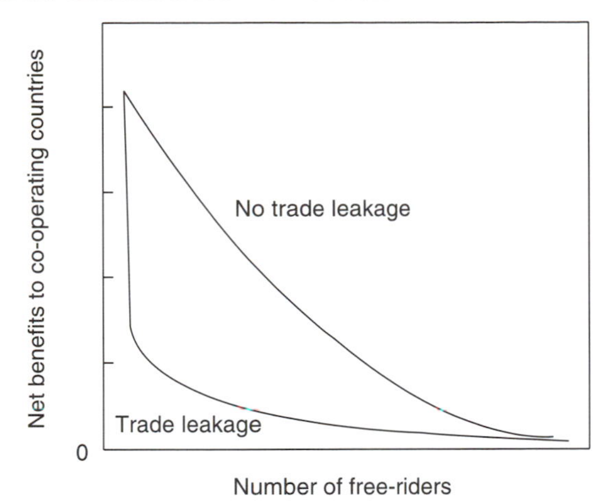

Figure 9.7 Net benefits to co-operating countries with, and without, trade leakage

example of this effect in the case where a treaty is agreed between a large number of homogeneous countries. The benefits of co-operation are plotted as the number of free-riders increase, and the remaining co-operators re-optimise their emission levels. Though the benefits of co-operation are markedly decreased by trade leakage they do not go negative, so there is no stable equilibrium caused by competitiveness effects. This is because the marginal cost of leakage falls with abatement and so co-operating countries can always re-optimise their emissions downwards to achieve a positive level of net benefits (benefits of co-operation minus the benefits of non-cooperation).

With a heterogeneous group of countries there may be several possible equilibrium points, and the equilibria will depend on the order of defection. The effect of a country with a strong iron and steel industry leaving the agreement will be very different to that of a country with a similar GDP and carbon emissions but most of whose income comes from service industries and high value added manufacturing; however, it is very difficult to generalise which type of countries will defect first. Small, high cost of abatement countries have an incentive to defect because they will gain competitiveness compared to low cost of abatement countries, who will be abating most in an efficient co-operative agreement. On the other hand, if the number of free-riders is already large the low cost countries, who will probably experience the most trade leakage, have the biggest incentives to free-ride. This is because they abate most, suffer the most from trade leakage, but receive the same benefits as the other co-operating countries (all other things being equal except abatement costs), therefore their gains from co-operation are smaller.

To simplify analysis of these issues we use our usual model of the three OECD regions to see if there would be a stable trade leakage enforced equilibrium between these coalitions, and whether leakage to the non-

OECD would break down any agreement. Using three actors allows us to model all the different permutations of defection and accession, but removes the many (>10) player aspect of the stability game, which underlies many of the assumptions used in the above theoretical analysis. These results should therefore be seen as indicative of forces undermining, or strengthening, the possibility of a lasting agreement as there are many other strategic processes to consider in a full analysis; including the issues of side payments, heterogeneous benefit functions, order of commitment and institutional design which are discussed in Chapters 8 and 11.

Figure 9.8 shows the change in OECD incomes caused by terms-of-trade effects, expressed as a per centage of OECD GDP, under a range of taxes; all figures are undiscounted summations over the simulation period 1995–2030.

Two different trade models are used, the standard estimated model in EGEM (LBS 1993), and an enhanced model where the export elasticities in each country have been doubled and the effects of industrial relocation included. The model for industrial relocation is based on the one described above to calculate CO_2 leakage, the value of industries moving abroad is based on the share of energy intensive industries in each economy which is scaled by the proportion of CO_2 leakage. The relocation calculated in each country is added to its import demand, and these increased imports are divided among countries not in the treaty. As before, leakage to OECD countries who do not impose carbon taxes is assumed to be proportionately higher than that to non-OECD countries, due to their more advanced technology. Additionally when an OECD country defects from an agreement it is assumed that some of the leakage which would have gone outside

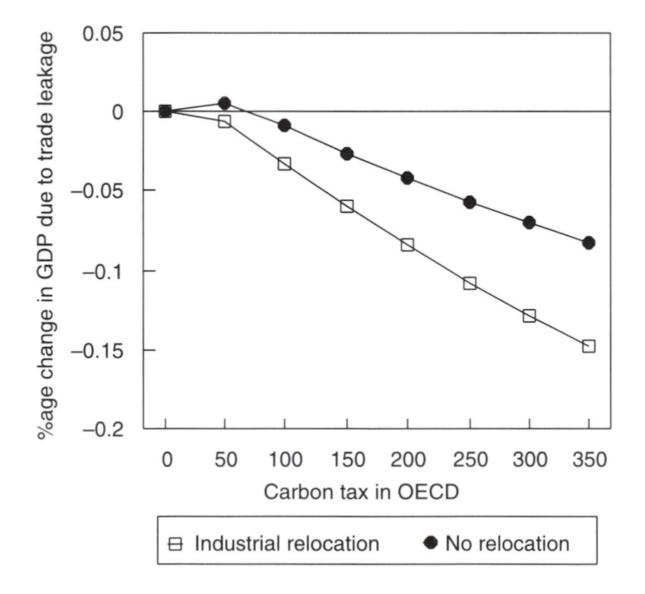

Figure 9.8 Trade leakage from the OECD with, and without, industrial relocation

the OECD is diverted to them, so gains to the rest of the world are reduced.

A tax of $350/tC causes a GDP drop of 0.6 per cent in the OECD over the simulation period, so it can be seen that even with the existing model the effects of trade leakage are substantial; accounting for a 0.08 per cent drop in GDP (13 per cent of the GDP drop from taxes). With higher trade elasticities and industrial relocation this effect almost doubles to 0.14 per cent of GDP (24 per cent of the impact of imposing a tax). This is a significant proportion, but is not large enough to outweigh the environmental benefits from abatement. Figure 9.9 shows the extra amount of trade leakage to committed countries, over and above that for an OECD-wide tax, when other members of the OECD defect, but without the effects of industrial relocation and higher trade elasticities.

It can be seen that the effects are very small, well under 0.1 per cent of the co-operating countries' GDP, and when North America defects they are positive. This is because America has a higher demand for imports when it does not impose a carbon tax and this income effect totally outweighs the price response. The income effects are smaller when Japan and Europe defect because they suffer proportionately less from the tax (see Chapter 10 for details) and the price response gives a net increase in trade leakage. Figure 9.10 shows the trade leakage if only one country, America or Europe in this case, stays in the treaty; the results do not

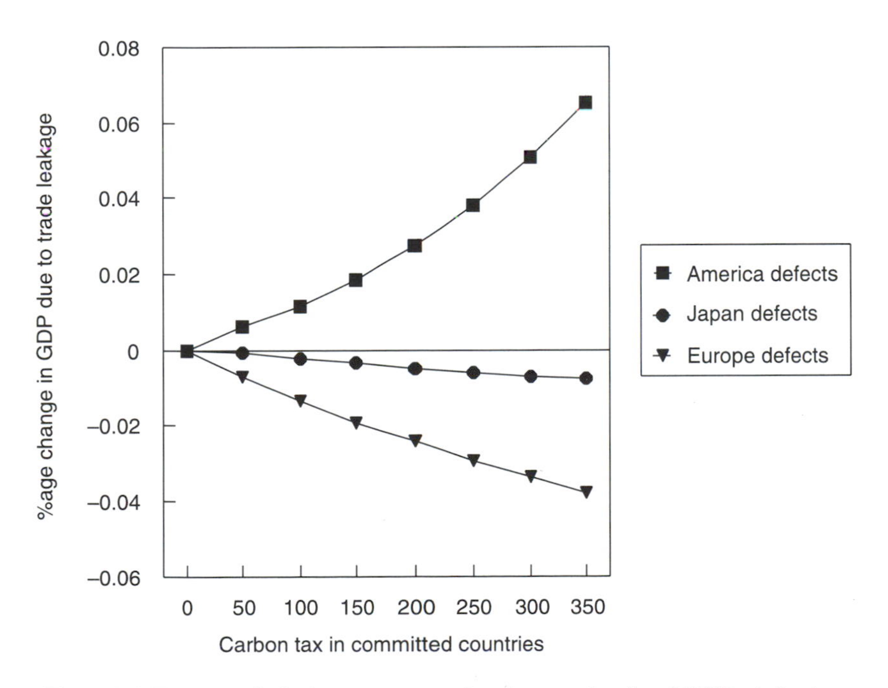

Figure 9.9 Extra trade leakage to committed countries for OECD defection

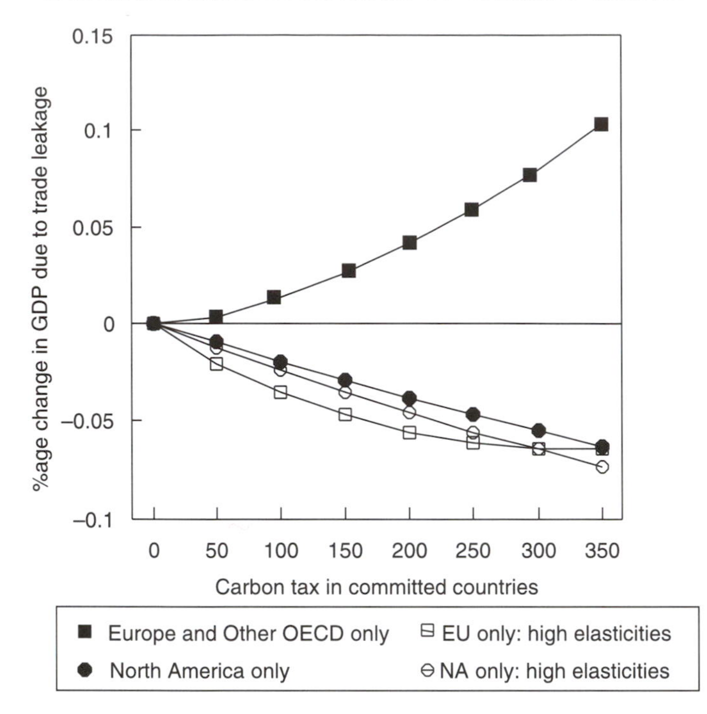

Figure 9.10 Trade leakage with unilateral abatement in North America and Europe

include the effects of industrial relocation, but show the original and doubled trade elasticities.

Comparing Figures 9.9 and 9.10 it can be seen that the additional defection of Japan makes little difference to the trade leakage in Europe and America, but doubling the trade elasticities has a very large effect on leakage from America, making it robustly negative. The most dramatic change, however, is seen when industrial relocation is included in the consequences of OECD members defecting. This is shown, along with the results for the original model, in Figure 9.11.

With industrial relocation allowed for, when North America defects there is a large amount of extra trade leakage from Europe and Japan, up to 0.4 per cent of GDP at the highest tax rates. Table 9.3 summarises these results, expressing the trade leakage in this case as a proportion of the GDP drop experienced in the co-operating countries when they were in an OECD-wide agreement, at the same level of carbon tax. If we assume that the consumer surplus from preventing global warming, that is, how much the total benefits outweigh the total costs, is likely to be in the region of 2–3 times, despite the obvious uncertainties previously discussed in Chapters 7 and 8, then the trade leakage from America's free-riding is of the right order of magnitude to outweigh the environmental benefits of abatement to Europe and Japan. This raises the possibility that an OECD-wide agree-

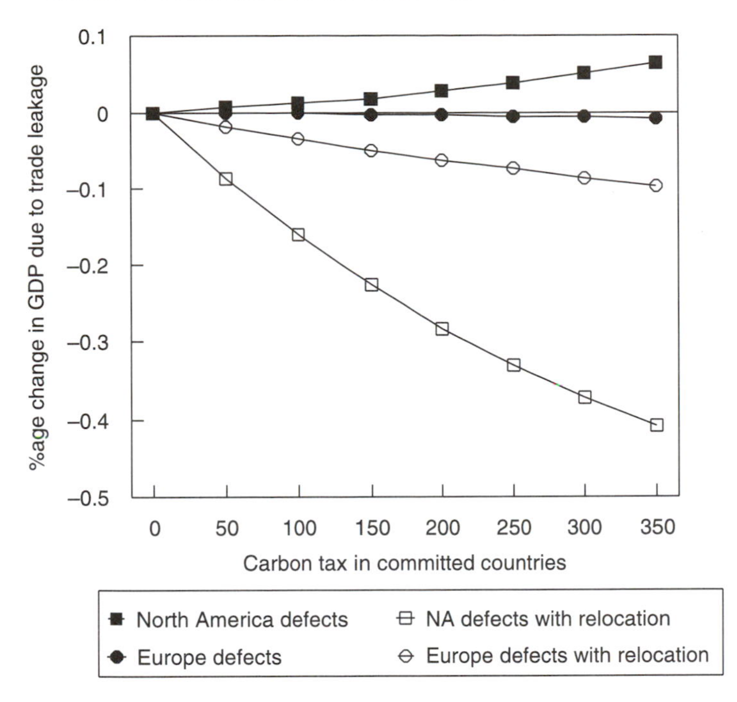

Figure 9.11 Regional trade leakage with, and without, industrial relocation

ment could contain a trade-leakage equilibrium, where a credible threat of agreement breakdown can be levied on the USA if it wishes to leave. However, there is no similar threat that could be directed against either Europe or Japan, because the additional trade leakage caused by their defection is too small.

The ability of high trade leakage to re-enforce an agreement depends on whether the remaining co-operators can renegotiate the agreement at a lower level of abatement, and still gain benefits compared to not co-

Table 9.3 Trade leakage costs with, and without, industrial relocation

	Percentage of GDP drop caused by trade leakage						
Carbon tax	*50*	*100*	*150*	*200*	*250*	*300*	*350*
OECD-wide Tax							
No relocation	−4.25	4.29	8.58	10.81	11.94	12.41	12.48
Industrial relocation	5.52	14.98	19.47	21.32	22.48	22.63	22.36
North America defects							
No relocation	−12.08	−11.22	−11.58	−12.34	−13.23	−14.15	−15.08
Industrial relocation	174.11	157.26	141.22	127.07	114.72	103.88	94.29
Europe defects							
No relocation	5.02	5.70	6.02	6.17	6.21	6.19	6.13
Industrial relocation	13.32	14.90	15.64	15.97	16.06	16.03	15.92

operating. If the marginal cost of trade leakage fell quickly as abatement efforts slackened, indicating a threshold effect where trade leakage is small, then renegotiation could be possible and a trade leakage equilibrium would not hold. In the case of North America defecting, the marginal cost of trade leakage remains virtually constant at ≈ $300 per tonne of carbon abated, for all levels of abatement. If the marginal benefits of co-operation rise as abatement falls (indicating falling marginal abatement costs, and rising marginal benefits from abatement), it is possible that a re-negotiation point does exist where Europe and Japan would remain in an agreement.

Using the damage model developed in Chapter 8, the optimum levels of abatement in the remaining co-operating countries were calculated for the unilateral defection of one OECD region; both with and without industrial relocation. The co-operating countries are assumed to maximise their net benefits from abatement (which includes that produced by any relocation), assuming that emissions in the rest of the world are unchanged (no leakage), but changes in competitiveness increase the cost of abatement. This myopic objective function is the scenario most comparable to the theoretical model in Mabey (1995a). Table 9.4 gives the results of these simulations showing the gross reduction in regional and OECD emissions from the business-as-usual case, and the net change in global emissions *including* leakage; all values summed over the simulation period 1995–2030.

These results show that consideration of the costs of relocation does tend to lower abatement relative to the case where only aggregate trade flows are affected, but that this is a small effect relative to the impact of defection on abatement levels. The small number of participants in the agreement means that defection by anyone greatly lowers the benefits of abatement, and the renegotiated optimum is at a much lower level of co-operative abatement. Leakage via relocation does affect the global levels of abatement but as calculated before this only amounts to 10–15 per cent of total co-operative abatement. The strangest result is probably that including relocation effects *increases* abatement when Japan defects from any agreement. This small effect results from changes in terms of trade between

Table 9.4 Optimal abatement levels with defection: percentage change in CO_2

	OECD-wide		*NA defects*		*JP defects*		*EU defects*	
Region	*Trade*	*Relocate*	*Trade*	*Relocate*	*Trade*	*Relocate*	*Trade*	*Relocate*
OECD CO_2	−21.5	−20.5	−5.1	−3.2	−17.7	−18.6	−18.1	−15.9
World CO_2	−5.3	−4.3	−1.3	−0.5	−4.4	−4.1	−4.0	−2.9
NA CO_2	−26.2	−28.3	0.1	−0.04	−18.9	−21.8	−28.52	−25.7
JP CO_2	−8.7	−11.8	−3.9	−12.0	0.1	0.5	−4.0	−0.1
EU CO_2	−15.8	−10.7	−17.7	−7.3	−21.9	−18.9	0.06	0.25

Europe and North America as abatement levels alter, and thus is a complication not considered in the discussion above.

Using the optimising structure above shows that it is unlikely that the costs of trade leakage will enforce a more complete equilibrium agreement between the regions of the OECD, because the effect is small compared to changes in the co-operative benefit function caused by defection. If North America defects from an agreement it lowers OECD abatement by 75 per cent of its co-operative level. Trade leakage reduces abatement by a further 40 per cent from this very low level; though if carbon emissions leakage is also included it drops by nearer 70 per cent. Trade effects may therefore only be significant to equilibrium abatement if larger numbers of countries/regions are considered, or if some countries are considered to have proportionately small damages, as this will minimise the amount of co-operative benefits lost by defection, but maximise the amount of trade leakage.

Trade leakage and the FCCC

The model of treaty stability analysed above involves countries optimising their utility using well defined cost and benefit functions; as Chapter 8 explained, this is not obviously the case in the climate change negotiations as the damage costs from carbon dioxide are very uncertain. Justifications put forward for considering optimising behaviour in these circumstances are that it shows the long run full information equilibrium, or that in the presence of uncertainty countries will act as if they were maximising expected utility by assigning subjective (and most likely spurious) probabilities to potential outcomes. In the presence of hard uncertainty, however, this description of country behaviour may break down and risk minimisation may be a more appropriate analytic framework. If the developed countries are truly concerned by the risk of climate change, and wish to influence the future behaviour of developing countries because their emissions will eventually cause the most damage, they may try to keep to stabilisation targets even if one party defects from the treaty.

Table 9.5 shows the GDP loss and average cost of abatement for stabilising OECD emissions over the simulation period, both for an

Table 9.5 OECD emissions stabilisation, with industrial relocation

Changes in co-operating countries	*OECD-wide tax*	*North America defects*	*Europe defects*	*Japan defects*
ΔGDP (%age of OECD)	−0.625	−0.641	−0.580	−0.684
$\Delta GDP/\Delta CO_2$ ($/tC)	250.0	700.9	237.1	272.8
ΔCO_2 (MtC total)	25224	10538	24254	25304
Tax levels ($/tC)				
North America	275	NA	438	302
Japan	275	741	438	NA
Europe	275	485	NA	302

OECD-wide treaty and with various countries having defected. The figures for North American defection reflect the fact that with trade leakage there is no feasible set of taxes which allows Europe and Japan alone to stabilise OECD emissions, because emissions in North America rise too much. Therefore, taxes are set to stabilise each region's emissions over the period, rather than emissions in the whole OECD.

From the results the importance of North America again becomes apparent, because when Japan defects average abatement costs for stabilisation rise by only ≈ 10 per cent, while they fall by ≈ 5 per cent when Europe defects. This fall results for two reasons: firstly, because taxes are equal in all countries, the marginal abatement cost in Europe is much higher than in North America or Japan, so Europe's removal from the agreement mirrors the process of sharing abatement efficiently between countries; secondly, Europe's defection diverts trade leakage from the rest of the world and the resulting increase in output boosts the economies of Japan and America via trade in non-energy intense goods, reducing the overall output loss.

Unlike the optimisation case, if countries agree to set a target for CO_2 emissions, because of uncertain information or the need to set an example to developing countries, trade leakage may well provide a credible threat to the defection of North America. This will occur if lower commitments will not persuade the developing world to commit to self-funded abatement measures, and so lower abatement is virtually useless in terms of achieving long run minimisation of the risks of climate damage. The size of North American emissions and their higher rate of growth means that in the long run they must join any treaty which has a chance of success, whatever the domestic politics of restricting the automobile culture!

Summary

Losses in competitiveness, or industrial relocation to countries which do not impose carbon taxes, may strengthen or weaken a multilateral treaty to control emissions. If such 'trade leakage' losses are smaller than the benefits of co-operation they will just make abatement more expensive, and so lead to higher equilibrium levels of emissions. If the costs are larger than net environmental benefits then they have the potential to destroy the whole agreement. However, this potential to destroy the agreement can be used constructively by countries that wish to co-operate because it implies there is a minimum number of signatories to the treaty that will make it worthwhile. This minimum number allows a credible threat of non-agreement, or breakdown of agreement, to be levied at countries that try to free-ride or do not wish to join any agreement.

The results from EGEM suggest that trade leakage to the rest of the world from a unilateral OECD agreement will be small, even if significant amounts of industrial relocation are assumed to occur. Trade leakage is unsurprisingly higher when other OECD countries free-ride, but with no industrial relocation it is partially or completely offset by the non-trivial effects of income reductions on trade. When industrial relocation is considered, only free-riding by North America produces trade leakage large enough to destroy an agreement, but this effect is smaller than the loss in co-operative benefits caused by America's defection. If countries aim to reach a target (as in the FCCC) rather than optimise costs and benefits then trade leakage increases the costs of abatement in Japan and Europe to a point where they cannot feasibly stabilise OECD emissions on their own. Therefore, American participation is vital for any substantive abatement to occur but trade leakage is merely re-enforcing other motivations towards including North America in any agreement.

As the number of self-funding abators in the FCCC grows the trade leakage argument may become more relevant, as industrialising countries will face great pressure to join Annex I before they gain too large a competitive advantage. This may well be linked to membership of other multilateral organisations such as the OECD, thus giving strong negotiating power to the developed world. If these multilateral institutions fail to include economies such as India and China, then the threat of agreement breakdown because of trade leakage may become credible again but this will depend on the circumstances surrounding the debate.

CONCLUSIONS

The leakage rate is important in policy for two reasons: firstly to assess how effective a set of OECD policies would be in reaching *global* CO_2 emission targets; and secondly in terms of the incentives for countries to join a treaty. The most important factor affecting treaty stability is the net loss of industry or competitiveness to abating countries, rather than the shift towards less energy intensive industry. The net loss in aggregate competitiveness may be nearly zero with appropriate tax recycling policies, but energy intensive industries will suffer disproportionately and if they relocate to countries outside the treaty both the costs of carbon taxes and the amount of emissions leakage will increase.

The long run values of critical parameters used to model these effects, such as trade elasticities and oil market supply elasticities, are not well-known and almost certainly vary over time. We have therefore attempted to use the best available estimates from the literature, along with the econometrically estimated energy demand model for the OECD, to assess the magnitude of the various avenues for leakage.

It appears that oil price leakage effects are small, which, despite the obvious uncertainties in modelling world oil markets, appears a fairly robust conclusion. The main proviso over these results is the assumption that gas and coal markets do not become sufficiently internationalised to contribute their own effects. At higher carbon tax rates the oil price effect does increase more than proportionately, indicating its potential importance in the future. A major limitation of the model is that fuel markets outside the

OECD are simplistic, including no interfuel substitution. Chapter 4 showed that coal and gas prices inside the OECD are often partially determined by oil prices, as well as extraction and delivery costs, and so a fall in oil prices could also increase consumption of higher carbon content coal and lower carbon content gas. The immaturity of energy markets in the important coal consuming areas, especially China, India and the former Soviet Union, makes the type of econometric analysis of price reactions we performed in the G9 countries inappropriate here, so the size of these effects can only be 'guesstimated' at the present time.

Industrial relocation by energy intensive industry seems potentially to be a more important source of leakage, especially at medium taxation levels, but it is an area that needs a lot more empirical investigation. At low taxation levels it is likely that industry would not relocate as transaction costs would be too large. With a large tax all of the most energy intensive industries may move away and so proportionately the influence of relocation will decline compared to oil price and income effects. The highest rates of leakage from relocation will therefore occur at medium taxation levels which means incremental movement to higher taxes may encounter strong, but temporary opposition, and a more long term view of the consequences would have to be taken by responsible politicians.

The model derived here attempts to give a probable upper bound, by using high values for trade elasticities, but limiting the effect to the most energy intensive industries. However, it should be noted that if process efficiency of the relocated industry in non-OECD countries remained much lower than that in OECD, leakage would be higher than modelled here. Industrial relocation of energy intense industry over the simulation period may occur even without energy taxes as demand for basic raw materials, such as steel and bulk chemicals, grows in the developing world. Therefore, the extent of potential leakage from relocation will be heavily dependent on structural changes in developed and developing economies, and the tariff policies of the various trading blocks.

Reductions in OECD incomes due to carbon taxes are shown to have very significant effects on leakage and can completely remove it at higher tax levels. OECD income losses may be small ($\approx$ 1 per cent) for stabilisation of emissions, but these have larger than proportional effect on trade and thus emissions outside the OECD. These GDP reductions may not take place if climate change damage in the base case severely impacts production, as well as welfare, but we do not model this eventuality. Another weakness of our modelling in this area is that it does not reflect how changing patterns of world-wide consumption will affect trade, making non-OECD countries less dependent on exports for growth. Changes in the pattern of OECD consumption due to energy taxes will also affect energy use outside the OECD, helping to stimulate technological innovation, and possibly causing a shift away from energy intensive processes. New theories of how countries innovate and imitate at different stages of development suggest that, even with no world-wide agreements to limit emissions, any low energy technologies that are produced will be disseminated world-wide as they become cheaper, and 'standard' products incorporate previously state-of-the-art technology (Grossman and Helpman 1992). In fifty years' time the OECD may be importing renewable energy technologies rather than cars from South-East Asia! To model such shifts numerically is impossible, but the potential for such synergistic effects, which have been repeatedly observed in the past, should be borne in mind when interpreting the likely biases in any quantitative results.

From a policy perspective, it appears more useful to target the energy intensity of a country rather than emissions as such. It is for this reason that this chapter has concentrated on leakage excluding income effects, despite the latter being very important in terms of actual emissions and target achievement. In the long term, to address carbon leakage or global stabilisation effectively, non-OECD countries must face some emission restrictions, preferably aided by technology transfer to ensure their energy intensity is kept low. This may include anything from efficient power generation and consumption technologies to effective biomass and renewable energy plant. Implementing efficient market incentives for carbon abatement, via some form of joint implementation, or 'producer taxes' on energy (particularly oil) that are then partially redistributed to consuming countries, could also play their part in stabilising developing country emissions without restricting economic growth. In the shorter term however it appears that while carbon leakage may reduce effectiveness of carbon taxes, particularly if adopted by one country unilaterally, a tax throughout the OECD would lose a relatively small proportion of its abatement through leakage, at greatest of the order of 10 per cent, which is certainly not sufficient to justify abandoning abatement efforts in the developed countries.

The income effects of reduced OECD competitiveness will also be important, but our estimates show that even with quite pessimistic estimates for industrial relocation these will only amount to 25 per cent of abatement costs at high taxation levels. This is because the effect of reductions in OECD income on world trade balance out some of these effects, though more detailed modelling of exchange rate reactions would also be needed to investigate this fully. Income effects become most significant however when there is only partial agreement inside the OECD. This is because the competitive advantage of countries is smaller and so changes in relative prices have a proportionately larger impact than with non-OECD nations. If North America failed to join a treaty the relocation of energy intensive industries (iron, steel, mining and chemicals) from Europe and Japan would double or triple the marginal cost of abatement, depending on the stabilisation level chosen. This effect is potentially large enough to wipe out all the benefits of abatement, though in simulations using the benefit function defined in Chapter 8 this effect was swamped by the change in the co-operative benefits due to North America's defection.

This effect was more relevant when considering a stabilisation treaty between the OECD countries as it made unilateral OECD stabilisation by Japan and Europe virtually impossible. This implies that there is no incentive for North America to free-ride from an existing OECD-wide agreement as Europe and Japan would probably also leave, removing any advantage from defection. Conversely Europe and Japan could agree to a binding and credible ratification level for an abatement agreement which stipulated America's involvement. However, there is no such sanction on the unilateral defection of either Europe or Japan, because they cause much lower trade leakage and effects on total benefits. An agreement only containing America and Europe, or America and Japan, would still give positive co-ordination benefits, and so these countries could not credibly enforce treaty compliance by threatening the free-riding country that they will stop abatement in response to its actions.

The potential for industrial relocation therefore provides another strong motive for ensuring North America participates in any international abatement treaty, and this is re-enforced by the fact that the macroeconomic cost of abatement is lowest there and there is no chance of stabilising OECD emissions without America's participation. In Chapter 11 we look in detail at how different policy instruments can contribute to forging complete agreement, recognising that any efficient distribution of abatement may also have profound effects on trading patterns inside the OECD.

10

THE DOMESTIC POLITICAL ECONOMY OF CARBON TAXES

INTRODUCTION

International agreements are negotiated between countries which are assumed to be acting in their own self-interest, and in formal modelling (such as in Chapters 8 and 9) governments are assumed to have well defined, if not objectively derived, cost and benefit functions on which to base their decisions. In reality this is not the case; even if all the uncertainty as to the impacts of climate change and the cost of mitigation were eliminated, governments would still have the problem of accurately representing the diverse views of their populations. The views of the population (or their domestic representatives) are important in international negotiations, because they define both the actual range of compromise solutions which are acceptable, and the perceived willingness of a country to participate in co-operative action (Putman 1988). Countries known for having legislatures which often reject proposals negotiated by their executive branch (for example, the United States), can use this fact to strengthen their bargaining position, without resorting to face-to-face confrontation at the conference table.

The literature on social choice (for example, Fishburn 1973) shows that it is very unlikely that any voting system will conform to the conditions necessary to produce a ranking of different actions consistent with an economically-based social welfare function. However, participatory democracy is the commonest public decision making system in use, and so we must alter our economic analysis to take this into account, and not vice versa. Therefore, the perceived costs and benefits of different abatement commitments and policy instruments will have to be seen to benefit the majority of the population, without shifting the costs of compliance on to a significant minority.

The domestic economic impacts of carbon taxes are composed of direct welfare effects, macroeconomic effects and benefits from energy tax recycling. Each of these factors is distributed differently: direct welfare losses from reductions in energy consumption proportionately damage the poor more (Fankhauser 1995), macroeconomic effects are relatively evenly spread (except for employees in energy intense industries which lose competitiveness) and tax recycling options differentiate between the employed and unemployed workers. The timing of any taxation scheme is also important as different types of taxation and recycling can shift the cost of mitigation further into the future thus penalising future generations, but benefiting current voters!

To introduce these problems the first part of this chapter gives a basic outline of a general theory of environmental/energy taxation and its welfare, employment and output impacts. Modelling approaches to these issues are then discussed highlighting the strengths and weaknesses of CGE and econometric models, especially in the modelling of the labour market.

EGEM is then used to quantify the magnitude of the different effects in the G7, concentrating on the impact of tax recycling on employment and income distribution, and the potential reactions of workers to falling real incomes. The benefits of carbon tax recycling, and reciprocally the cost of public funds, will be one of the main influences on the international acceptability of tradable permit schemes, which are currently supported by UNCTAD (1994). The implications of EGEM's results on this issue, and other political debates, are therefore discussed in detail at the end of this chapter and in Chapter 11.

A GENERAL THEORY OF THE ECONOMIC IMPACTS OF ENVIRONMENTAL TAXATION

Over the last few years political interest in using economic instruments, such as Pigouvian taxation, to control environmental problems has increased. One of the driving forces behind this interest is the idea that switching the tax burden from economic 'goods' (Labour and Capital) to environmental 'bads' will not only improve the environment but also increase the efficiency of the productive economy by removing existing tax distortions. There have been several moves by governments to perform such revenue neutral reorientation of the tax system, including the imposition of a carbon tax in Sweden and the recent landfill tax imposed in the UK. The most ambitious proposal to date was the European carbon/energy tax proposal put forward by the European Commission (CEC 1992); the main argument in this proposal was that revenues from the energy tax could be used to reduce employers' labour taxes, which in turn would reduce the high levels of structural unemployment in the European Union (EU).

Given this political interest it is unsurprising that economists have moved to analyse the potential and size of any employment and output effects from tax shifting; however, this research has resulted in conflicting and confusing conclusions. Researchers using empirically derived economic models have often found significant positive employment and output benefits from tax shifting (e.g. Barker 1994, Barker and Gardiner 1994); on the other hand, general equilibrium analysis, both numerical and theoretical, has generally concluded (except for Capros *et al.* 1994) that there is little scope for increases in output or employment (e.g. Bovenberg and Goulder 1994, Carraro and Soubeyran 1994). These diverging results derive from both semantic problems in defining the positive effects of tax shifting, and substantive structural differences between the models used (Bohm 1995). These structural differences basically concern the modelling of the labour market, and whether it is in or out of equilibrium, and demand or supply constrained. Another important strand of disagreement is the modelling of wage setting, and resultant inflationary, monetary and fiscal impacts from tax shifting. This divergence between empirically based and theoretical results causes problems for policy makers who have to try to assess the likely impact of tax reforms. The following sections try to define the essential constituents of the problem, and detail the important differences between theoretical and empirically based models which are driving these disagreements.

Single, double and triple dividends: what are they, and where do they come from?

The semantics of this debate have revolved around the concepts of multiple 'dividends' from imposing revenue neutral environmental taxation. Different researchers use different nomenclature, thus adding to the confusion of diverging results, but here we use the following categories:

Single dividend Imposing environmental taxation increases total welfare by removing a negative externality from the economy.

Double dividend Shifting the tax burden from labour to environmental inputs raises employment by reducing the relative price of labour.

Triple dividend Reductions in taxation distortions will increase direct economic output *above its level in the base-case.*

It is immediately obvious that these three 'dividends' constitute very different economic objective functions. The single dividend involves straightforward aggregate social welfare optimisation; the triple dividend is looking at the usual macroeconomic indicator of gross economic activity in the economy; the employment dividend has no strict economic significance – and is subsumed into consideration of economic output – unless the social welfare function explicitly contains the level of employment as a public good. Given the connections made between unemployment, poor health and crime this would seem to be a very reasonable assumption to make. However, none of the general equilibrium analyses includes employment in its basket of public goods, and so mainly concentrates on whether tax shifting can increase overall economic activity, thus making environmental taxation a completely free lunch.

The analysis below will show that this polar case of the triple dividend is somewhat of a straw man to disprove, unless the environmental externality is significantly depressing production; though an economically credible case can be made for its existence inside a second-best public finance framework. The more credible case, which is more commonly assumed in the policy environment, is that using environmental taxes to replace other distortionary taxes makes complying with environmental targets cheaper than if revenues were returned through lump sum transfers or income tax cuts. This is particularly important in the case of climate change, where the magnitude of environmental damages is uncertain and will occur far into the future. Policy makers are therefore looking for reassurance that a precautionary mitigation strategy will not impose massive costs on the economy in the short to medium term.

Public goods and public finance: the general equilibrium case

In order to discuss the numerical modelling of these dividend effects, we need to outline the principal determinants of the economy, in both production and consumption. The general utility function of the representative consumer can be given by (for clarity public goods are shown in bold):

$$U = u(I, F, E_C, \mathbf{N}, \mathbf{P})$$

$$I = Y - \text{Input Taxation}$$

where I is real wage income, Y is total output, F is leisure or free time, E_C is energy directly consumed, **N** is direct consumption of the basket of environmental goods which will be affected by the proposed taxation and **P** is a vector of all other public goods, including equity and unemployment. The productive side of the economy is given by:

$$Y = f(L, K, E_P, \mathbf{N})$$

where L is labour, K is capital, E_P is energy used in production and **N** is the same basket of public goods as in the utility function, but here they are being used as an input to production. In the case of climate change, examples of these goods could be: forestry productivity, agricultural productivity, parasitic disease rates, flood and storm damage costs, etc. The utility and production functions are linked by monetary income flows Y, total energy use $E_P + E_C$ and the externality functions below:

'Ecology' Function: $\mathbf{N} = e(E_P + E_C),\ d\mathbf{N}/dE < 0.$

'Equity' Function: $\mathbf{P} = p(L, T),\ \partial\mathbf{P}/\partial L > 0,\ \partial\mathbf{P}/\partial T > 0.$

where T is total government taxation revenue. Increasing energy use therefore adversely impacts on the availability of environmental goods/inputs; for simplicity the stock nature of the environment is not considered – tax setting in this case is covered by Wirl (1994). Assuming a second-best world, where there are no lump sum taxes available, government finance for supplying public goods is raised by proportional taxation on private inputs and profits/income. The available taxation instruments are:

Income Tax: $T_I = t_I*I$

Input Taxes: $T_L = t_L*L$, $T_K = t_K*\Delta K$, $T_E = t_E*E_P$

Energy Consumption Tax: $T_C = t_C*E_C$

where ΔK is the yearly investment stream and I contains both wages and investment income. The structure of the model means that income taxes and consumption taxes on goods other than energy are equivalent in the distortion they will place on the economy, so the only explicit consumption tax considered is on the polluting good, energy.

In the initial state, the environmental externality N has not been recognised as a problem and so the optimal steady state solution[1] of setting second best tax rates is to equalise the apparent marginal welfare burden on all available taxes, excluding the environmental externality. The optimal revenue raising tax rates are given by the solution of:

$$\partial U/\partial t_I = \partial U/\partial t_L = \partial U/\partial t_K = \partial U/\partial t_E - \partial U/\partial \mathbf{N}.\partial \mathbf{N}/\partial t_E = \partial U/\partial t_C - \partial U/\partial \mathbf{N}.\partial \mathbf{N}/\partial t_C$$

Obviously this is a sub-optimal position because the externality has not been considered, and so the inclusion of the externality into the optimal solution will unambiguously raise aggregate welfare, giving the first dividend. This must be true because for the same amount of public funds raised[2] the energy tax will be higher, all the other taxes will be correspondingly lower, and as all the taxes have marginal burden curves which are monotonic ($\partial U/\partial t < 0$, $\partial^2 U/\partial t^2 > 0$), aggregate welfare will rise.

Having recycled the revenues from the carbon tax so that the economy has moved to the true optimal taxation position,[3] where the marginal social burden of all taxes is the same, we can analyse the potential for the other two dividends by looking at the changes in factor demand and output between the two scenarios.

The first point to notice is that the optimal tax on the polluting good can be lower than the Pigouvian tax level; that is, lower than the marginal social cost of environmental damage.[4] This is a well-known result (see Bovenberg and De Mooij 1994) and reflects the existence of production externalities involved in reducing energy use. Reducing energy use involves using distorting taxation which lowers the productivity of other inputs, and thus the welfare increases from environmental protection are partially offset by reductions in production. If energy were a pure consumption good, or there were no production externalities unseen by individual decision makers, the tax would be set at the Pigouvian level; unless increasing consumer taxes decreases the labour supply by reducing the after-tax wage.

The output (as opposed to welfare) effects of imposing the tax are ambiguous, because the environment is a positive externality in the production function, and not just a direct consumption good. Therefore, decreasing pollution will increase production, but the reduction in energy use will correspondingly decrease productivity. If the economy was originally at an optimal productive equilibrium (including no labour market inefficiencies), and the environment only plays a small part in production, then output will unambiguously decrease with the introduction of the energy tax; these assumptions implicitly underlie most general equilibrium analyses of resource taxation.

The effect on employment of tax shifting is harder to characterise than the welfare or output implications, because there are interactions on both the demand and supply sides of the economy. The different influences are:

- Labour Demand will alter with any change in output. Output may decrease if the productivity drop from using less energy is greater than the increase in productivity from an improved environment, and vice versa.
- Labour Demand will increase if labour is substituted for energy in production, due to re-optimisation of the factor mix and lowering of labour input taxes (t_L).
- Labour Supply may decrease as increased consumption taxes (t_C) mean a drop in the real after-tax wage, causing a substitution effect into leisure in the utility function; this drop in wages cannot be completely offset by tax recycling through labour or income taxes if overall productivity in the economy has fallen.
- Labour Supply may decrease if energy and leisure are substitutes; that is, decreasing direct energy consumption increases leisure use, but remains unchanged if leisure and energy are independent.
- Labour Supply may increase if the environment and leisure are substitutes; that is, increasing environmental amenities reduces the need for leisure, but remains unchanged if leisure and the environment are independent.

Therefore, in a simple general equilibrium economy, where environmental taxation is introduced in response to a newly realised externality, there will always be a single dividend, and there may be the potential for double or triple dividends. If environmental goods do not appear in the production function there will never be a triple dividend, as output will always decrease. If all goods in the utility function are independent, then the existence of a double dividend depends on the relative strength of reactions in labour supply and demand to factor substitution and tax shifting.

Double and triple dividends in a non-equilibrium economy

The above analysis was predicated on the assumption that the economy is growing along an optimal steady state path, apart from the influence of the environmental externality. In the real world there are two major imperfections in the economy which will radically affect the impact of an environmental tax: existing non-optimal taxation and labour market imperfections.

Non-optimal taxation

In the base-case above, because public goods must be funded the output of the economy is below its maximum productive potential due to existing tax distortions. These distortions are optimal, because more utility is gained by the purchase of public goods than is lost through reductions in money income. However, in the political debate it is often assumed that current taxation is non-optimal (marginal burdens are not equal), and this argument has prompted tax reforms such as the imposition of a carbon tax, and cuts in marginal direct taxation, in Sweden (Bohm 1995). Most economists have argued that such a tax re-optimisation, and implied triple dividend, cannot be said to be a by-product of environmental taxation, because such actions would be optimal even if no externality existed. In this scenario there is no credible reason – barring incompetence and ignorance – why the energy tax could not be imposed, and therefore the current taxation distribution must *by assumption* be optimal. In this type of simple analysis the future welfare improvements from tax shifting would only be forgone if they did not outweigh the immediate costs of implementation (when discounted). The role and choice of discount rates in determining these effects have been covered in part in Chapter 7, and they remain an important determinant on the political feasibility of any changes.

However, even without discounting effects non-optimal taxation could exist if there are public good externalities from energy taxation for which the environment is a good substitute, but income is not. Considering the utility function given above, P is a basket of public goods which includes equity; the distribution of final consumption among the population. Equity is bought at an efficiency price of funding transfers through taxation, though there are also positive production externalities, such as improved health, associated with it. Energy taxation is a regressive tax, as the rich spend a lower proportion of their income on energy than the poor. As such, in the absence of an environmental externality, introducing energy taxation reduces equity, and compensates for the loss of this public good through higher output. Higher output will, at best, be redistributed equally among the population, but will probably further increase inequalities. Therefore, in the absence of changes in the progressive nature[5] of the taxation scheme, a revenue raising energy tax would be sub-optimal because it reduces equity. However, if the environment – being another public good – is a strong substitute for equity in income, which is quite likely, raising energy taxes would become optimal and this rise would be accompanied by increasing output and employment. In this case a triple dividend has been achieved even if the environment does not enter the production function, and is uniquely dependent on the existence of an environmental externality.

Labour market imperfections

Despite the possibility of existing non-optimal taxation, the existence of non-equilibrium in the labour market is more important from a modelling viewpoint, because it is driven by fundamental structural assumptions about economic behaviour. Explanations for the prevalence of high unemployment rates in many countries are diverse and conflicting, and there is no obvious consensus position about the main causes (Layard *et al.* 1991). Some of the more popular theories are:

- Maintenance of high (non-market clearing) wages rates due to union power, and insider–outsider effects.
- High levels of voluntary unemployment due to high benefit levels, or an inappropriate taxation structure which leads to widespread poverty traps.
- High level of long term transitional unemployment, and hysterysis in employment, after macroeconomic/technical shocks; caused by skill shortages, job search costs and immobility of labour (often linked to sunk housing costs).
- Mismatch between the dynamics of investment and capital accumulation and labour supply, due to inappropriate macroeconomic (monetary) policy.

These types of imperfections have ambiguous effects on the existence of double and triple dividends, and are too complex to cover comprehensively in this chapter. The limitations of aggregate macroeconomic modelling mean that the only mechanisms which can easily be investigated are the effects of high real wages, and the dynamics of investment and monetary policy; as we are considering steady state changes the monetary policy issues are less relevant so we will concentrate on wage setting. On the other issues, Bovenberg and Van der Ploeg (1994) consider the effects of search costs with energy tax recycling inside a theoretical model, and Allen and Nixon (1995) construct a model where the 'natural rate' of unemployment depends on capital accumulation and supply side interactions.

If wages are set by a bargaining process between employers and local unions then there is potential for those in employment to bargain the average wage ('going rate') up to a level where the labour market does not clear; this benefits those in employment at the expense of those left outside the labour market. If this is the case, then any decrease in direct labour taxes has the potential to increase both output and employment, because the initial equilibrium is sub-optimal. If it is assumed that increasing employees' after-tax wages, by lowering direct taxation, will not increase the amount they work or coax more people into the labour market, then the only way to reduce unemployment and raise output is to lower labour taxes imposed on employers (t_L). Given the size of energy tax revenues needed for carbon emission stabilisation, this may result in an employment subsidy to employers.[6]

People in employment before the imposition of the energy tax will initially see a fall in their real incomes if taxes are recycled through employers' labour taxes, rather than income taxes; even though in the medium term aggregate income will probably be higher. This fall will be compensated in part by increased utility from lower unemployment, and decreased social transfers, but it is still possible that recycling through t_L will leave original employees worse off. If the money were recycled through direct income taxation original employees would regain the bulk of their lost earnings from the energy tax. However, employment would not increase as much and neither would output, so the long run outcome for original employees relative to employer labour tax recycling is ambiguous. If workers feel they will lose out from labour tax recycling they may try to bargain the average wage up again, to account for higher prices, and this will reduce and may eliminate both the second and third dividends.

The effectiveness of any energy tax will therefore depend on the precise nature of the bargaining procedures in the labour market, of which we know little for certain. Economists differ in their opinions about how different forms of taxation affect wages, and the dynamics of convergence to notional equilibrium positions. Results in this area will be driven by the theoretical assumptions underlying models, and many different solutions will match the observed data.

GENERAL EQUILIBRIUM AND ECONOMETRIC MODELLING OF THE THREE DIVIDENDS

The above analysis shows that inside both an equilibrium and a non-equilibrium framework there are several ways that all three dividends could occur, even if all past government behaviour has been strictly optimal. Whether the dividends actually exist will depend on the relative magnitude of the various effects, and cannot be conjectured a priori.

The existence of a triple dividend could depend on three different influences: environmental externalities in production, existing non-optimal taxation (perhaps from equity concerns over revenue raising energy taxation) and labour market disequilibria. To date no theoretical general equilibrium model has included any of these factors, so models have been structurally unable to produce output increases from tax shifting; this is a result of their modelling assumptions, not an observed feature of the real economy.

The possibility of a double dividend is harder to model, but in the absence of output increases it is dependent on factor substitution effects outweighing any decrease in labour supply brought about by lower after-tax wages. Numerical general equilibrium models tend not to show increases in employment because economic output always drops. They also tend to have weak factor substitution, and strong leisure/work trade-offs in a supply constrained labour market, which gives no scope for increasing employment through labour tax recycling. However, the actual results will depend on the parameters assumed in calibrating the model, and indeed Capros *et al.* (1994) produce output and employment increases in some countries.

In contrast, econometric models always model labour market disequilibria because this is a feature of the measured economy, and thus they have the potential to produce all three dividends. As real data on input taxation is used in an econometric model there is also no need to assume that the initial taxation scheme was optimal. Therefore, even when the supply side of the model is fully consistent (that is, estimated with restrictions to ensure rational, optimising behaviour by firms), shifting the pattern of factor taxation can plausibly give a rise in economic output; this is also true for numerical general equilibrium models if prices for traded factors are taken as exogenous.

In modelling the labour market most econometric models have no explicit household utility function, so there are no work disincentives from lower wages. This model structure is a *result* of the empirical evidence of this effect being very weak in a demand constrained labour market (Layard *et al.* 1991). However, econometric models include, and measure, several effects that are ignored in general equilibrium studies: for example, the effect of unemployment on employee bargaining power, how inflation levels affect long run wage expectations and the dynamics of wage and employment transitions after price shocks.

In summary, theoretical GE models have to date ignored the potential for both double and triple dividends in the structure of their models; numerical GE models on the other hand can give all three dividends if original taxation is sub-optimal. Given the importance of labour market disequilibrium, econometric models such as EGEM provide the best tool with which to measure the potential for a double or triple dividend; however they suffer from the lack of an explicit utility function with which to optimise tax setting behaviour. Both types of modelling have so far failed to address the influence of environmental damage on productivity and how this may also produce a triple dividend. This is especially important in the case of a carbon tax, where secondary benefits from reductions in congestion and local pollution will produce large measurable gains on top of reduced climate change damage. However, most studies are not actually looking for a real triple dividend, but aim instead to measure the cost of reaching a certain environmental target, the primary benefits of which are probably poorly quantified. Thus, not modelling productive environmental effects reduces the studies to producing estimates of the *cost-effectiveness* of different policies, and not the type of cost analysis needed to assess optimal taxation levels.

MODELLING TAX RECYCLING, EMPLOYMENT AND DISTRIBUTION EFFECTS IN EGEM

EGEM was used to investigate the output and employment effects of carbon taxes utilising the endogenous technical change supply side developed in Chapter 5. A range of flat rate taxes, from $50/tC to $350/tC, were imposed in the G7 economies between 1995 and 2030. This reflects a reasonable range for future taxation levels as the model stabilises carbon dioxide emissions over this period with a tax of $275/tC. Assuming welfare

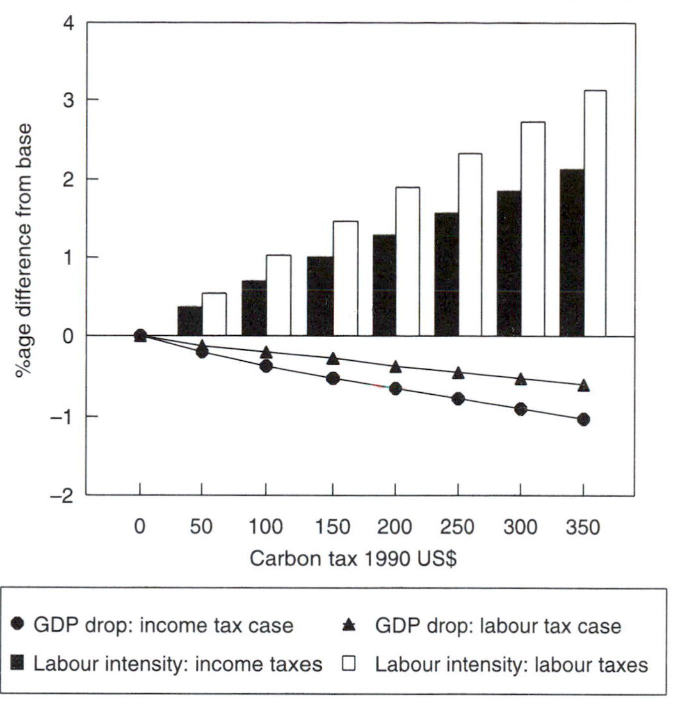

Figure 10.1 Change in output (Y) and labour intensity (L/Y) in the G7

effects are roughly proportional to taxation levels, and there are no large income effects, the changes in regional impacts caused by different forms of taxation recycling can be compared by just considering the macroeconomic costs. This simplifies analysis as it removes the need for assumptions about the utility functions of consumers in different countries, without affecting the accuracy of the results. Of course, welfare effects are significant when the impact of taxation in different regions is compared with a view to equalising, or sharing, the abatement burden and this process is discussed in Chapter 11.

Figure 10.1 shows the changes in economic output (Y) and labour intensity (L/Y) in the G7 when carbon tax revenues are recycled through employers' labour taxes or income taxes; in this and all following figures monetary values are summed over the simulation without discounting, unless otherwise specified.

There is no triple dividend as economic output is unambiguously lower with carbon taxation in both recycling scenarios (despite some short term dynamic gains which are discussed in Chapter 5), but there is a strong double dividend as employment intensity rises faster than output drops. The fall in output derives from the need to invest in energy efficient capital in order to reduce emissions; this investment does not increase the productivity of the economy, but just reduces the amount of energy needed to maintain current productivity levels. In addition the investment in energy

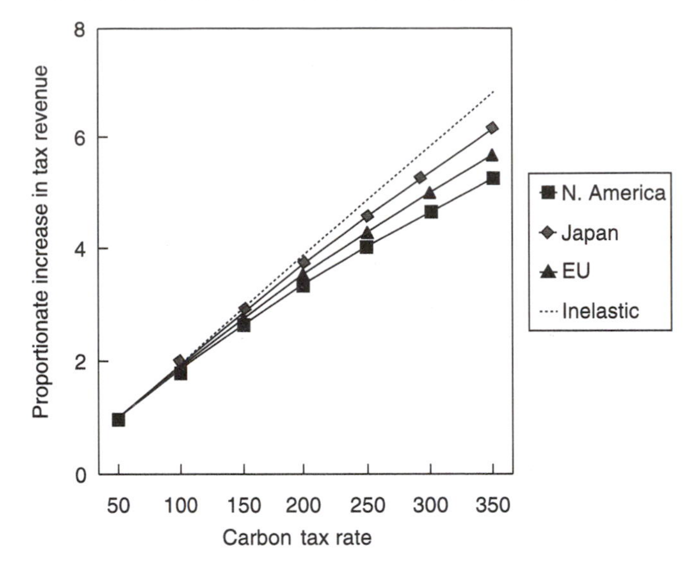

Figure 10.2 Regional increases in carbon tax revenues as tax levels rise

efficiency is assumed to be at the expense of investments to augment labour productivity, which correspondingly falls when the tax is imposed. Without labour tax recycling the rise in labour intensity in the economy comes from both a direct substitution from energy to labour, due to the change in prices, and the fall in labour productivity.

Though energy use does significantly reduce with taxation, Figure 10.2 shows that demand is inelastic enough in the medium term to produce large revenue flows from the tax. Recycling these revenues through employers' labour taxes, instead of reducing income taxes, reduces the drop in output by over 40 per cent.

Companies see labour as a cheaper input to production and so increase employment, which gives the subsequent rise in output. These new jobs will not be as productive as those in the rest of the economy so average labour productivity will drop, even though aggregate output rises. From Figure 10.1 the similarity of the changes in labour intensity in the two recycling scenarios shows that in EGEM employment is more influenced by energy/labour substitution than by this range of reductions in producer wages.

Figure 10.3 shows that the majority of output gains from labour tax recycling come in the first 10–20 years of imposing a tax, and the difference between the two forms of recycling narrows over time. This is because emissions are being stabilised, and so carbon tax revenues are virtually static in real terms; therefore, revenues are falling as a proportion of real wages, which carry on growing along with the economy. The effectiveness of labour tax recycling in creating new jobs and raising output therefore falls over time.

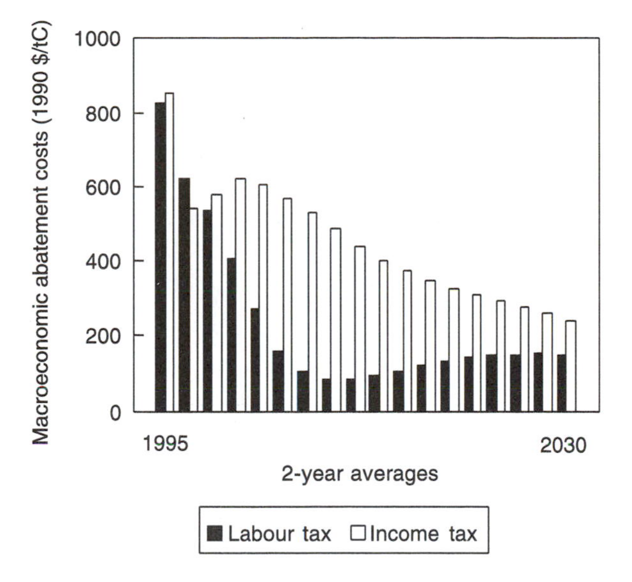

Figure 10.3 Macroeconomic costs of abatement in the G7 with $250/tC tax

Figure 10.4 displays the regional[7] variations in economic output. With income tax recycling the largest percentage drop in GDP is in the European Union, but with labour tax recycling North America becomes the hardest hit region. This is because the high initial rates of unemployment in Europe mean that significant new employment can be created, without stimulating higher inflation from over-tightening the labour market. Output losses are much smaller in Japan than the other two regions due to its low fossil fuel intensity, but the average macroeconomic cost per unit of carbon saved is comparably high in Japan. What is not shown in Figure 10.4 is that, when recycling through income taxes, unemployment in Japan will rise slightly at some carbon tax levels, despite an overall increase in labour intensity.

The large output gains from labour tax recycling mean that this policy will be superior to income tax recycling for both the employed and the unemployed. Table 10.1 shows that the amount of money going to the unemployed[8] and formerly unemployed nearly doubles compared to the no tax case at high carbon tax levels, while the fall in employees' disposable income is partially offset by reduced social transfers.

Despite the similarity in employment intensity for the two recycling methods, the unemployed increase their income by nearly three times as much with labour tax recycling. This is because part of the employment increase is outweighed by a drop in output; the higher output in the labour tax case means that more of the additional jobs created are actually reducing unemployment relative to the base case. These figures are undiscounted but, as Figure 10.5 shows, in both North America and Europe income tax cuts increase disposable income above the labour tax case in the

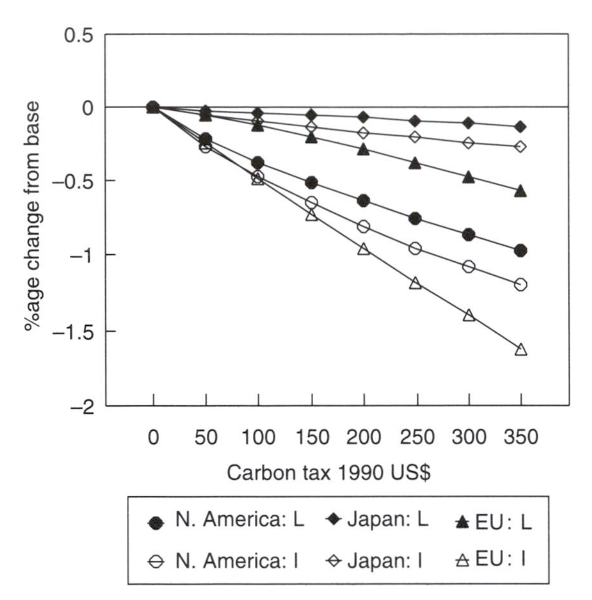

Figure 10.4 Changes in output: labour (L) and income (I) tax recycling

Table 10.1 Real disposable employee wages and income to unemployed in the G7

	Carbon tax $/tC						
%age change from base	*50*	*100*	*150*	*200*	*250*	*300*	*350*
RPDI[a]							
Income tax recycling	−0.66	−1.19	−1.69	−2.19	−2.68	−3.18	−3.69
Labour tax recycling	−0.38	−0.68	−0.96	−1.26	−1.58	−1.91	−2.27
UI[b]							
Income tax recycling	5.14	10.02	14.83	19.61	24.38	29.15	33.93
Labour tax reccycling	16.47	31.34	45.17	58.21	70.60	82.46	93.85

a Real Personal Disposable Income = (Compensation − Taxes)/(Employees*Consumer Prices).

b Unemployed Income = (Real Welfare Payments + RPDI*New Employees)/Baseline Unemployed.

first 6–10 years after they are imposed. This is because the benefits of lower producer wages take time to filter into the labour market and output, while income taxes immediately inflate demand.

In the European Union, if employees discount future consumption at 4 per cent per year they will consider the RDPI loss from income and labour tax recycling to be identical; though this level of discounting does not equalise the overall change in output – including the income of the newly employed. In North America the corresponding figure is rather high at 12 per cent, while in Japan labour tax recycling is superior for the whole

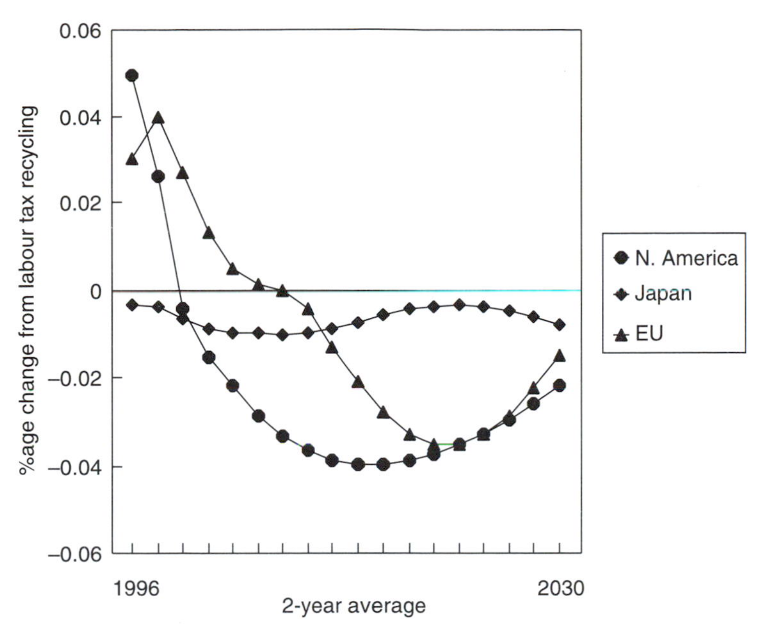

Figure 10.5 Changes in RPDI between income and labour tax recycling ($250/tC)

simulation period. From a political point of view therefore there could be a great temptation to use the revenues of environmental taxation to produce short term economic gains for the majority of the electorate. The acceptability of this type of fiscal shift will partly depend on the distributive effects of carbon taxes, and whether the lower levels of employment from recycling through income taxes damage the least well-off, enough for this to impinge on the political choices of the employed majority.

Distributive effects of carbon taxation

Carbon taxes produce both macroeconomic costs and welfare costs. If, for example, car production reduces in response to higher fuel prices the costs of this structural shift are measured by the decrease in total macroeconomic *productivity*, as modelled in EGEM, not by the change in the industry's production *level*. Analysing the welfare impacts on households is not so straightforward. For very low income groups, especially the elderly, reductions in energy use may severely affect health and general well-being. Not only is this morally wrong, but it will increase costs in other areas such as healthcare and nursing home provision. By definition aggregate macroeconomic models cannot easily model the efficiency consequences of such distributional concerns, but these effects must be taken into account when designing policy.

In most countries carbon/energy taxes disproportionately impact on lower income groups. Table 10.2 shows the relevant bias in the taxation

Table 10.2 Distribution of carbon tax costs by income group

Country	*Energy*[a] *%age of expenditure*	*Carbon tax/expenditure: normalised to lowest value*			
		1st quartile	*2nd quartile*	*3rd quartile*	*4th quartile*
USA[b]	2.99	1.45	1.41	1.25	1.00
Japan[c]	2.65	1.42	1.24	1.12	1.00
Germany[d]	3.80	1.33	1.24	1.14	1.00
France[d]	3.87	1.27	1.17	1.08	1.00
Italy[d]	3.87	1.00	1.07	1.04	1.03
UK[d]	3.87	2.33	1.52	1.25	1.00

a OECD 1994, includes non-fossil energy.
b Proportions from Poterba 1991.
c Proportions from Japanese National Accounts 1994.
d Proportions from Smith 1992.

incidence for energy directly consumed by households. It should be noted that these figures do not measure the distributive effect of energy taxes on inputs into production, and how that will change the price of manufactured goods. Obviously the distribution of changes in product prices will change markedly if taxes are recycled through labour taxes, as labour intensive goods will be proportionately cheaper, but there has been little detailed analysis of these effects (see Symons *et al.* 1994 for a treatment of the UK).

Charges based on carbon and energy content are less regressive than an analysis of gross energy expenditure would suggest, because higher income households use greater proportions of petrol which is expensive, inefficient and polluting. Thus, whereas the ratio of energy expenditure shares between the top and bottom quartiles in Germany is 1.64, the difference in carbon tax burdens is 1.33. These figures must be interpreted with some caution however because, as Poterba (1991) notes, households move in and out of the income quartiles and so the full distributional impact of a carbon tax should be measured over their complete lifetime. Taking this into account lessens the regressive effects of carbon taxation for each individual household. Unfortunately, finding survey data measuring shifts between income groups is difficult, so snapshot figures are used here.

Instead of returning carbon tax revenues through general income or labour taxation, they could be used to make the policy expenditure-neutral for low income groups. Mechanisms for this include recycling through low band income tax reductions, targeted benefits and pension increases. The proportion of carbon taxation revenue raised directly from the bottom quartile of consumers is 12.9 per cent, 20.9 per cent and 18.6 per cent respectively for North America, Japan and Europe. Assuming, due to lack of data, that lower income groups do not purchase more energy intense goods than the well-off, the proportion of total carbon taxation imposed on the lowest quartile in each region is 11.6 per cent, 18.6 per cent and 17.1 per cent. Therefore, a simple policy to reimburse this income group completely would be to divert 20 per cent of tax revenues into direct

Table 10.3 Effects of tax recycling and equity programmes

	Gain in economic output from labour tax recycling per $ of carbon tax revenue raised			*Gain in total unemployed income per $ of carbon tax revenue raised*		
Amount recycled	*100%*	*80%*	*Loss*	*100%*	*80%*	*Loss*
North America	0.07	0.03	0.04	0.18	0.14	0.04
Japan	0.16	0.14	0.02	0.34	0.27	0.07
European Union	0.39	0.30	0.09	0.36	0.29	0.07

transfers to the poorest groups. Table 10.3 shows how such a strategy reduces the gains from labour tax recycling, by giving the difference in economic output and total unemployed income (as defined in Table 10.1, multiplied by the number originally unemployed) between the labour tax recycling and income tax recycling cases, expressed as a return on tax revenue raised. This changes slightly with the size of tax levied and the results here are averages over the entire range of taxes modelled.

Recycling through labour taxes gives a good return for the aggregate economy, especially in Europe where it increases output by 40 per cent of the revenue stream. That these output gains are progressively distributed is shown by the fact that the income of the unemployed increases by a much larger proportion than their very small share of total income would suggest. Of course, this aggregation hides the fact that the increased income only affects a specific group of families who gain new jobs, and the rest are still dependent on benefits.

In Japan and North America the figures seem to show the unemployed receiving so much of the gains from labour tax recycling that employees would either gain nothing or actually lose out. This is misleading however because, as is shown in Table 10.1, employee disposable incomes also rise. The reason for this is that, while per capita before tax wages fall as more employees are hired, due to a drop in the marginal product of labour, this is fully compensated by reductions in direct taxation as unemployment, and so transfer payments, fall. Also the gain in aggregate output from labour tax recycling is smaller than the gain in wages, because the reduced level of unemployment raises inflation and consequently interest rates, leading to a lower rate of investment than in the income tax recycling case.

The diversion of 20 per cent of funds to poverty reduction has a nearly proportional impact on output and uncompensated unemployed income, as the marginal value of a recycled dollar is virtually constant. The efficiency cost of poverty reduction is therefore quite small, 2–9 cents in the dollar depending on the region being considered. The unemployed lose up to 7 cents in the dollar as fewer jobs are created, but as the direct compensation is 20 cents per dollar this will wipe out the effect in the aggregate.

Of course if the creation of new jobs may be seen as inherently good, diversion of revenue to poverty programmes reduces the gains from the tax more dramatically. This could be avoided by recycling all energy tax revenue through labour taxes and raising income taxes to fund the compensation programme. If the distortionary effects from income taxes are lower than the losses reported in Table 10.3 this would reduce the costs of providing equity, but at the expense of increasing the taxes on the employed again. However, this may be politically unfeasible, and again raises the issue of how sub-optimal taxation can exist if the government is acting optimally in raising energy taxes. If, as argued above, the existence of an environmental externality is needed to justify the imposition of energy taxes to the electorate, then employees may wish these revenues to pay all the costs of the tax, including the costs of maintaining equity. Otherwise, income taxes could be raised in order to lower labour taxes without consideration of environmental issues.

Direct transfers will not be the best *environmental* solution to the redistribution problem if they just produce increases in fossil fuel demand in these low income groups. What is needed is a shift to more efficient appliances, which produce improved energy services at a lower level of carbon emissions. This may have to involve direct intervention to provide such devices, which usually involve high upfront investment, because it is well known that low income groups, especially those in rented accommodation, have very high effective discount rates when considering energy efficiency investments. These arise for several reasons including informational failures, lack of tenure security, inability to borrow at low interest rates due to poor credit ratings and the short-termism which comes naturally from living at near subsistence income levels. Though direct programmes of investment and energy monitoring will probably be needed for the non-working poor and elderly, the increases in employment from tax recycling will give the largest boost to low income working households.

The funding for these programmes is problematic if it is seen as an increase in public expenditure by the Exchequer, as it will then be subject to the same demand cycle pressures as other spending programmes. Revenue from carbon taxes is fairly constant through the economic cycle, and so is a prime target for diversion into other areas. This raises the whole problem of hypothecation and how the revenue neutrality and equity of carbon taxes can be guaranteed to the electorate. Of course, there is no economic reason why carbon taxes should be revenue neutral, but if this is not the case it will probably be harder to find domestic support for any binding international abatement commitments.

WAGE EXPECTATIONS, INFLATION AND CARBON TAXES

The above results from EGEM show a rather optimistic picture for carbon taxes. Not only will creative fiscal policy greatly reduce the macroeconomic impact of controlling emissions, but the regressive nature of the tax is greatly lessened by the increase in employment and much of these gains are preserved even with ample direct compensation to the lowest income groups. In all countries both the employed and the unemployed gain by not recycling revenue through income taxes, unless they value the present much more than the future.

However, these results depend on workers accepting the immediate real income drop produced by carbon tax increases, and not trying to restore their real purchasing power through wage negotiations. Of course, if carbon taxes were recycled through income taxes this is unlikely to be a problem as pay packets immediately increase and the drop in workers' incomes, compared to the base case, comes slowly as the economy moves to a less efficient mixture of factors. This is only a small incremental effect which will be lost in the general trend increase in wages, and so is unlikely to affect employee behaviour substantially.

It is a very different story if revenue is recycled through labour taxes because employees see an immediate and sizeable reduction in purchasing power, with no compensating increase in take home pay. If markets are perfectly competitive the price of labour intense goods will drop to reflect the lower producer cost of labour, but instead companies may try to take the cost reductions as larger profits. Perhaps more importantly in the short to medium term workers may expect this to happen, and because a large increase in the price of a few goods is more noticeable than a slight reduction in the price growth of a wide range of goods, the chances of an inflationary spiral are high.

In EGEM the wage setting procedure is modelled by making wages tend towards the marginal productivity of labour in the long run; the shift of factors brought about by energy taxation therefore decreases average wages as well as raising manufacturing prices. The tendency of trend wages to equal labour productivity is counteracted by the fact that decreasing unemployment gives workers more bargaining power, and thus the ability to demand higher wages. It is this bargaining factor that produces a disequilibrium labour market in the model, otherwise it would tend towards the general equilibrium case of full employment, as long as the level of interest rates allowed demand growth to match labour productivity growth. This structure means that wages are unaffected in the long run by the 'wedge' between producer prices and retail prices, which consists of consumption taxes, import prices and input material prices including energy.

To model the potential for inflationary spirals in these economies this model was augmented; it was assumed that employees could negotiate inflationary wage increases to make up for 25 per cent and 50 per cent of the extra money being spent on energy. In most countries the 50 per cent level approximates to the proportion of taxes paid by consumers on directly purchased energy, and so is the most visible portion of the price increase.

Figure 10.6 shows the impact of this type of wage inflation on output in the G7 over the whole range of taxation, by plotting the increase in aggregate output from recycling through labour taxes, relative to income tax recycling. Clearly, there is a reasonably discrete point where all the

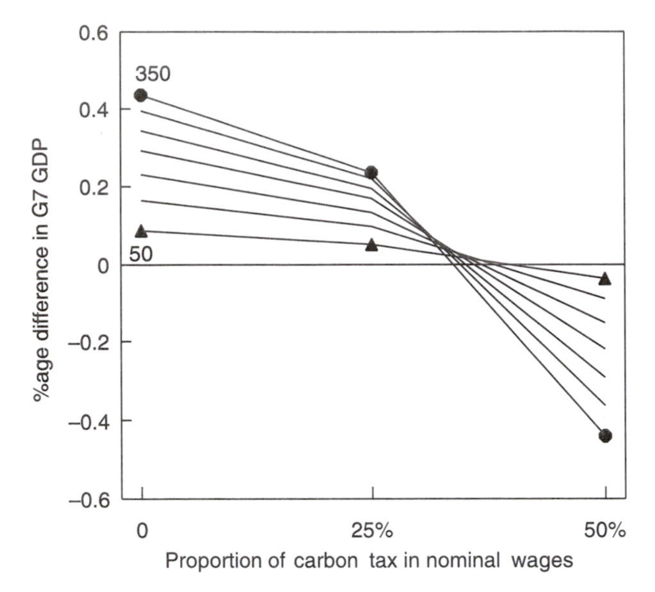

Figure 10.6 Difference in G7 GDP between recycling options with wage inflation

benefits of labour tax recycling are wiped out, and it becomes more efficient to return revenue via income taxes. The output losses are caused by the rise in interest rates needed to control higher inflation, depressing investment and in some countries consumption as well. This slow down in investment affects both short run demand and long run growth, because labour productivity depends on an accumulating capital stock in the supply side of the model. This connection between the demand and supply side of the model makes these monetary and fiscal interactions more important than in other models.

The regional distribution of this effect is given in Figure 10.7, which shows, as in Table 10.3, the average difference in output caused by a unit of revenue recycled through labour taxes instead of income taxes. Japan is the country most vulnerable to inflationary problems, because investment plays such a large part in economic growth. On the other hand, even with a wage claim for 50 per cent of tax revenues labour tax recycling still has high benefits for the European economies, because they gain relatively more from the added employment.

Despite this overall drop in output, labour tax recycling is still more effective at creating jobs than income tax recycling. Table 10.4 shows the average amount of carbon tax revenue per year needed to produce an extra job in each scenario. In any region, even if average wages are bargained up by 50 per cent of revenues the cost of creating a job is less than two-thirds of that in the income tax case.

The potential for inflationary pressures puts a degree of uncertainty on the advantages of labour tax recycling in Japan and North America, though

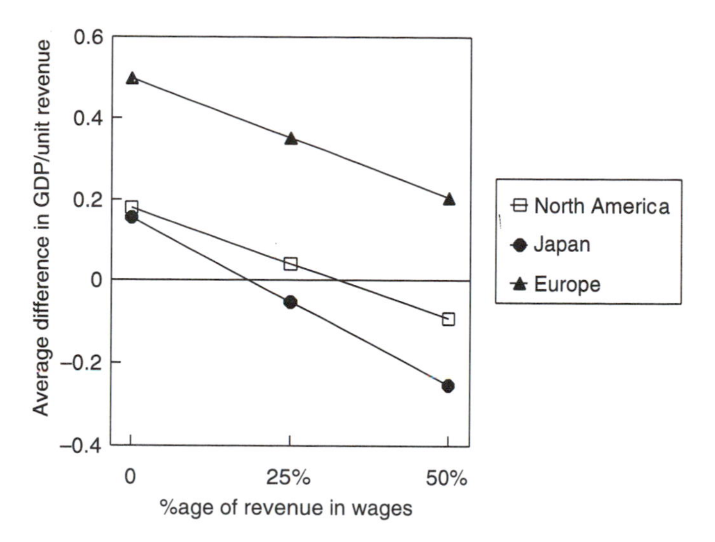

Figure 10.7 GDP difference/revenue between labour and income tax recycling

Table 10.4 Effect of inflationary wage claims on employment change from base

	Carbon tax revenue per job created (1990 US$)			
Recycling mode	*Labour taxes*	*Labour taxes + 25% wage claim*	*Labour taxes + 50% wage claim*	*Income taxes*
North America	20924	26128	34394	48593
Japan	28547	31951	36121	457011
European Union	16978	19361	22273	45215

the positive effects on employment and income distribution, which are robust to inflationary pressures, will mitigate part of this downside risk. In practice it is very difficult for governments to anticipate the reaction of the economy to the large tax increases needed to stabilise emissions; the formation of wage expectations will depend on the point of the economic cycle at which the taxes are imposed, the credibility of the government and the perceptions of employees and unions of price behaviour. However, these effects are clearly of the same order of magnitude as the macroeconomic costs of supply side shifts caused by carbon taxes, and so will weigh heavily on the minds of decision makers when they consider signing international agreements.

SUMMARY OF MODELLING RESULTS

EGEM's simulations indicate that the macroeconomic costs of stabilising G7 carbon emissions at, or even slightly below, their 1990 levels over the next thirty-five years are quite small; even when using international flat rate taxes which are an inefficient policy instrument. Overall, losses from a \$350/tC tax are under 1.1 per cent of output summed over the whole period, with a range of 0.3–1.6 per cent between the different regions; by the end of the simulation period this tax has reduced carbon dioxide emissions by 37 per cent compared to the base-case. If tax revenues are recycled through employers' labour taxes these losses reduce to 0.6 per cent in aggregate, with a range of 0.14–1.0 per cent.

These cost estimates are small compared to some other modelling exercises, and this is due to the structural assumptions underlying EGEM. By modelling innovation as an evolving and endogenous process, increased energy prices stimulate the development of energy efficient technology which can be applied costlessly into the future. This contrasts with the conventional approach where energy saving is assumed to involve continuous direct substitution into labour, technology is unchanged, and the effects on capital accumulation are ambiguous, or are assumed a priori. In EGEM the evolution of technology results in unit abatement costs being highest when taxes are introduced, due to structural inertia and dynamic effects, and then, as the economy adjusts and technological substitution starts to outweigh behavioural changes, unit costs decline to a fairly steady level.

A \$350/tC international tax raises revenues amounting to 2.7 per cent of G7 output over this period; with the percentages in each region being 3.9 per cent, 1.1 per cent and 2.7 per cent for North America, Japan and Europe respectively. Therefore, in both political and economic terms, the use and incidence of the tax revenues will be of the same importance as the effect on macroeconomic productivity which has received most attention in economic studies to date. The value of recycling these revenues through labour taxes is high, both in output and employment terms. In Europe each ecu gives a 0.39 ecu increase in output compared to recycling through income taxation; the returns are less impressive in Japan (0.16) and North America (0.07), but they are still large enough to offset 40 per cent of the macroeconomic costs of abatement in the whole G7. Most of these gains only occur in the medium term however, because with stabilised emissions carbon tax revenues fall relative to wage costs, and so become less effective at creating employment over time.

Unsurprisingly, imposing such a significant carbon tax causes a large shift in the economic structure of these economies: with a \$350/tC tax cumulative carbon emissions drop by 27.5 per cent and average labour intensity per unit GDP increases by 3.6 per cent (regional range 1.3–4.5 per cent) in the G7. This rise in employment nearly doubles the average income of previously unemployed workers, as unemployment is reduced by on average 35 per cent. However, the remaining 65 per cent who do not find new jobs, and other low income groups such as pensioners and the working poor, will suffer a large drop in purchasing power. In every country except Italy, the bottom 25 per cent of earning households spend a higher proportion of their income on fuel and power than richer households, and future growth in personal vehicle use will exacerbate this feature. To combat inequality carbon taxes could be made purchasing power neutral to these households by diverting under 20 per cent of the tax revenue to increased benefits, pensions and lower income taxes. The economy-wide efficiency cost of diverting this money ranges from 10 per cent to 45 per cent of the money transferred, depending on the region and size of tax, and because fewer jobs are created low income workers specifically will lose 20–35 per cent of the transferred amount. The cost of such a compensation programme to low income groups could mean that it should be funded from another source, such as income tax on high earners, but the practicality of this will depend on the politics surrounding the implementation of the tax.

Though in an ideal world recycling carbon taxes through employers' labour taxes seems to have many advantages, there are several political and economic drawbacks. When summed over the whole simulation period labour tax recycling is generally superior, but if consumers prefer consumption now, rather than in the future, the efficiency gains from slowly increasing employment become less compelling. In Europe even a low consumption discount rate of 4 per cent would make income tax recycling more attractive to employees; of course, the unemployed would still prefer higher employment. This problem is less pronounced in North America where a discount rate of 12 per cent would be needed, and does not exist given the dynamics of consumption in Japan.

A larger problem is the possibility that, when faced with no immediate compensation in the form of income tax cuts, employees respond to higher energy taxes by bidding up wages. This results in an inflationary spiral, higher interest rates, lower investment and consequently lower growth. In Japan, recycling through labour taxes leads to lower growth, compared to income tax cuts, if employees claim more than 20 per cent of the tax back in their wage packets, in North America the critical figure is 30 per cent and in Europe it is unlikely to happen before 75 per cent is reclaimed. Since the proportion of carbon taxation directly falling on consumers, rather than on inputs to production, is around 45 per cent, this type of reaction seems reasonably likely in Japan and America, and so these results call into question the benefits of labour tax recycling there. However, it should be pointed out that this is probably the worst case scenario for inflationary effects as the tax is imposed in a discrete increase, and a smoothly rising tax rate would reduce the initial effects. Even with such inflationary spirals cutting labour taxes still significantly increases employment, and so on balance may still be a politically acceptable strategy in Japan and North America.

CONCLUSIONS

Deciding to impose a carbon tax and electing what to do with the revenues are matters of political economy not 'pure' economics, because they involve balancing multiple objectives which cannot be measured in the same units. Employment, environmental quality and economic output will all be affected, and from EGEM's results there seem to be no permanent win–win–win situations where all three are improved. These would exist if the costs of climate change had a major impact on production, rather than welfare, but at the moment the distribution and composition of these costs are too uncertain to be used in a cost/benefit analysis, though the large secondary benefits of CO_2 control are better defined and seem to be of a similar order of magnitude. Therefore, the decisions on whether to tax and by how much will be made by a non-formal assessment of the relative importance of the three factors, subject to standard democratic decision making mechanisms. These mechanisms are not economically perfect, but then such institutions cannot exist outside a 'perfect' dictatorship. They are, however, the only way to deal *legitimately* with questions involving efficiency, distribution and public goods.

All types of taxation on inputs and outputs from production introduce distortions into an economy, even ones that are aiming to internalise an environmental externality. 'Ecological' tax reform is often promoted as a way of producing employment, at no or small net cost to the economy, by using green tax revenues to lower employers' labour taxes. An economist's reply would be that if there were potential for such benefits from lowering labour taxation, this could also be done by raising other more conventional taxes. Such employment benefits can only be uniquely associated with carbon taxes if there is no other politically feasible, or economically efficient, way to raise the revenue.

In the real world this may well be the case given current world-wide aversion to income and capital based taxes, and the regressivity of consumption taxes. The unique feature of energy taxation is that the associated environmental benefits mitigate the distributional and efficiency effects, thus making it, on net, less distortionary than other available instruments. An agreement to stabilise G7 CO_2 emissions at 1990 levels over the next 35 years would give regional income streams of $\approx$1–4 per cent of GDP with which to perform such tax reform. The resulting tax shift would reduce unemployment by 35 per cent and lower output losses by 40 per cent compared to recycling funds through income taxes; though these gains will reduce proportionately as the economy grows.

Arguments against such taxation schemes have been many: especially that it would be costly and ineffective and would hurt the poorest groups in society. Our analysis shows that, even when ignoring the environmental benefits from lower CO_2 emissions and other pollutants, the macroeconomic costs would be small, generous relief for the poorest is affordable, and if past behaviour can be considered a good guide the economy will adjust to increased prices with significantly and permanently lower energy consumption.

On the other hand this analysis has shown that the demand side impacts of taxation and recycling are of a similar size to the supply side effects which have been the sole focus of previous, more long run, studies. A significant reorientation of the tax system could lead to an inflationary spiral as workers bid up wages to take account of higher prices. The consequent higher interest rates needed to maintain macroeconomic stability could wipe out all the output gains from labour tax recycling, even though the employment gains would largely remain.

However, probably the largest problems associated with ecological tax reform, and carbon taxes in particular, are in the political dynamic of implementation. The majority of the electorate will immediately lose from imposition of a tax, and in energy intensive industries there will be large job losses. These industry and labour constituencies will be far more vocal than the unemployed that find work due to lower producer wages, or the present and future population who benefit from cleaner air, less traffic congestion and reduced risk of adverse climatic change. Even if the environmental benefits become clear and garner mass support, the temptation to channel revenues through income tax reductions will be great, because in the short term this will often provide more benefits to the majority of the electorate than reducing labour taxes.

The choice of taxation levels and instruments is therefore far more complex than merely equating marginal costs and benefits, whether domestically or internationally. Institutions aiming to co-ordinate international action to reduce emissions will have to recognise the constraints imposed by domestic politics and economics, and these will form a large component of any substantive negotiations on sharing the abatement burden. While North–South interactions on this issue are currently driven by ideas of equity and opportunity for development, negotiations between the industrialised countries will be based on economic power, relative macroeconomic costs and the impact of taxes on domestic welfare in all its facets. The ability for disagreement is high and, as the next chapter will show, it is not always the most efficient treaty that will guarantee a long lasting agreement.

11

OECD CO-OPERATION UNDER THE FRAMEWORK CONVENTION ON CLIMATE CHANGE

INTRODUCTION

In previous chapters we have investigated the economics of some major questions lying at the heart of the global warming debate, notably: by how much should we reduce GHG emissions, and when should we start this reduction? can the developed countries effectively control their emissions unilaterally, and will this effect their international competitiveness? will electorates agree to reductions in fossil fuel use, and what are the distributional impacts of such a policy?

The first question is prescriptive, in that it asks what should be done, while the second and third are descriptive in that they aim to analyse the consequences of actions, without balancing costs and benefits in any formal way. This chapter is a mixture of the two approaches, and in this way follows closely in the spirit of the Framework Convention on Climate Change (FCCC).

As described in Chapter 2, the FCCC has a mixed bag of objectives; it aims to prevent dangerous anthropogenic climate change, to avoid irreversible – if uncertain – damage, and to do this without damaging economic growth! Chapter 7 laid out in detail the reasons for this multi-objective framework, and our present inability to transfer it into a simple cost/benefit problem using the traditional tools of economic welfare analysis. The overall aims of the FCCC will be determined politically, by a process that balances costs and benefits to different countries, linkages to other issues (e.g. aid and trade policies), and the perceived common threat from climate change. It is unlikely that a global co-operative solution, where countries internalise the notional global damages caused by emissions, will be implemented in the near or medium term. The international community has a bad record of achieving substantive co-operation, unless there is an imminent crisis. Non-cooperative outcomes will be most likely if some major emitters of GHGs assume that they will suffer negligible, or positive, effects from a changing climate. In this type of mixed co-operative/non-cooperative negotiation, the distribution of abatement and damage costs will do much to determine the political stability of international agreements. Side payments may well have to be paid to major polluters in order to encourage them to abate, and this has severe equity implications which could destroy any agreement. If most welfare damage is suffered in developing countries, and they do not have the financial resources to compensate richer polluting countries for carrying out abatement, non-cooperative results will be very likely and too much climatic change will occur.

The interactions of many of these issues were explored in Chapter 8, which showed how important the scope of international co-operation was in achieving substantial abatement commitments. The declining OECD share of global CO_2 emissions means that the current structure of the FCCC can never achieve its stated aims of stabilising GHG concentrations, unless the developing world agrees to accept controls on its own emissions, or the developed world massively subsidises emissions reductions in these countries. The developed country response to the challenge of the FCCC can therefore be seen as a signal to the South that they take climate change seriously, and are prepared to co-operate globally to prevent it. To this end, the OECD must agree to abatement levels which are *greater* than they as a co-operative group would agree unilaterally. Only this extra abatement (that is, the amount above the optimum level if not co-operating with the South) has the potential to stimulate progress in setting future global targets. The illustrative figures in Chapter 8 suggested that, using monetary valuations of climate damage, such commitments may imply the OECD reducing gross CO_2 emissions by 22 per cent from 1995 to 2030, and this is approximately equivalent to stabilisation at 1990 emissions levels. However, if equal financial value is placed on non-market damage in developing countries, the optimum could increase to a 37 per cent gross reduction; that is, nearly a 25 per cent drop from 1990 levels. This tougher target could also be seen as a payment by the developed countries for their past use of the global atmosphere as a pollution sink, because this implies they have fewer rights to pollute now, and so should be subject to higher abatement commitments than a simple balancing of current costs and benefits would suggest.

With some mechanism for transferring abatement commitments between countries – either Joint Implementation (JI) or a global tradable permits scheme – these OECD commitments could be actually carried out in the developing world. However, it is unlikely that the North would agree to commitments which imply *de facto* zero emissions from its economies, unless obligations have been agreed elsewhere. Therefore, we cannot expect technology transfer or investment from the North to solve the climate change problem on their own.

Taking stabilisation at 1990 levels, and 25 per cent below 1990 levels, as OECD abatement regimes which could represent possible pre-conditions for any future global agreement, we aim to look in detail at the potential for OECD co-ordination in the short to medium term. Therefore, this chapter concentrates on analysing the achievement of *cost-effective* and *permanent* international abatement, inside the present framework of the FCCC. This assumes that the most probable medium term target will be to stabilise developed country emissions at 1990 levels over the next thirty-five years, but there is a reasonable likelihood of a tougher target of 25 per cent reductions over the same time period.

Following the main themes introduced in Chapter 2, we will consider how the choice of international co-ordination instrument – emissions targets, taxes or permits – will affect the costs of meeting international commitments, and the stability of these agreements over time. The analysis also incorporates the conclusions of Chapters 9 and 10, by looking at the effect of leakage and industrial relocation on policy effectiveness, and considering how the domestic valuation of employment effects and choice of recycling policy could affect the outcome of international negotiations.

POLICY QUESTIONS

Assuming that a level for OECD emissions stabilisation has been agreed as a medium term objective of the FCCC, what type of co-ordinating policy mechanisms should be put into place to achieve this goal? There are three possible candidates: emissions targets, international emissions taxes and tradable permit schemes. Targets are currently implied, if not enforced, in the FCCC; tradable permits are being seriously considered (UNCTAD 1994); international taxes seem perhaps less likely because of the sovereignty issues involved. The other main policy tool which is already included in the FCCC is Joint Implementation, where developed countries pay to reduce emissions in developing countries (or other developed countries) and this abatement is then credited to them. Though quantitative analysis of JI is beyond the scope of EGEM at the moment, Chapter 6 demonstrates a methodology which could model this process as an influx of investment capital into developing countries, and gives some results for India which are illustrative of the future potential for JI.

In EGEM the only way to reduce CO_2 emissions is to impose carbon taxes inside countries, so the role of the different co-ordination mechanisms we are considering is to determine the distribution of the tax burden among countries, which will in turn decide the efficiency and stability of any agreement. In reality targets and tradable permits also allow governments to use non-price mechanisms, such as regulation and direct stimulation of energy efficient technology, which may producer smaller macroeconomic distortions and trade effects. Though EGEM, like most other macroeconomic models, cannot quantify these advantages, they must be borne in mind in the final policy analysis and may well play a significant part in the choice of policy instrument.

Classic economic analysis of the different instruments would argue that applying equal targets in each country (at either a level of emissions or an equal percentage reduction) is the least efficient way of meeting a common stabilisation level, if emitters have different unit costs of abatement. Emissions taxes and tradable permit schemes will be more efficient co-ordination instruments, as they should allow the marginal cost of abatement to be equalised across all sources of pollution. However, these arguments, while valid inside individual countries, cannot be simply extended to analysis of international agreements, because of the complexities caused by non-continuous macroeconomic and taxation regimes.

Facing an international emissions tax countries will reduce emissions until the marginal cost of abatement equals the tax level, but this is only efficient if technology exists to remove emissions directly, and there are no existing tax differentials between countries. In the case of CO_2 neither of these conditions holds: carbon dioxide cannot be efficiently 'scrubbed' as a combustion by-product (unlike sulphur dioxide), and so emission reductions are produced by lowering fossil energy use. Fossil fuels, especially in transportation, are already subject to differing tax regimes in each country; when a carbon tax is added on top, fuel use will be reduced until the marginal cost of energy conservation equals the total price of fuels. Therefore, the marginal cost of abatement will differ between countries because of the existing fuel tax regimes. Assuming all taxation costs are seen by fuel consumers, efficiency could be ensured if the total tax increment on each fuel was harmonised to a minimum level in each country. However, this will cause problems if the tax revenues are needed to fund international side payments, because the existing energy taxes are already raising government revenue which would have to be replaced from elsewhere at a higher macroeconomic cost. If there are significant macroeconomic costs not seen directly by consumers, attached to fuel taxation – for example, inflation and productivity effects – or energy use markets are imperfect, then even a harmonised tax will not necessarily guarantee the least cost distribution of abatement. Of course, if a central authority had perfect information about each country's economy it could impose the second-best optimal tax rates so as to minimise the total cost of reaching the communal target. Such a scheme is impractical due to lack of information, and the reluctance of countries to devolve tax raising powers to supra-national bodies.

Tradable permit schemes allow countries to trade emissions reductions until the marginal cost of abatement is equal in each. This bypasses the need for a central authority to set optimal tax rates, and allows governments flexibility as to how they achieve abatement inside their territories. These schemes are efficient if the agents involved are companies which face the same marginal costs of raising revenue with which to buy permits (i.e. the relevant current cost of capital). However, if governments finance the buying and selling of permits, and they can only raise revenue through taxation, this imposes a marginal productivity burden on the economy. Countries will only buy permits when the cost of abatement in their country is greater than the permit price *plus* the cost of raising the revenue to buy it. Conversely, at equilibrium countries will offer to sell permits at the macroeconomic cost of abatement *minus* the benefits of receiving a unit of revenue from outside the country. In this market sellers will subsidise their goods because selling reduces the domestically raised tax burden.

The equilibrium position found by permit trading will, therefore, not be at the point where the marginal costs of abatement are equal in all countries, but will depend on the source of revenue used to fund permit purchases. There is therefore an efficiency cost for using tradable permits if all emissions reductions are produced by energy taxes; if other non-tax instruments are used this efficiency cost will be reduced or disappear. In extreme cases, where the costs of raising revenue vary highly with the amount raised, there will be a different equilibrium mix of abatement for each distribution of permits (see Appendix 11.1 for proof). In this case, finding the optimal distribution of a system of tradable permits to ensure the lowest aggregate abatement cost requires as much centralised information as an optimally distributed, second-best international taxation scheme, but has higher transaction costs.

An efficient distribution of abatement between countries, however this is achieved, does not guarantee that all will gain from participating, and side payments may be needed to ensure all countries are satisfied with the agreement. With no limits on transfers (that is, zero internal costs of raising revenue), and small income effects on the marginal utility of payments, there are always enough benefits to redistribute so as to make every country better off from co-operating. However, some policy instruments facilitate this process better than others: national targets, or internationally set taxes which are collected domestically, provide no easy mechanism for calculating or distributing side payments; if taxes are collected internationally such a redistribution is possible, and the same effect can be achieved with the initial distribution of tradable permits. Of course, the marginal costs of raising revenue to fund side payments will complicate the negotiations for an internationally collected tax, in the same way as for the tradable permits equilibrium defined in Appendix 11.1.

Therefore, macroeconomic effects, and different existing government policies, complicate the a priori assessment of what will be the best policy instrument. If raising the money to buy permits has large macroeconomic costs, and the heterogeneity of direct abatement costs between countries is small, then an internationally set tax (at a flat rate or possibly harmonised), which is collected domestically, is probably the best instrument. However, the results of any particular tax level will be hard to predict, and so this method risks missing environmental targets. Alternatively, if equity, flexibility in choosing domestic policy instruments and certainty of achieving FCCC conditions are seen as important, international targets may be the best choice even though they may incur some efficiency losses. If costs are very heterogeneous between countries, and the costs of raising revenue are small, then tradable permits are probably the best solution. Permits also satisfy the equity, operational and environmental advantages of international targets, if at a higher transaction cost. The need for side payments to ensure adequate co-operative rewards for each county further complicates the choice of instrument, and depends on the perceived benefits of abatement in each country – currently an unknown quantity. Instruments that do not easily allow side payments, such as national targets, may prevent some countries from exercising their negotiating power and appropriating much of the benefits from co-operation. However, this may cause those countries to leave the agreement, making everybody worse off, and a more flexible instrument would be more useful.

In the next sections we model the different policy instruments in EGEM, and assess the difference in costs between them, to see if the quantifiable effects of better efficiency seem likely to outweigh some of the non-quantifiable effects; such as the flexibility to use non-price instruments, size of transaction costs and the likely accuracy of hitting environmental targets. Having considered questions of efficiency, we attempt to model the stability of agreements by inferring the possible damage costs countries have from their current behaviour when negotiating the FCCC, and then modelling interactions as full gaming solutions, with and without side payments. In the next section we outline the modelling methodology underlying these simulations, and discuss general results from EGEM, so as to put the optimisations into the context of the overall simulation properties of EGEM.

DEFINING AND MODELLING THE COSTS OF CO_2 ABATEMENT

The costs of controlling emissions using carbon taxes consist of macroeconomic distortions caused by changed input prices, and welfare costs caused by reductions in the direct consumption of energy by consumers. There may also be terms of trade effects and competitiveness externalities, and these are discussed more fully in Chapter 9. The scope of the fully detailed economic models inside EGEM means that only the G7 countries can be investigated completely, and their results are aggregated inside three regional negotiating coalitions: North America (USA and Canada), Japan and the Core European Union (Germany, France, UK, and Italy). These G7 countries account for 88 per cent of total emissions associated with the countries in Annex II of the FCCC which have committed to stabilisation of GHG emissions at 1990 levels by the year 2000 (that is, excluding economies in transition).

EGEM has two different supply side structures which can be used to measure the macroeconomic distortions: a conventional three factor production function, and one that incorporates endogenous technical change in labour and energy productivity. In all the simulations reported here we use the endogenous technical change model whose simulation properties are given in Chapter 5. The main feature of this model is that it assumes that costless, price-induced technical improvements occur in the energy sector in the long run. Macroeconomic costs are further reduced by recycling carbon tax revenues through employers' labour taxes, reducing unemployment and raising output relative to recycling through income taxes.

Welfare costs result from a change in consumers' purchasing behaviour caused by increased prices. As EGEM has no explicit welfare function for consumers these changes have to be approximated, but they will have vital effects on the outcome of abatement sharing between regions. Welfare losses from reducing energy use will be partially compensated by increased utility from other non-energy intensive goods, but there will an overall decrease in utility if we assume rational, optimising behaviour by consumers. The welfare loss can be approximated by calculating the change in consumer surplus from the energy demand equations in EGEM,[1] and then positing that a fixed proportion of this consumer surplus is lost in the shift between consumption bundles.

Welfare costs will be proportional to taxation revenue, which will always be larger than macroeconomic costs in this model. For example, a tax of \$350/tC (1990 real dollars) raises revenue equivalent to 2.7 per cent of G7 output over the period 1995–2030, with a regional range of 1.1–3.9 per cent; while macroeconomic costs amount to under 0.7 per cent of aggregate output, with a range of 0.14–1.0 per cent. Table 11.1 shows the average total costs (macroeconomic + welfare) of abating a tonne of carbon over the simulation period for different sized taxes, and under different assumptions as to the amount of consumer surplus forgone. The first three rows show the regional average macroeconomic abatement costs per unit of saved carbon, without including welfare effects. The next rows then show the total costs in Europe and Japan under different assumptions about the proportion of consumer surplus lost from shifting consumption patterns. These total costs are expressed as a percentage of the abatement costs in North America for a \$350/tC tax. As a unilateral tax in North America of \$570/tC would stabilise G7 emissions at 1990 levels, this presentation indicates the potential for distributing emissions between the regions in order to achieve an efficient agreement.

Table 11.1 Total abatement costs with different assumptions on welfare costs

Regions		*Average macroeconomic costs \$/tC*			
Tax levels \$/tC		*50*	*150*	*250*	*350*
North America		167	170	181	194
Japan		228	220	244	274
European Union		104	155	199	243
Distribution of welfare costs between regions		*%age difference in total abatement cost compared to the cost of a \$350 tax in North America*			
America 0%:	Japan 0%	17.8	13.6	26.1	41.2
	Europe 0%	−46.5	−20.2	2.8	25.6
America 30%:	Japan 30%	93.7	91.3	100.0	111.0
	Europe 30%	−14.8	4.4	21.5	38.4
America 30%:	Japan 10%	10.9	2.4	6.5	13.0
	Europe 20%	−31.0	−14.8	−0.2	14.5

Looking at the macroeconomic costs of abatement it can be seen that over this range of taxes North America is by far the cheapest place to abate CO_2, except for low tax levels in Europe. This is unsurprising because existing energy prices are much lower in North America, and so we would expect the marginal productivity of energy to be lower. At low tax levels recycling revenues through employers' labour taxes produces significant output gains in Europe, offsetting the costs of saving energy, but these gains become proportionately smaller at higher tax levels. The other striking result is how slowly macroeconomic costs fall as tax rates fall, as in a simple analysis we would expect the average cost of abatement to be below the tax level. The reasons for this are discussed in Chapter 5, but they mainly stem from the demand side (inflation, interest rates and trade) parts of the model. These have a proportionately higher impact than supply side substitution, especially through capital accumulation, at lower taxation rates.

If consumers in all countries are assumed to lose 30 per cent of their consumer surplus as they re-optimise their consumption mix, then total Japanese abatement costs rise to over twice that produced by an equal $350/tC tax in North America. This stems from the fact that both Europe and Japan have lower implicit energy elasticities than North America, and so if equal proportions of consumer surplus are lost in each region, they will incur proportionately higher welfare losses. Therefore, the greater the proportion of consumer surplus lost, the higher these regions' costs will become relative to North America's.

This assumption of homogeneity between countries' consumption habits is misleading however, as it supposes that they are all choosing from the same consumption bundles, with the same utility curves, but are just starting from different initial price levels. In reality, the reason why energy taxes are lower in America than in either of the other regions is because it is politically unacceptable to impose them, owing to cultural preferences, a harsher climate and higher dependence on personal transport in North America. This implies that welfare losses from reducing energy use are proportionately higher in America, and so utility curves are very different in each region; shaped by evolutionary developments which give European and Japanese citizens more attractive substitution options away from direct energy use than North Americans. In the last rows of Table 11.1, the proportion of consumer surplus lost in each region is 30 per cent, 10 per cent and 20 per cent for North America, Japan and Europe respectively. In contrast to just comparing macroeconomic abatement costs, this distribution reduces Japanese costs relative to North America, and reduces the differential between American and European costs. Unlike the case when all regions lose 30 per cent of consumer welfare, in this case there is the potential for significant shifting of abatement between America and Europe to produce efficient CO_2 reductions in a co-operative agreement.

As explained above, these interactions will also be influenced by the cost of raising revenues to buy permits, or the benefits of receiving foreign revenues which allow domestic tax distortions to be reduced. In Chapter 10 the average macroeconomic benefit of recycling carbon tax revenues through employers' labour taxes, rather than income taxes, was calculated in EGEM as 0.07, 0.16 and 0.39 of the revenue stream, for North America, Japan and Europe respectively. Therefore, the cost of North American abatement will be ≈ 30 per cent higher than European abatement at a tradable permit equilibrium, leading to efficiency losses relative to a 'perfect' international taxation scheme. The actual decision variables used by governments may, of course, be multi-objective, including equity, employment and special interests (e.g. energy intense industries), as well as aggregate output and welfare. This will further complicate the theoretical elegance of economic policy instruments.

Modelling the economic efficiency of policy instruments

Using numerical optimisation and control algorithms we can simulate the different policy instruments inside EGEM. The carbon taxes consistent with meeting a G7-wide target with an equal international tax or regional stabilisation targets can easily be computed, but those for optimal taxation and tradable permits are harder to define. In order to minimise the welfare costs of abatement, taxation levels must be traded-off between countries without compromising the overall G7 abatement target. However, defining such a constraint directly is hard to do inside EGEM because its numerical optimisation involves balancing two different functions, not a constraint and a function; for example, the simulations in Chapter 8 using the costs and benefits of abatement. In order to define a workable endogenous constraint on G7 emissions, carbon taxes in Europe were redefined as the residual needed to ensure the G7 target was met, once taxes in America and Japan had been set; this was accomplished by defining a feedback rule for European taxes inside the model. Simulations involved minimising the G7 costs of abatement using the tax levels in Japan and America as independent control variables. In tests this gave determinant optima, and allowed reasonable precision (±1 per cent) in hitting abatement targets. This modelling methodology meant that it was easier to model Europe as a whole, rather than as separate countries, implying that abatement would be met by a EU carbon tax which was undifferentiated between countries. Given the likelihood of harmonised taxation inside the EU in the future this is not an insupportable assumption, but is still an approximation to a country-based permit or taxation scheme.

The tradable permits scheme produced another modelling difficulty in defining the amount of money transferred between countries. This involves calculating the number of permits bought and sold, as well as the clearing price of the permit market. In a dynamic model, with important investment and technological innovation effects, defining an instantaneous clearing price for permits is very difficult. The quarterly marginal cost of abatement (or running average) will not account for the potential benefits of forward-looking investment by governments, which are characteristic of the energy demand equations in EGEM. In addition the instantaneous macroeconomic cost of abatement includes economy-wide externalities, such as interest rate and inflation effects, and the benefits of tax recycling. All these processes have significant lags and interlinkages, making the determination of a permit price which is not forward-looking very difficult. EGEM can be solved using forward-looking rational expectations about endogenous variables, but these simulations proved to be computationally unwieldy given the scope and complexity of the model. Therefore, as a working approximation the instantaneous clearing price for permits was taken as the emissions weighted carbon tax level in the G7; this is not the theoretically correct answer, but should give appropriate marginal signals for optimisation. Therefore, in each quarter fiscal flows are calculated between the regions based on their permit allocations and current emissions, priced at this aggregate taxation level.

Carbon taxes, costs and abatement in EGEM

Figure 11.1 shows G7 emissions over the simulation period 1995–2030, for levels of international carbon taxes (i.e. a single tax rate applied across all countries) ranging from \$0 to \$750/tC. Though average annual emissions

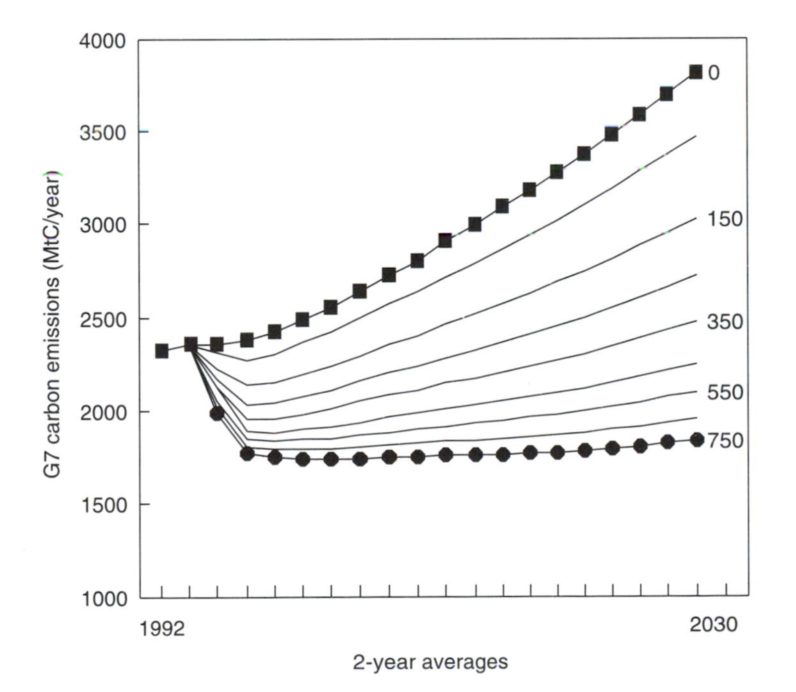

Figure 11.1 G7 carbon emissions for different taxation levels, 1995–2030

stabilise at 1990 levels with a $275/tC tax, it can be seen that demand is still rising at the end of the simulation period. Emissions are only truly stabilised, in that they stop growing, when tax levels reach $750/tC, and if this is applied at the same rate from 1995 stabilisation occurs at about 72 per cent of 1990 emissions. The level at which emissions are stabilised will depend on when the tax is introduced, and how it changes over time, because increased taxes principally reduce emissions by altering the growth rate of energy efficiency, though the level of energy use is affected directly by a small price elasticity (see Chapter 4 for an analytical definition of this equilibrium).

Figure 11.1 shows that energy demand becomes significantly less elastic as taxation levels rise above $350/tC, implying that the total costs of abatement rise quickly after this point as well. The average cost of abatement (macroeconomic + welfare) for the whole G7 over a wide range of international taxes ($50–850/tC) is shown in Figure 11.2, plotted against the percentage of CO_2 saved from the baseline. These results are disaggregated in Figure 11.3 into the three principal regions of the G7, and unit costs plotted against the carbon tax rate for comparability.

Abatement costs rise slowly, albeit from an initially high level, for taxes below $350/tC, but then increase rapidly after this. Figure 11.3 shows that much of this rise comes in Japan, though this is exaggerated by the influence of falling import demand in North America, which produces

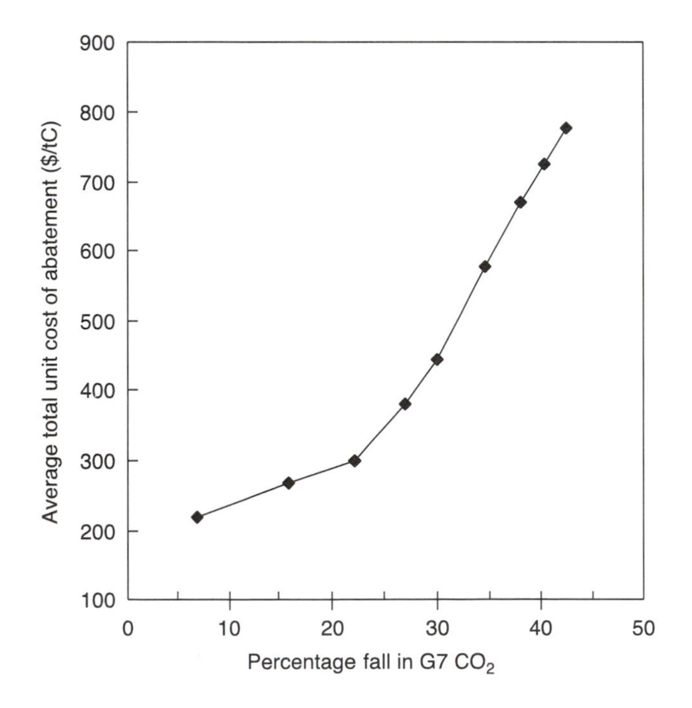

Figure 11.2 Average unit abatement costs in the G7 (equal regional tax rates)

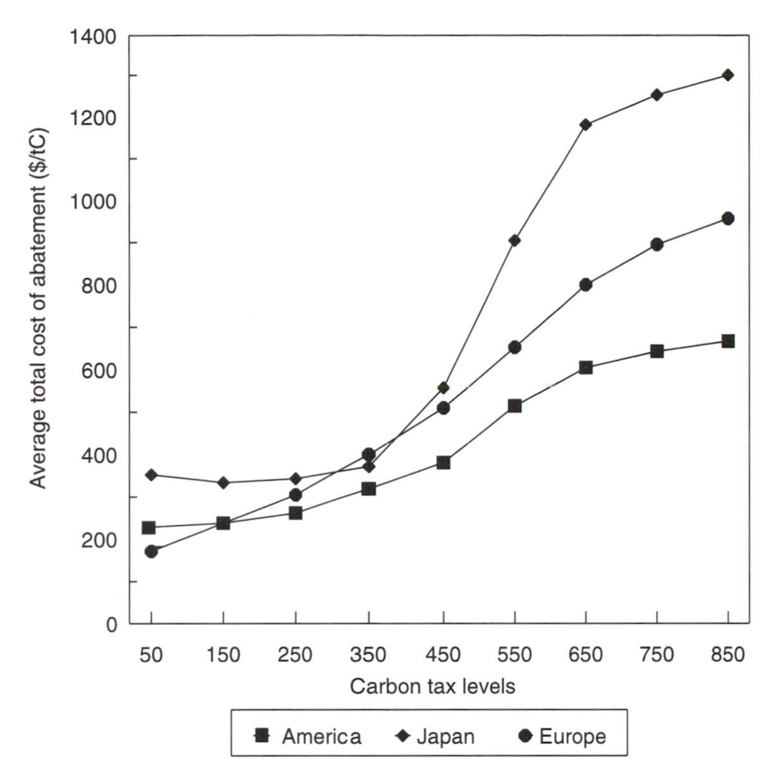

Figure 11.3 Average unit regional abatement costs over a range of carbon taxes

up to 30 per cent of the change in Japanese output. This change is permanent because real exchange rates are fixed inside EGEM; the welfare costs of declining imports due to demand effects are therefore represented by GDP losses in *exporting* countries, as are relative price driven changes in the terms of trade.

The results in Figure 11.3 suggest that the extent of cost saving possible when aiming to stabilise emissions at 1990 levels is highly limited, because unit costs do not vary much between regions, or with differing tax levels. The optimal distribution of abatement will therefore be sensitive to factors such as the allocation of tradable permits, and the specific objective function of each region, but a simple optimisation will probably not include any abatement in Japan. For the stricter target of a 25 per cent cut from 1990 emissions levels (equivalent to a $675/tC tax in all regions) there is a far greater potential for cost savings by using efficient policy instruments, and the resulting optima will probably be more robust to changes in permit distribution and so on.

STABILISATION AT 1990 EMISSION LEVELS: EFFICIENCY, EMPLOYMENT AND CO-ORDINATION

EGEM was used to investigate the efficiency of the various policy instruments in reaching the modest target of stabilisation of G7 emissions at 1990 levels, averaged over the simulation period 1995–2030. This formulation of the target understates the cost of compliance because under-emitting in early periods can be balanced against over-emitting at the end of the period. The macroeconomic costs of control are not discounted as this has already been taken into account inside the supply side, and further discounting tends to cause anomalous end-effects by producing high tax levels in the final periods where their full macroeconomic costs will not be felt.

Table 11.2 compares the effectiveness of the co-ordination instruments which imply no side payments – regional emissions targets (1990 levels), a flat rate international tax and a harmonised tax – with the first best taxation scheme which would be imposed under conditions of full information. The tax rates shown are calculated from the model by dividing carbon tax revenues by carbon emissions, and thus represent the actual price increment on aggregate fuel seen by consumers, which is especially important for the harmonised taxation case. As was expected from studying Figure 11.3, the different instruments do not lead to markedly different results, with only a 14 per cent saving in total abatement costs separating the use of regional stabilisation targets from an optimal regional taxation regime; even though the taxes imposed in each case are very different. The optimal abatement distribution implies zero abatement in Japan, with virtually equal taxes in Europe and North America. From Figure 11.3 we would have expected proportionately higher taxes in America, but the loss in trade from higher relative export prices into Japan marginally raises American costs in this case.

One surprising feature of the results in Table 11.2 is that harmonised energy taxes do reduce welfare losses compared to a flat rate international tax, but they incur higher macroeconomic costs than all the other instru-

Table 11.2 Stabilisation at 1990 emission levels: no transfers

Changes from base-case	*Regional targets*	*International flat tax*	*Harmonised tax*	*Optimal taxes*
ΔWelfare (%)	−0.87	−0.80	−0.77	−0.75
ΔOutput loss (%)	−0.56	−0.49	−0.53	−0.45
ΔEmployment (%)	2.53	1.89	1.62	1.77
Total cost ($/tC)	352.34	317.24	325.11	305.84
Output loss ($/tC)	221.80	194.20	210.92	179.19
Tax rates ($/tC)				
North America	182.4	276.3	316.4	298.9
Japan	741.1	277.5	194.9	0.3
Europe	485.2	275.0	97.8	289.2

Table 11.3 Stabilisation at 1990 emission levels: regional welfare losses

	Change in welfare as percentage of GDP			
Policy instrument	*Regional targets*	*International flat tax*	*Harmonised tax*	*Optimal taxes*
North America	−0.90	−1.35	−1.42	−1.47
Japan	−0.78	−0.32	−0.23	0.06
Europe	−0.86	−0.51	−0.43	−0.57

ments except regional targets. This is a reflection of the supply side distortions found in real economies; this undermines the simple application of harmonised taxes, because their efficiency is based on the assumption of a general equilibrium holding in the economy. By harmonising taxation the effective tax rate in Europe falls markedly because of existing high taxes on oil, and taxation is more effectively targeted on to untaxed coal use. This reduces the funds available for recycling through employers' labour taxes, and thus the gains from removing existing tax distortions. In Europe every unit of tax revenue that is recycled gives a 0.39 unit output gain, so this drop in European tax rates removes a lot of recycling benefits. However, as would be expected, harmonising taxes reduces the welfare cost of the tax by increasing abatement in North America which has a relatively high energy elasticity, and the fall in welfare costs outweighs the rise in supply side costs in the aggregate.

Table 11.3 disaggregates the above results into regional welfare losses, showing that in the most efficient agreements North America loses over 60 per cent more welfare than when it merely has to stabilise its own emissions. Given the scale of this abatement shift, it would seem likely that some form of transfers from Europe and Japan would be needed to compensate this loss, before America would agree to an optimal taxation scheme.

Though regional costs differ markedly, the total cost of G7 stabilisation is quite similar for all these policy instruments. However, the amount of employment created is reduced, with harmonised taxes creating only 64 per cent of the employment increase associated with regional stabilisation targets. Employment shows constant returns to scale in EGEM, so relative differences in output are not large enough to explain this shift. Variations in employment relative to the base case therefore must come from the tax recycling policies, and distribution of substitution effects in the model. Table 11.4 gives the changes in G7 labour intensity – that is, labour used per unit GDP – and unemployment (from the baseline level of 6.7 per cent of the workforce) for the three non-optimal tax regimes; with results given for recycling tax revenues through employers' labour taxes, and through income taxes.

Presenting these changes as labour intensities normalises the influence of

Table 11.4 Changes in G7 labour intensity and unemployment with recycling

	Percentage change in G7 labour intensity		
Tax recycling policy	*Regional targets*	*International flat tax*	*Harmonised tax*
Labour tax recycling	3.43	2.71	2.41
Income tax recycling	2.70	1.90	1.53
	Percentage change in unemployment		
Labour tax recycling	−38.8	−28.3	−24.5
Income tax recycling	−21.1	−11.3	−7.6

output changes caused by the different taxation and recycling schemes, and allows the influence of price substitution and recycling effects to be examined. The results show that harmonised taxes give lower employment increases through substitution of labour for energy (i.e. in the income tax recycling case). However, the difference in labour intensity between the two recycling methods shows that harmonised taxes are most effective at producing employment through lowering employers' labour taxes.

In the end it is the balance of substitution, recycling and output effects that determine the net impact on unemployment of the different co-ordination instruments. Table 11.4 shows that this is much more finely balanced than the changes in labour intensity. G7 unemployment falls by 39 per cent with labour tax recycling, but this drop is cut by 36 per cent if harmonised taxes are used. With recycling through income taxes the differences are even greater, and harmonised taxation reduces the fall in unemployment by 70 per cent, compared to that produced by regional targets. Therefore, if some carbon tax revenues have to be diverted away from recycling through employers' labour taxes, in order to neutralise the regressive nature of the tax on the poorest groups of consumers, then regional targets will create proportionately more jobs than the more efficient instruments. In Chapter 10 we argued that diverting 20 per cent of the tax revenues would adequately perform such compensation, but this proportion may have to be increased if other interest groups, such as small business, successfully argue for specific tax reductions – for example, in the business rate – to offset higher transport and fuel costs. Such political sweeteners are economically unnecessary, as the reduction in employers' labour taxes should be adequate compensation for higher energy prices, but the politics of imposing energy taxes is never that simple!

As was explained in Chapter 10, one of the main arguments for imposing carbon/energy taxes immediately is that they will not only reduce local pollution and the threat of global warming but also stimulate employment by reducing existing tax distortions. The potential for such a 'double dividend' has mobilised wide political coalitions in support of increased energy taxation, or environmental tax reform as it is often known; especially in Europe which has endemic problems of high unemployment.

However, the figures in Tables 11.3 and 11.4 show that a trade-off exists between the efficiency of abatement policies and their ability to reduce unemployment. If a harmonised tax scheme were agreed at the international level in order to give efficient abatement, it could inadvertently destroy domestic political support to carry out abatement commitments, because of the lower impact on employment. This seems to be especially true if a significant part of carbon revenues are diverted away from relieving employment related tax distortions. Therefore, even for this modest abatement target, the decision regarding which policy instrument to use is a non-trivial problem of political economy, given the different domestic objectives – efficiency, job creation and equity – that each government must consider, in order to gain support for tax driven reductions in energy use.

Tradable permits and multi-objective decision making

Given the problem of building and maintaining domestic support for carbon taxes, tradable permits become far more attractive as policy instruments. This is because international negotiations just have to focus on their distribution and the aggregate level of abatement; each government can then decide whether or not to trade, given its own assessments of the merits of gaining abatement efficiency or creating jobs. Therefore, the scope for disagreement at the international level is smaller, and domestic governments are given maximum flexibility with which to ensure political support for international commitments. To investigate these effects we model the results of using tradable permits to reach stabilisation when countries aim to minimise the welfare costs of abatement, and when they also include unemployment reduction inside their objective functions.

Table 11.5 gives the results of using tradable permits to stabilise emissions at 1990 levels with no weight given to unemployment effects; the results for optimum taxation are repeated for comparison. Three permit distributions were modelled: permits allocated based on average G7 carbon

Table 11.5 Stabilisation at 1990 emission levels: tradable permits

Changes from base-case	*Optimal taxes*	*CO_2/GDP TPs*	*Per capita TPs*	*Energy use TPs*
ΔWelfare (%)	−0.75	−0.80	−0.80	−0.79
ΔOutput loss (%)	−0.45	−0.50	−0.49	−0.47
ΔEmployment (%)	1.77	1.83	1.95	1.84
Transfers (% of G7 GDP)	0.00	0.33	0.20	0.07
Total cost ($/tC)	305.84	316.43	316.55	313.68
Output loss ($/tC)	179.19	198.62	194.81	185.44
Tax rates ($/tC)				
North America	298.9	245.2	235.8	297.2
Japan	0.3	133.9	228.7	39.2
Europe	289.2	343.1	399.0	277.9

intensity (CO_2/GDP) multiplied by each country's GDP; allocation of an equal number of permits per capita; permits allocated hinged on baseline energy emissions. Allocating permits based on carbon intensity benefits carbon efficient countries such as Japan and France, per capita allocations benefit carbon efficient and poorer countries, and energy-based allocation subsidises energy inefficient countries such as the USA and Canada. Therefore, the first two permit distributions represent objectively 'equitable' solutions, where countries are not penalised for having saved energy in the past, or are rewarded for past profligacy. Distributing permits based on energy use could only be regarded as a fair solution if North Americans are assumed to value energy use more highly than the other regions, thus implying that this allocation equalises welfare losses from abatement.

All the permit systems incur greater costs than optimal taxation without transfers, as was expected from the analysis in Appendix 11.1; though they are always more efficient than regional targets, if not harmonised international taxes. The permit distribution which involves the least transfers near the optimum tax levels, that founded on baseline energy use, is the most efficient. The other two distributions result in far too much abatement in Japan and Europe. With per capita permits both Japan and Europe gain on aggregate from transfers; while distribution based on carbon intensity results in Japan's gaining at the expense of the other two regions. Japan abates more with a per capita distribution because the transfers it gains from selling permits are smaller, and so the marginal value of recycled revenue is higher (for any given abatement level) than with permits based on carbon intensity.

Table 11.6 gives the regional distribution of welfare losses when using tradable permits. Though net transfers caused by buying and selling permits are a reasonable proportion of overall welfare losses (up to 40 per cent), there is a similar dispersion in results as for the different taxation regimes in Table 11.3, though the taxes in each region are very different. This goes against the simple theoretical analysis which, for example, would set American abatement to the same level as with optimal taxes and no transfers. When permits are allocated by either carbon intensity or population, America would still have to buy permits abroad at this abatement level thus increasing the dispersion of welfare losses between the regions. However, this does not happen here because America gains little from recycling

Table 11.6 1990 stabilisation with tradable permits: regional welfare losses

	Change in welfare as percentage of GDP			
Policy instrument	*Regional targets*	*CO_2/GDP TPs*	*Per capita TPs*	*Energy use TPs*
North America	−0.90	−1.36	−1.32	−1.47
Japan	−0.78	−0.21	−0.32	−0.02
Europe	−0.86	−0.62	−0.56	−0.64

revenue (under 7 per cent of revenue recycled), compared to Japan (16 per cent) and Europe (39 per cent). These regions are therefore prepared to subsidise the purchase of permits they hold, and abate at higher unit costs than North America. It is therefore optimal for America to purchase permits abroad, rather than abating at home until it reaches the optimal taxation level. Thus, while these distributions of tradable permits produce slightly more expensive stabilisation, they do not seem to exacerbate the welfare losses to North America arising from participation in an efficient agreement.

Table 11.7 gives the same results as in Table 11.5, except that here each region sets its tax level based on minimising the welfare costs of abatement plus the benefits of creating new jobs. The value of new jobs in the objective function is set at 1.5 times the unemployment benefits each country pays; benefit payments being calculated as a constant proportion of real wages (see Chapter 10 for replacement ratios).

Including job creation in the regional objective function marginally increases the cost of stabilisation (≈ 5 per cent), but surprisingly *decreases* the total increase in employment in the G7. This is because net job creation becomes skewed towards Japan, and away from Europe, with North America being virtually unchanged; hence taxation levels are higher in Japan and lower in Europe. Fewer jobs are created across the whole G7, but Japan values each job much more highly than the other regions, and so abates more in order to maximise the inflow of permit revenue, and energy/labour substitution. These interactions are summarised in Table 11.8, which shows the percentage change in labour intensity for regional stabilisation targets, and each distribution of permits, when the objective function contains just welfare costs (W), and both welfare and unemployment costs (W + U).

Europe suffers a large drop in additional labour intensity given this valuation scheme for unemployment, while Japan has a complementary increase. Of course, these results will be sensitive to the value placed on

Table 11.7 Stabilisation at 1990 levels: optimisation with unemployment costs

Changes from base-case	*Optimal taxes*	*CO_2/GDP TPs*	*Per capita TPs*	*Energy use TPs*
ΔWelfare (%)	−0.77	−0.84	−0.82	−0.78
ΔOutput loss (%)	−0.45	−0.53	−0.52	−0.47
ΔEmployment (%)	1.75	1.81	1.94	1.82
Transfers (% of G7 GDP)	0.00	0.37	0.21	0.06
Total cost ($/tC)	306.33	332.88	325.93	310.61
Output loss ($/tC)	180.36	212.52	207.73	185.31
Tax rates ($/tC)				
North America	305.9	270.8	242.9	290.9
Japan	0.0	387.0	717.1	48.0
Europe	271.2	245.7	252.7	291.4

Table 11.8 Stabilisation at 1990 levels: changes in labour intensity

	Percentage change in labour intensity (L/GDP)						
Permit system	*Regional*	*CO_2/GDP*		*Per capita*		*Energy use*	
Objective function	*targets*	*W*	*W + U*	*W*	*W + U*	*W*	*W + U*
North America	2.09	2.42	2.64	2.40	2.46	3.32	3.19
Japan	2.53	1.61	2.42	1.29	2.80	−0.15	−0.11
Europe	4.48	3.14	2.24	3.69	2.47	2.87	2.88

extra jobs, and the inflationary effects of reducing unemployment by labour tax recycling investigated in Chapter 10. If Europe valued employment creation proportionately more than the other regions it would impose higher abatement taxes, and thus gain from both additional permit revenue and greater labour energy substitution. This would tend to move the optimum nearer the regional targets position, where the largest increase in employment intensity occurs in Europe.

Summary

Stabilisation at 1990 emission levels over the next thirty-five years is not a very ambitious target, and this is reflected in the low cost of abatement for reaching it (0.87–0.75 per cent of G7 GDP). Unit abatement costs do not differ vastly between regions, or with abatement level, over the range in question, and so the ability to lower costs by using efficient policy instruments is limited. Switching from a system of regional stabilisation targets to an optimally differentiated tax gives only a 14 per cent drop in welfare losses.

However, achieving this small efficiency increase involves a large shifting of abatement responsibilities, mainly away from Japan and on to North America. Welfare losses in America increase by 60 per cent compared to when it merely stabilises it own emissions over the period. Efficient abatement also reduces by over 30 per cent the amount of employment created from carbon tax recycling and labour/energy substitution. Therefore, the side effects of efficiency from both the shift in abatement and lower employment creation could undermine domestic political support for carbon abatement – with North America needing transfer payments to make up for its extra effort, and Europe wishing to impose higher taxes in order to stimulate employment.

Tradable permits allow individual governments more flexibility in balancing the conflicting goals of efficiency and employment, without having to discuss these issues explicitly in an international forum. In contrast, if an optimal taxation regime were to be negotiated under the FCCC all these factors would complicate the division of abatement effort, and the setting of overall abatement targets. However, because buying and selling permits involves differing dead-weight taxation losses to economies, most distributions of tradable permits will not result in the same equilibrium as an optimal tax. Distributing permits based on population or carbon intensity results in a 7 per cent efficiency loss, compared to optimal taxation, which – though small – is half the saving gained by not simply imposing regional stabilisation. Distributing permits grounded on baseline energy use in each country results in an equilibrium very close to the optimum, and very little buying and selling of permits. This shows that with macroeconomic distortions the only sure way to minimise costs is to allocate permits based on optimal abatement levels, if these can be centrally calculated.

If regions use the flexibility of tradable permits to balance increasing employment against abatement efficiency, the results seem to be highly sensitive to the relative valuation of jobs created. Using the *ad hoc* assumption that governments value a new job at 1.5 times the unemployment benefits it saves, overall G7 employment in EGEM actually falls, but much more is created in Japan where benefits are relatively high. The efficiency cost of targeting employment means that for some permit distributions G7 abatement costs are nearly the same as for regional stabilisation, though more unevenly spread among the regions.

The interaction of all these issues suggests that, for such a modest abatement target, the expense of constructing a full tradable permit regime will not be compensated by either a marked increase in abatement efficiency or a greater ability to balance international commitments with domestic agendas on employment and equality. Therefore, in the medium term an extension of regional stabilisation targets beyond 2000 will be almost as efficient (especially if JI is used), easier to agree on and cheaper to co-ordinate than a tradable permits system.

STABILISATION BELOW 1990 EMISSIONS LEVELS: STRUCTURAL SHIFTS IN PRODUCTION

From Figures 11.2 and 11.3 it can be seen that stabilising emissions well below 1990 levels is a qualitatively different proposition than reaching the 1990 target. Forcing a 25 per cent reduction in CO_2 emissions from 1990 levels stimulates so much technological change in the energy sector that fossil fuel demand virtually stops growing. Increasing taxes markedly above this level would make the growth in fossil energy saving technology outstrip the growth in demand, leading to no CO_2 emissions in the long run. Therefore, EGEM has an implicit backstop technology, priced at the tax levels consistent with this size of emissions reduction, and indicative of a marked structural change in the economy where fossil fuels are progressively eliminated from production. This process is not costless, but it is also not limited by the a priori assumption behind most supply side models (without explicit backstops), that fossil energy must always remain a contributor to the productive process. With fixed price elasticities which are lower than one (as is common in the CES production structure used in most climate change models – see Chapter 3), progressive reduction of energy use brings proportionately higher and higher costs to the economy. However, just as the elimination of land from production models was caused by the industrialisation of the last century, so a vigorous response to global warming should eliminate fossils fuels from production in the long run.

Reaching this target requires very high tax rates, and consequently unit abatement costs increase markedly. Large differences also appear between costs in each region, suggesting that using efficient economic instruments will produce far larger gains than before. Table 11.9 gives the results for stabilisation at 25 per cent below 1990 levels, a 40 per cent drop in emissions from the base case, using regional targets, a flat tax, a harmonised tax and optimal taxation without side payments.

Though emissions have only dropped from 23 per cent to 40 per cent of baseline emissions, the unit cost of abatement is approximately double that of achieving the 23 per cent reduction; so total welfare costs are around three to four times as great. The majority of this increase comes not from welfare losses but from changes in output which now make up around 80 per cent of costs, as opposed to around 50 per cent in the earlier case. This is because in the long run the implied energy elasticity approaches (and may surpass) unity, and so welfare costs remain virtually constant per unit of abated carbon. However, a more sophisticated treatment of welfare losses may not show this feature, and so total costs may be understated here.

Imposing stabilisation targets at the regional level is far more expensive than the other instruments, and using an optimally differentiated tax reduces total costs by 24 per cent; this saving is equal to the total welfare cost of hitting the target of 1990 emissions. Perhaps surprisingly, the harmonised, flat and optimal taxes give proportionately more similar results than for the lower abatement target; though, as before, the differences in output costs are greater than those for welfare costs.

Table 11.10 shows that for an efficient agreement regional welfare losses are mainly in North America; though when shifting from regional stabilisation to optimal taxes, they do not increase by as high a proportion as when stabilising at 1990 levels. If regional stabilisation targets were considered an

Table 11.9 Stabilisation at 25 per cent below 1990 emission levels: no transfers

Changes from base-case	*Regional targets*	*International flat tax*	*Harmonised tax*	*Optimal taxes*
ΔWelfare (%)	−3.51	−2.79	−2.74	−2.71
ΔOutput loss (%)	−2.83	−2.14	−2.24	−2.05
ΔEmployment (%)	3.81	3.04	3.23	2.78
Total cost ($/tC)	861.79	686.73	674.66	668.49
Output loss ($/tC)	695.73	527.46	550.17	506.23
Tax rates ($/tC)				
North America	478.1	675.9	808.7	876.3
Japan	1305.6	675.9	689.5	126.1
Europe	1220.0	676.0	602.2	448.1

Table 11.10 Stabilisation at 25 per cent below 1990 emissions: regional welfare losses

	Change in welfare as percentage of GDP			
Policy instrument	*Regional targets*	*International flat tax*	*Harmonised tax*	*Optimal taxes*
North America	−3.25	−4.15	−4.63	−5.23
Japan	−1.84	−0.92	−0.89	−0.15
Europe	−5.66	−2.84	−2.54	−1.89

equitable way to meet the aggregate target, then the larger gains made possible by setting optimal taxes allow more liberal use of side payments to ensure agreement.

Of course, an equitable distribution of abatement costs is one of the motivations that can underlie the allocation of tradable permits between regions, and many would argue that using carbon emissions per unit output is the most fair approach. Table 11.11 shows that though this permit allocation alters the optimal carbon taxes levied in each region, the aggregate output loss is very similar to an optimal tax regime with no side payments. In fact the differences between optimising instruments are proportionately lower than for stabilisation at 1990 emissions levels. This indicates that the influence of dead-weight taxation losses on decisions to buy permits has decreased as unit abatement costs have risen, making tradable permit equilibria more robust to different initial permit distributions. This is a result of both the increased costs of abatement and the decreasing marginal benefits from tax recycling. The latter stems from both the falling marginal productivity of labour and the increased inflationary pressures on wages bought about by decreased levels of unemployment.

The tradable permit simulations in Table 11.7, where the objective functions explicitly include the benefit of creating new jobs, were repeated

Table 11.11 Stabilisation at 25 per cent below 1990 emissions: tradable permits

Changes from base-case	*Optimal taxes*	*CO_2/GDP TPs*	*Per capita TPs*	*Energy use TPs*
ΔWelfare (%)	−2.73	−2.94	−2.81	−2.74
ΔOutput loss (%)	−2.05	−2.29	−2.15	−2.06
ΔEmployment (%)	2.78	2.76	2.78	2.86
Transfers (% of G7 GDP)	0.00	0.60	0.22	0.27
Total cost ($/tC)	673.49	720.48	689.86	678.54
Output loss ($/tC)	506.23	559.65	528.67	509.53
Tax rates ($/tC)				
North America	876.3	764.8	744.5	860.5
Japan	126.1	328.9	345.0	157.2
Europe	448.1	574.3	607.7	463.1

Table 11.12 Welfare costs of creating new employment: both targets

	Yearly welfare cost per new job created (1990 $)				
Policy instrument	*Regional targets*	*International flat tax*	*Harmonised tax*	*Optimal taxes*	*CO_2/GDP TPs*
1990 emissions	8573	9691	10482	9711	9993
25% below 1990	21145	21121	19486	22568	24514
percentage change	146.6	117.9	85.9	132.4	145.3

for this emissions target; however, there was virtually no change in the optimum tax rate for each permit distribution (under 1 per cent change). This is a very different result than for stabilisation at 1990 levels, where including a premium for new employment skewed abatement so as to create many more jobs in Japan. Again the reason for this is that the cost of reaching the abatement target is significantly higher than before, and so achieving cost efficiency completely outweighs employment considerations; given the value placed on new jobs of 1.5 times the existing benefit level. Table 11.12 clearly shows this change by comparing the macroeconomic cost per job created for the two abatement targets.

With welfare losses per job created up to 2.5 times that for the lower emissions target, even with extensively more carbon tax recycling (revenues increase by ≈ 100 per cent for most policy instruments), job creation is quite rightly lower on the list of governments' priorities. The higher welfare costs are in part caused by the marginal productivity of new jobs falling markedly, and so lowering the importance of creating employment – as opposed to saving direct abatement costs – in the objective function.

To illustrate this effect, Figure 11.4 plots the average marginal productivity of *new* employment (change in welfare/change in employment), when different proportions of carbon tax revenues are recycled through employers' labour taxes. All of the simulations are for the optimal tax needed to stabilise emissions at 25 per cent below 1990 levels. The remaining tax revenue is deducted from income taxes, which gives no direct employment effect in EGEM. Therefore, the figure only shows the value of jobs created by lowering the real wage as seen by employers, because energy/labour substitution is the same in each case. Productivity falls by over 30 per cent compared to the first jobs that are created, though the number of jobs created per unit of revenue recycled is virtually constant for all proportions of recycling.

Abatement and trade links

Such high taxation levels are likely to have non-marginal effects on the terms of trade between countries, especially if energy intensive industries relocate to outside the OECD. In Chapter 9 these effects were modelled

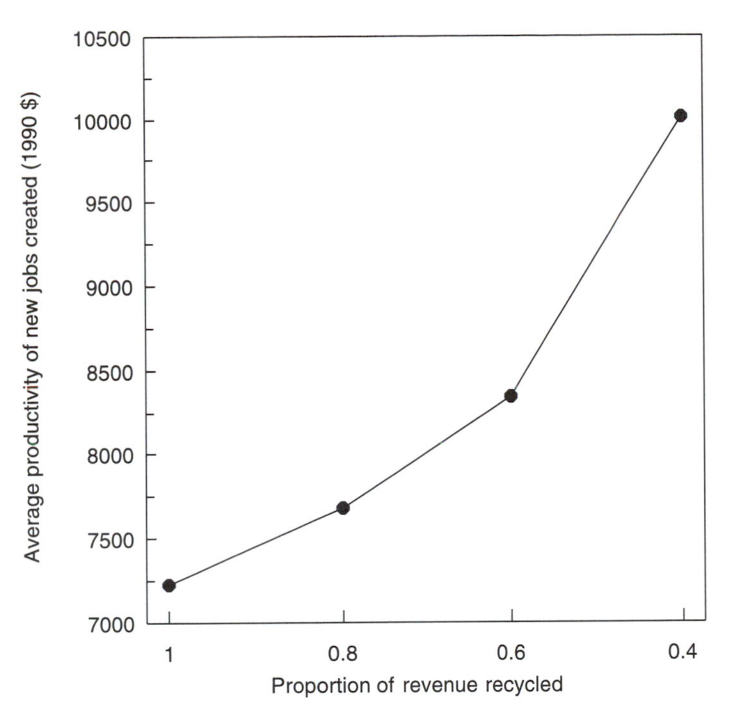

Figure 11.4 Average productivity of new jobs for different recycling proportions

inside EGEM by increasing the price elasticity of trade and by defining an industrial relocation equation which calculates the proportion of energy intensive industry which will leave at any particular tax level. With a $350/tC tax approximately 25 per cent of the GDP loss in the G7 came from these additional trade effects, as opposed to 14 per cent in the standard model as estimated from the data. Real exchange rates are fixed so relative price changes remain as debits to the balance of payments, but do not affect domestic interest rates.

Table 11.13 gives results for regional stabilisation, and the optimal tax instruments, when the potential for industrial relocation is included in the model. Perhaps surprisingly, the cost of abatement using regional stabilisation targets only increases by 6 per cent, and with optimal taxation instruments the change is even lower. The increased relocation from higher taxes seems to have been virtually completely balanced by the output drop in the G7, depressing imports and regaining a balanced trade position even with no exchange rate movements. However, carbon leakage becomes larger at 15–19 per cent of G7 abatement, depending on the policy instrument used, compared to under 2 per cent for the 1990 target. Leakage has increased through not only industrial relocation but also and especially the oil price effect, which makes emissions from the rest of the world proportionately less sensitive to fluctuations in G7 output.

As well as increasing the cost of abatement, this added relocation factor

Table 11.13 Emissions at 25 per cent below 1990 levels: industrial relocation and trade effects

Changes from base-case	*Regional targets*	*Optimal taxes*	*CO_2/GDP TPs*	*Per capita TPs*	*Energy use TPs*
ΔWelfare (%)	−3.72	−2.77	−3.08	−2.91	−2.81
ΔOutput loss (%)	−3.06	−2.09	−2.40	−2.23	−2.12
ΔEmployment (%)	3.15	2.76	2.60	2.66	2.78
Transfers (% of G7 GDP)	0.00	0.00	0.61	0.19	0.30
Total cost ($/tC)	919.11	682.85	748.83	711.32	691.56
Output loss ($/tC)	756.40	514.14	583.75	544.49	522.25
Tax rates ($/tC)					
North America	478.1	930.6	909.3	921.6	931.6
Japan	1305.6	221.8	458.4	471.6	225.5
Europe	1220.0	344.9	322.0	304.6	342.7

forces all the optimal instruments to a similar equilibrium position, where tax rates are the same for all permit distributions and optimal taxes. The main reason for this convergence is that Europe suffers more heavily from relocation leakage, especially to the developing world, and so there is a large incentive for taxes to decrease there. This involves Europe having to buy permits from Japan and America, for all distributions, whereas previously the high macroeconomic cost of not recycling revenues domestically had kept taxes high. In this way the non-marginal effects of industrial relocation at these higher tax levels become a major feature in governments' objective functions. However, all the caveats given in Chapter 9 surrounding the derivation of this part of the model still hold, and more detailed research into trade and welfare effects under floating exchange rates is needed.

Summary

From these results it seems that the political economy of carbon taxes changes as the abatement target becomes more stringent. If major reductions in CO_2 levels are needed, the importance of 'double dividend' effects and of increased employment is reduced as the marginal cost of raising revenue from energy taxes increases rapidly and the marginal value of new jobs falls. Much of this fall in the value of new jobs comes from inflationary pressures as the labour market tightens, and this effect will be exacerbated if employees do not expect aggregate price levels to be virtually unchanged after the carbon tax is imposed. The non-environmental benefits of ecological taxation, or environmental tax reform, are therefore highly dependent on the amount of macroeconomic slack in the economy and on the expectations of workers and unions. The extent of a free-lunch re-optimisation of the tax system, where shifting a higher proportion of the tax burden on to resource use gives compensatory macroeconomic benefits, is limited. It will probably not be a major factor in long run resource use decisions, but may play a limited short to medium term role in stimulating broad agreement over reaching relatively modest abatement targets, while the costs of climate change still remain highly uncertain.

In contrast to the declining importance of employment effects at these abatement levels, trade and carbon leakage issues, which were very small when targeting 1990 emissions levels, exert a far more important influence here. Carbon leakage rises markedly, to 15–19 per cent of G7 abatement, though the *net* output effects of industrial relocation are proportionately not as strong as before. With strong competitiveness effects the use of optimal instruments becomes even more critical, with savings rising to 26 per cent of the cost of imposing stabilisation of emissions in each region separately. As costs are potentially higher so the effect of revenue raising concerns of tradable permit equilibria is smaller, and, unlike the other cases using tradable permits, it produces the same distribution of abatement as an optimally set international tax regime. Therefore, the transaction costs of setting up a full tradable permit regime, which were probably not worthwhile when stabilising emissions at 1990 levels, will be easily recouped with this more stringent target. The optimality of a permit system will not be so undermined by the costs of raising revenues, and, if trade leakage is as bad as predicted in EGEM, the flexibility of trading permits will be very important for governments needing to minimise their costs of compliance.

TREATY STABILITY AND THE USE OF POLICY INSTRUMENTS

The results from EGEM have shown that when a significant cut in emissions is needed, the total cost of abatement is highly dependent on the type of policy instrument used. The use of tradable permits does produce savings over a system of regional targets, but the costs of raising revenue reduces these gains compared to the hypothetical equilibrium found by an optimally differentiated international tax. However, such an international tax is unlikely to be used because of sovereignty issues, and the results for a harmonised tax show the difficulty in calculating the appropriate taxation levels in a second best world with market distortions. The corollary of the sovereignty issue is that using permits allows governments to construct their own objective functions, and so weigh the benefits of increasing employment, and reducing tax regressivity, against abatement cost reductions. This increased flexibility is likely to make domestic political agreement with international policy easier to achieve, as it allows political tradeoffs between different internal constituents.

However, even with the flexibility to use different domestic policy instruments and to optimise the political acceptability of abatement commitments, there is no a priori reason why countries should co-operate together to reduce GHGs emissions. The optimisations carried out in Chapter 8 showed how important co-operation, in the sense of all countries internalising the global cost of their pollution, was to achieving significant abatement commitments (given the existence of a well-defined damage function). However, the current progress of the FCCC does not seem to agree with an abstraction of the negotiation process as moving towards such perfect co-operation. So far the developed countries have agreed to a vague emissions target, and are committed to discussing more concrete aims in the near future, despite the inability of the Conference of Parties to define procedural rules for making non-consensual decisions.

These new aims will involve the defining of an aggregate target for developed country emissions, and the distribution of abatement efforts. The scope of co-operation is most important when deciding abatement targets, as it will determine the valuation placed on damages in non-OECD countries, especially non-market damages in the developing world. Still, given that substantial cost savings can come from moving abatement between countries, agreeing the distribution of abatement will also have an important influence on the aggregate target which is agreed. The results from EGEM suggest that North America seems to be the cheapest place to abate GHGs, so any cost-effective agreement will involve significant changes in behaviour in both the USA and Canada. It is not obvious whether these obligations would be politically acceptable, or how much compensation would be required in order to make them so. These questions must be answered as they will determine the potential for reaching agreement on substantive targets inside the FCCC process, or conversely the stability of any agreement formed under these institutions.

Co-operation and agreement stability with side payments

The stability of coalitions made up of actors who are working together to achieve common benefits, but who have an incentive to cheat unilaterally on any agreement, has been extensively studied in the economics literature; especially in the context of cartel formation (for example, Donsimoni *et al.* 1986). However, many of these treatments concentrate on the theoretically tractable case of homogeneous actors, and empirically based analysis on the formation of lasting coalitions between asymmetric actors are rare. Botteon and Carraro (1995) use a simple model to analyse the stability of coalitions of up to five regions, covering most of the world's population, working together to abate GHGs. They argue that asymmetry between regions, in terms of marginal abatement costs and climate damages, seems to allow larger stable coalitions to be formed than in the symmetric case; where the maximum number of parties is usually three, if co-operation can bring substantive benefits. The equilibrium agreements defined in such multi-player, asymmetric games are often highly sensitive to a priori assumptions about the order of action, information available to each party and the institutional arrangements within which they work. There is often no natural hierarchy of equilibria, which rather diminishes the predictive power of modelling, and makes the transparency of assumptions paramount when explaining results.

Working coalitions in the real world will differ markedly in size, abatement costs and perceived climate damage. Inside the OECD the USA stands out as both the largest economy and by far the biggest emitter of CO_2. The results given in Chapter 9 showed the importance of American participation, because there is no feasible way of even stabilising OECD emissions at 1990 levels if fossil fuel demand does not decrease there. This fact, coupled with the reluctance of the US Administration to raise energy prices so as to reach current FCCC targets, and an unjustified attitude of optimism towards the likely consequences of climate change, seem to form the largest barrier to a binding agreement to limit developed country emissions.

America could be encouraged to participate if side payments were available to compensate it for its abatement effort, as long as these transfers do not outweigh the benefit it brings to any existing agreement on accession; that is, the other regions all see an improvement in their welfare from its accession to the agreement. Such transfers between coalitions can be split into two types (from Botteon and Carraro 1995):

- Profitability payments, which ensure all countries benefit from agreement.
- Stability payments, which ensure no country can gain by defecting from the agreement.

A stable treaty must be profitable by definition, but the reverse does not hold. It is possible that there is no way of distributing the benefits from co-operation which both leaves all parties better-off and dissuades them from free-riding.

The profitability of co-operating for each country depends on its non-cooperative alternatives, and the distribution of the abatement burden inside any agreement. Ensuring the profitability of a treaty usually comes from having a rule concerning how to share the surplus produced by co-operation; for example, a Nash Bargain where each party negotiates so as to produce a Pareto distribution of abatement and transfer payments, given the assumed reaction functions of the other parties. The simplest Nash Bargain case between two equal sized parties, with unlimited transfers, has an equilibrium where the gains from co-operation are split evenly (Rasmusen 1989). However, as was demonstrated above, providing transfers has a significant macroeconomic cost, and so the equilibrium bargaining position will be affected by the form of revenue recycling, and the distribution of any tradable permits between regions.

While Nash Bargains tend to provide an equitable sharing of the benefits of co-operation, they do not necessarily reflect the dynamics of accession to the treaty, and the negotiating power of the different coalitions. These are captured by different concepts, one of the most popular being the Shapely Value, which measures the added value each participant brings to a coalition by their accession. Obviously this value depends on the order of accession, and the overall negotiating power of any particular party is calculated by summing the added value it provides over all different possible permutations of accession to an agreement. Using the Shapely Value to determine side payments means that benefits will be more unequally distributed between the co-operating regions, biasing towards those that contribute most. This is more likely to give a stable agreement than distributing transfer payments based around a Nash Bargain, but there is no theoretical reason why this distribution will be stable. In fact there may be no way of producing a stable agreement, using side payments, which produces a Pareto improvement to all parties compared to a less extensive agreement. If this is the case then it is optimal for some countries to free-ride, and for the remaining parties to co-operate just among themselves.

Stability in the absence of side payments

Even if side payments are not available, due to funding constraints, or are not sufficient to deter free-riding, there are other mechanisms which can stabilise coalitions of countries. These are determined by structural aspects of the game, especially the amount of emissions caused by each country, and so are particularly important when assessing the ability of developing countries to participate in effective co-operation, when side payments are not available.

Countries have an incentive to free-ride, which does not necessarily disappear if co-operation is profitable, because by defecting they save their abatement costs but still benefit from the abatement of other countries. Incentives to free-ride will be greatest if a country has low damage costs or is small, so when it leaves the treaty the remaining countries will abate almost as much as before. This does not hold in the case of low damage cost countries if they are contributing a non-trivial proportion of abatement (that is, they are very large and/or have very low abatement costs relative to other regions), because the subsequent rise in abatement costs may radically reduce abatement in the remaining co-operating countries. As was mentioned before most researchers who have investigated this type of interaction suggest that the maximum size of coalition it can support is three, though this is a numerical and not a theoretical result (Barrett 1993, 1994a). Therefore, without side payments there could be a stable agreement between the G7 regions, as long as each regional coalition is stable – which is likely given their free-trade links – and North America does not have a very low valuation of climate damages.

The basic intuition behind stability with no side payments is that defection by one region causes the whole co-operative structure to break down, or at least reduces co-operative abatement to a non-substantive level. If this holds, there exists a credible threat to potential defectors that if they leave the treaty they will be no better-off than in the non-cooperative case, and so rationally they will not defect. The credibility of the threat lies in the remaining countries not being able to negotiate a feasible and substantive equilibrium after the defection, which would leave them better-off than the non-cooperative case. Chapter 9 investigated whether this equilibrium could be enforced by competitiveness externalities between the OECD regions (Mabey 1995a), and a similar effect can occur if abatement benefits are highly non-linear.

If the benefits of abatement are monotonic, then perfect co-operation implies that if a large country, such as the USA, decides not to co-operate because its damages are low – and feasible side payments unsatisfactory – then the remaining countries should continue abating a similar amount of GHGs. Total abatement would fall dramatically, of course, but abatement in each co-operating country will only change by the amount that the *marginal* benefits of abatement rise, or fall, because of higher global emissions, and the exclusion of American damages from the collective damage function. By definition American damages in this case are only a small proportion of total damages, and the biggest change in abatement results from alterations in marginal damage costs due to changes in total GHG concentrations. Increased atmospheric concentrations may raise or lower marginal damage costs, so abatement in co-operating countries may rise or fall due to defection. This is a logical result because countries can still avoid some damage by abating, even though total damage costs will be markedly greater.

The history of climate negotiations seems to argue against this type of interaction. The only unilateral attempt to implement substantive abatement – the proposed carbon/energy tax in Europe – failed because other OECD countries refused to participate. As was argued in Chapter 9, this could have been because of fears of reduced competitiveness, or carbon leakage, from unilateral abatement, or it could be that Europe did not think unilateral abatement to be of any practical use. This approach is also observable when developing country abatement is considered. There seems to be a strong belief that if the developing world does nothing to prevent climate change, then the developed world may as well not bother to abate since even a 50 per cent cut in OECD emissions will only reduce cumulative global pollution by 12 per cent from 1995 to 2030.

In the standard co-operative model, with convex functions, no abatement is only a credible and logical response to an incomplete agreement if marginal damages fall at high GHG concentrations. That is, there is a limit to the amount of damage rising temperatures can cause, and this will be easily reached if emissions continue as usual. If this holds, then defection by a large emitter will increase GHG concentrations, and so reduce marginal damages so much that remaining co-operators will not bother to abate. However, this view of future damages is not expressed by any scientific authority, and ignores the potential for catastrophic climatic effects. In fact many damage models assume damage grows faster than temperatures, so even though the radiative forcing caused by each unit of

GHG emissions decreases as concentrations rise, the marginal damage caused by emissions can rise with GHG concentrations.

If marginal damages rise with concentrations, zero abatement after a defection is only a rational strategy if the benefits of abatement are considered to reflect subjective valuations of the risk of climate change, and not expected damages as calculated in Chapter 8. If the probability of damage is not considered to be random, but to be non-linearly related to past emissions levels, then a small amount of abatement will not affect the probability of high damage. In this structure countries highly value risk mitigation, which involves substantive total abatement, but if abatement is too low really to affect atmospheric concentrations then it is not considered to be providing useful climate insurance. Another interpretation is that if the developed countries do not agree on co-operative abatement, then neither will the developing world and global emissions will be in effect uncontrolled. The profile of such a risk averse benefit function is shown in Figure 11.5, where defection causes total abatement to drop from A1 to A2, and then remaining co-operating countries react to the lower marginal benefit of abatement by re-optimising their abatement levels to zero.

Compared to a monotonic damage function, this type of risk averse behaviour raises the importance of co-operation, by making the price of non-cooperation higher. In the long run this effect should support agreement, because there is a minimum level of defection which will destroy any agreement, and after which no country can do better than by co-operating. However, in the short run it benefits low damage/high emitting countries

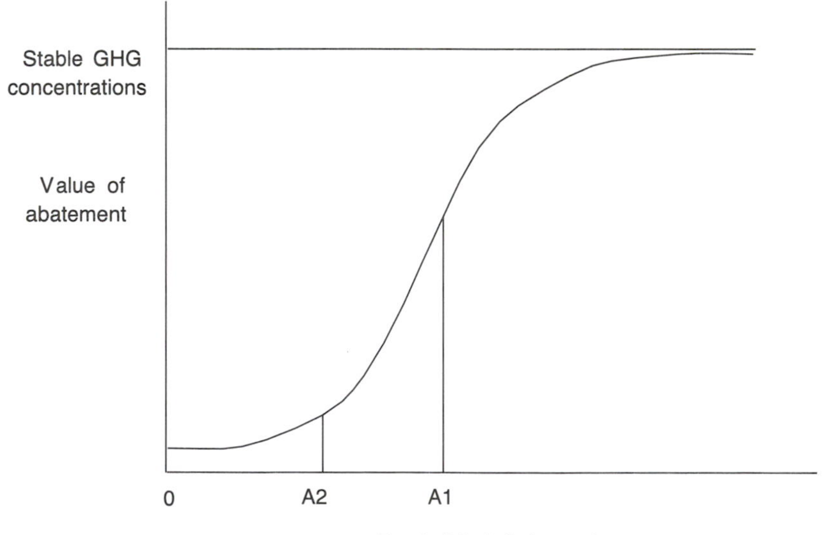

Figure 11.5 Schematic of a risk averse abatement benefit function

not to join an agreement, and so use their negotiating power to appropriate higher side payments.

To summarise, if countries face different damages from climate change and have different abatement costs, then even if they co-operate side payments will be needed to make the agreement profitable for each party. Side payments may also be needed to sustain a stable agreement, but this will not be possible in all cases. However, if agreement is to be reached between a small number of coalitions, it is possible that a treaty can be self-enforcing because defection by one party radically reduces abatement in the remaining co-operating regions. This equilibrium is especially likely if there are strong negative competitiveness effects from abatement, or if the marginal effectiveness of abatement at supplying greenhouse insurance falls rapidly at low levels.

The problem about modelling these types of interactions inside EGEM is that one side of the equation – the damage costs seen by each country – is unknown. Therefore, we need to hypothecate the likely distribution of perceived climate change damages between countries, and so incentives for agreement or dispute. In the following simulations we do this by assuming that the current agreements under the FCCC represent some form of cost/benefit optimisation by the parties, and then back-calculate the damage costs needed to justify emissions stabilisation in each region, given the costs of abatement inside EGEM. This is not meant to be an accurate reflection of reality, because no region really knows its potential damage costs sufficiently well yet, but implied distribution does have the intuitive attraction of positing low damages in North America, and higher damages in Europe, which seems to match current negotiating positions. Therefore, the following simulations should only be seen as illustrative of the potential for agreement, and not a description of what will actually happen under the FCCC. However, as was argued in Chapter 8, putting specific numbers on the problem allows clearer investigation of the salient factors involved in treaty stability, than merely considering general theoretical results.

Interpreting co-operative behaviour under the FCCC

In order to back-calculate each region's damage function from its commitments under the FCCC, the type of co-operation taking place must be defined. If countries truly co-operate, the negotiators in the Council of Parties of the FCCC can be considered to be debating the size of CO_2 reductions needed to maximise global welfare:

$$\max \left[\sum_{n} B_i(\Sigma A_i) - \sum_{n} C_i(A_i) \right] \tag{1}$$

where B_i, C_i and A_i are respectively the benefits of abatement, abatement costs and GHG emission reductions in each country i, out of a total of n countries. The benefit function here could be actual climate damages, or an appropriately risk weighted function of uncertain future damage costs, or could include the value of future co-operation by developing countries which may be stimulated by these actions. In this case the origin of these figures is unimportant because we are inferring their value back from an agreed target, not constructing it from raw data. The optimal abatement levels in each country j are therefore given by the simultaneous solution of (2) for all n countries:

$$\sum_{n} \frac{dB_i}{dA_j} - \frac{dC_j}{dA_j} = 0 \qquad (2)$$

This is an efficient solution because the marginal cost of abatement is equal in each country. Countries control emissions until the global cost of pollution is reached, and so those with low national damages, and low costs of control, will abate a lot but gain few benefits. To ensure profitability the other countries can give side payments which would make the country at least as well off as if it were not co-operating. With no side payments this simple model predicts that non-cooperative behaviour will occur, as the low damage/high abatement countries leave and set abatement using their own national damage costs, thus leading to lower abatement and higher damage.

The polar cases of perfect co-operation and non-cooperation could be used to define extremes of behaviour which could underlie the FCCC, but considering the different costs of control in each region an interpretation of perfect co-operation is rather unconvincing. An alternative solution concept is to consider the FCCC as being an agreement which is profitable to each region, has collective abatement above the non-cooperative solution, but no recourse to side payments. Such an equilibrium can exist if each country can observe abatement efforts in others, and abatement occurs simultaneously.

If abatement benefits in each country can be monitored, then a form of quasi-cooperation can exist where countries assume abatement to occur on a quid pro quo basis elsewhere; that is, if one abates a unit of CO_2 so will every other signatory to the treaty. This reduces the apparent unit cost of abating emissions, and for n homogeneous countries the optimality conditions for each country i become:

$$\frac{dB_i}{dA_i} - \frac{1}{n} \cdot \frac{dC_i}{dA_i} = 0 \qquad (3)$$

This equilibrium maximises each country's abatement benefits, so no side payments are needed to make sure all countries gain from co-operation; however, the treaty is not efficient as there will be different marginal costs of abatement in each country. Thus, there is a Pareto improvement that could be made by shifting abatement between counties, but only if the ability to make side payments exists. With heterogeneous countries the division of abatement effort could be based on per capita emissions, or emissions per unit of GDP, which would alter the calculation of how much 'matching' abatement has occurred. This type of quasi-cooperation there-

Table 11.14 Implicit damage costs: non-cooperative and quasi-cooperative cases

	Average marginal damage cost (1990 $/tC)			
Emissions target	*1990 levels*		*25% below 1990 levels*	
Damage region	*Non-cooperative*	*Quasi-cooperative*	*Non-cooperative*	*Quasi-cooperative*
G7 damages	2078	253	5096	828
North America	311	81	947	280
Japan	1054	73	2330	148
Europe	713	98	1819	400

fore corresponds to a regime of differentiated international targets, with no potential for Joint Implementation or emissions trading.

Table 11.14 shows the implicit damage costs needed to produce regional stabilisation at both 1990, and 25 per cent below 1990, emissions levels. To find these implicit costs EGEM was optimised against a multiple of the expected value damage function derived in Chapter 8. This function is based on the future damages caused by a unit of CO_2 over its lifetime in the atmosphere, when these have been discounted back to the period in which the CO_2 was emitted. The objective function being minimised is the summation of total macroeconomic costs and greenhouse damages over the simulation period 1995–2030, all values being undiscounted as before. The marginal damages reported in Table 11.14 are averaged over the simulation period, and in reality grow at around the same rate as output, from low levels in the beginning periods. As GDP and marginal damages grow at approximately the same rate discounting has little effect on the optimisation.

From Table 11.14 it can be seen that interpreting the FCCC as a non-cooperative agreement implies unrealistically high greenhouse damages, especially for the stricter target. Contrastingly, assuming quid pro quo abatement, or quasi-cooperation with no side payments, results in implied benefit functions which are much nearer those derived before, if slightly higher because of the inefficiency implicit in this distribution of abatement. Marginal damages are relatively equal in each region for the lower target, but are much more differentiated if higher values are used. This merely reflects the cost of meeting regional stabilisation, which as shown above is particularly costly in Europe; therefore, this type of back-calculation will imply that Europe suffers relatively high climate change damage. For all the following simulations the quasi-cooperative benefit functions were used, with the same regional distribution as in Table 11.14. To simplify discussion the benefit functions will be referred to as Low, for stabilisation at 1990 levels, and High, for the 25 per cent reduction.

Coalitions, abatement and stability

To investigate the influence of each region on optimal abatement levels each permutation of coalition was modelled inside EGEM, along with the non-cooperative case. The co-operating regions face the co-operative objective function as in (1), while the remaining region free-rides on their abatement efforts setting its abatement to its own regional marginal damages, but taking into account global abatement efforts. Therefore, there is a Nash equilibrium between non-cooperators and co-operators, and as the marginal damage costs of emissions change with abatement, so will the efforts of the regions. However, as the simulations in Chapter 8 showed, the linearity of marginal damage costs makes the importance of this second-order effect rather small in practice.

Table 11.15 gives the results for the different coalition combinations, and both high and low damage functions: G7 abatement levels including non-cooperative countries (but without trade leakage), and aggregate net benefits (Saved Damages − Abatement Costs) from the base-case expressed as a percentage of G7 GDP.

Though full co-operation does inspire some extra abatement above the quasi-cooperative level, it is more effective at increasing net benefits by distributing abatement efficiently, than by markedly increasing the amount undertaken. This is true for both benefit functions, but especially in the high cost case. Abatement falls compared to the quasi-cooperative case when fewer countries participate in a coalition, and this shows the ability of a simple organisational device such as monitoring emissions to stimulate substantive profitable abatement.

Table 11.16 shows the gains to each region when moving from non-cooperation to a two-region coalition, for both damage functions. For the low damage case neither North America nor Europe gains from forming a coalition with Japan, unless side payments are made; however, they both gain from cooperating with each other. Comparing these results with Table 11.15 shows that the free-riding region gains the majority of benefits when two-player coalitions form. For the high damage costs in Table 11.16 co-operation is more profitable, and the only side payments needed are to

Table 11.15 Abatement and net benefits for different sized coalitions

Coalition and co-operative mode	*Non-coop.*	*Quasi-coop.*	*Co-op.*	*America & Europe*	*Japan & Europe*	*America & Japan*
Low damage costs						
ΔEmissions (%)	−7.6	−23.1	−25.3	−14.2	−9.6	−11.9
ΔBenefits (% G7 GDP)	0.04	0.21	0.34	0.27	0.09	0.16
High damage costs						
ΔEmissions (%)	−16.7	−39.5	−43.1	−35.6	−19.5	−22.5
ΔBenefits (% G7 GDP)	0.36	2.32	2.74	2.61	0.97	1.8

Table 11.16 Changes in regional net benefits when moving from non-cooperation

Type of coalition formed	*Changes in regional net benefits (%age of G7 GDP)*			
	North America	*Japan*	*Europe*	*Sum*
Low damage				
America & Japan	−0.02	0.07	NA	0.05
Japan & Europe	NA	0.04	−0.02	0.02
Europe & America	0.03	NA	0.08	0.12
High damage				
America & Japan	0.12	0.37	NA	0.49
Japan & Europe	NA	0.12	0.02	0.14
Europe & America	−0.26	NA	1.62	1.36

North America when it co-operates with Europe, and the large net benefits of co-operation mean that these will be easily forthcoming.

As would be expected, for incomplete co-operation the greatest abatement, and highest net benefits, occur when North America and Europe co-operate, and are lowest when Europe and Japan co-operate. Between them America and Europe capture almost all the co-operative value in the high damage case, and over 75 per cent when using the low damage function. However, the even spread of damage costs between the regions means that, though Japan does little abatement itself, its presence in an agreement does foster markedly more abatement, because the benefits it gains can be redistributed to other regions so as to make their higher abatement efforts profitable.

The importance of Japan is shown by the Shapely Values, or relative negotiating power (normalised so that the weights sum to 1) for the low damage case, which are 0.36, 0.27 and 0.37 for North America, Japan and Europe respectively. Europe marginally outstrips America because of its low damage costs for small abatement levels, but the overwhelming impression is that the countries have remarkably even negotiating power by this measure, despite their very different emissions levels. For the higher target this spread of influence changes however, and the Shapely Values become 0.45, 0.18 and 0.37 respectively. Japan's importance is diminished by its proportionately low damage costs in this case, while Europe does not participate in abatement as fully. The fact that its marginal damages are nearly twice that of America keep its negotiating power high.

Interpreting Shapely Values as negotiating power is rather misleading however, as they merely reflect the value that countries bring to an agreement, and as we have seen this can be because they have either low abatement costs, like America, or high damage costs, like Europe. These values do not tell us who gains or loses from participating, or that a high damage cost country will only have influence if it is prepared to transfer side payments to low cost abators, so that co-operating remains profitable for them. Table 11.17 gives the net gains and losses to each region as co-

Table 11.17 Changes in regional net benefits when moving to full co-operation

Type of existing coalition	*Changes in regional net benefits (%age of G7 GDP)*			
	North America	*Japan*	*Europe*	*Sum*
Low damage				
America & Japan	0.057	0.111	*0.010*	0.178
Japan & Europe	*−0.014*	0.135	0.121	0.242
Europe & America	0.009	*0.041*	0.027	0.076
High damage				
America & Japan	−0.43	0.39	*0.98*	0.94
Japan & Europe	*−0.78*	0.60	1.95	1.77
Europe & America	−0.05	*−0.16*	0.34	0.13

operation moves from involving two to three regions. Results for initially free-riding countries are given in italics; positive values show that they would co-operate with no side payments; negative values imply that side payments will be necessary to stabilise a fully co-operative treaty.

Table 11.17 shows that for the low damage function there is only one case when a region loses out when co-operation expands from two to three regions, and this is when North America joins an existing partnership with Japan and Europe. However, Japan and Europe gain so much from America's participation that they can easily afford to fund side payments to ensure profitability and stability.

The situation is very different for results of simulations using the high damage function. In this case North America loses every time the coalition expands, though as would be expected this loss is greatest when it was previously free-riding. To guarantee America's participation in the full co-operative scheme a large proportion of the benefits from co-operation have to be transferred, and it must gain at around 50 per cent of the value that it brings to the coalition in order to prevent it from free-riding. Of the other regions, Europe's high damage function means that it always values full co-operation and needs no extra incentive to join; however Japan would prefer to free-ride on the abatement efforts of Europe and America, unless side payments were available.

Therefore, in order to ensure the participation of either Japan or North America, Europe will have to fund the side payments as it gains most from co-operation, even when it has been free-riding beforehand. Thus Europe plays a pivotal role in ensuring co-operation, because of its high damage function. In their work on global co-operation Botteon and Carraro (1995) suggested that India and China could also perform such a role, as they had the highest damage costs. These damage costs were mainly derived however from estimates of increased mortality, and so are subject to the problems of valuation mentioned before. Hence, regions with high damage costs can commit to support co-operative agreements unilaterally, but the result will only be stable if they have the funds to give actual financial side payments.

In the case of a coalition between developed countries, where we assume side payments are feasible, the problem is the more subtle one of how to fund such transfers from existing taxation instruments. There is an ≈ 40 per cent premium on carbon tax revenue which flows out of Europe, and is not recycled through employers' labour taxes. Even if the marginal return on recycling falls by 30 per cent, as is shown in Figure 11.4, it is unlikely that Europe would want to provide the transfers needed to entice Japan into a stable coalition, though it would probably wish to pay for America to join. If side payments are lump sum then it seems likely that the first best co-operative outcome can be achieved between the developed countries, and no region will wish to leave the agreement. However, designing a system of lump sum transfers is very difficult in practice as the optimal distribution of taxes must be known in advance, in order for the transfers to be calculated. The simulations above which investigated the efficiency of different instruments showed how the abatement equilibrium is highly dependent on the objective function of governments, and the distribution of tradable permits, especially for the lower target. Therefore, a priori calculation of lump sum transfer payments is both difficult as a technical exercise and likely to introduce unnecessary complications into international negotiations. The alternative is to impose side payments via the distribution of tradable permits, or a subsidy per unit of abatement from Europe to North America; however, this will be more expensive and will alter the equilibrium mix of abatement.

To investigate these effects in EGEM, the co-operative high damage case was re-run with side payments from Europe and Japan to North America; the transfer payments were set to increase American benefits just above the free-riding level. The model was solved in two ways: firstly, with lump sum transfers where America committed a priori to abate the optimal amount, and Europe rewards this with transfers until joining is profitable; secondly, in quasi-cooperative mode with Europe paying a subsidy for each unit of emissions saved in America in order to stimulate extra abatement. In both cases the total amount of funds transferred will be equal, if the model converges to the co-operative optimum distribution of abatement without side payments.

The lump sum payment to North America which just makes free-riding not worthwhile amounts to $245/tC at the optimum, though of course America could negotiate for a larger fee than this. The results in Table 11.18 show that this is a feasible outcome because the macroeconomic cost of providing transfers is virtually equal to their face value. This is not because there is no externality from transferring currency, but because Europe and Japan gain some trade benefits from stimulating the American economy with transfers. Hence America only sees a 0.5 per cent saving in GDP from the transfers, though domestic output increases by 0.9 per cent. Overall, the efficiency loss from providing transfers amounts to 0.2 per cent of G7 GDP which is small considering that the benefits of having North America in the treaty are 1.77 per cent.

Table 11.18 High damage costs and side payments

Changes from base-case	*Optimal taxes*	*Lump sum payments*	*Differences*	*Nash equilibrium*
Net benefits (%G7 GDP)	2.74	2.54	−0.20	2.42
Welfare costs (%G7 GDP)				
North America	−2.84	−2.34	0.50	−1.73
Japan	−0.07	−0.24	−0.17	−0.61
Europe	−0.65	−1.18	−0.52	−1.27
Total cost ($/tC)	777.7	820.3	–	787.3
Transfers (% of G7 GDP)	0	1.09	–	0.81
ΔCO_2 (%)				
North America	−54.2	−54.2	–	−48.6
Japan	−7.0	−7.0	–	−14.6
Europe	−31.5	−31.5	–	−34.4

When North America optimises non-cooperatively, and Europe and Japan provide a joint emissions subsidy, the benefits from co-operation diminish and the co-operative scenario gives virtually the same benefits as quasi-cooperative abatement. In a real negotiation however, the level of subsidy could become the control variable which is negotiated between the parties when abatement commitments are decided, but it will still not be as efficient as using lump sum transfers.

These scenarios have shown that, given this particular distribution of damage costs, it is possible to construct a stable agreement between the G7 regions if side payments are available. If a high target is to be agreed it is not obvious whether Japan is necessarily part of an efficient agreement, and it is unlikely that they would be offered side payments to induce them to join. Co-operation tends to rest on Europe's providing side payments, because it suffers the highest damages, and so gains the most from global abatement. The agreement could also be stable because of competitiveness effects, as modelled in Chapter 9, or the existence of a non-monotonic damage cost curve, but this has not been modelled here. The importance of transfers, especially when reaching a high abatement target, again argues in favour of using a tradable permit system to reach stabilisation, as this is the only policy instrument that can accommodate side payments, achieve efficiency and is acceptable to governments on sovereignty grounds.

CONCLUSIONS

The Framework Convention on Climate Change provides a basic structure of international law, within which countries can negotiate to their environmental and economic advantage. At present the aims of the treaty are relatively weak, but it is to be expected that they will be strengthened in the future. To fulfil its aims of preventing dangerous interference with the global climate, aggregate abatement targets will have to be a significant proportion of GHG emissions in the medium term, otherwise the marginal benefit of abatement – in terms of risk mitigation – will be very small.

Significant medium to long term abatement requires that the developing world reduce its emissions in the future, but this is unlikely to happen if the developed countries cannot agree to limit their own emissions, or provide the money and technology for abatement in other regions. Therefore, successful abatement in the developed world is a necessary precursor for wider global efforts. This implies that when setting abatement targets the OECD countries will have to take into account damage in the developing world, as well as their own regions, even though the developing countries bear no binding commitments. If these damage estimates are based on purely financial measures then it is likely that the majority of global damages, which seem to be non-market costs in tropical countries, will not influence OECD abatement levels. This is consistent with the OECD's expecting compensation for abating at a higher level than their own damage costs would justify, but not with co-operation based around mitigating *global* welfare losses.

When global welfare costs are not considered, an appropriate level for OECD abatement could plausibly be maintaining emissions at 1990 levels over the next thirty-five years. Including global welfare costs would probably push this target up to at least a 25 per cent reduction from 1990 levels over the same period. This more stringent target would cost 3–4 times as much to reach, even using the most optimal distribution of abatement, and recycling revenues, so as to relieve existing tax distortions. However, it seems unlikely that the developing countries will accept future restrictions on their emissions unless such a major effort was forthcoming, especially as the vast majority of current increases in GHG concentrations have come from economic activity which has benefited the developed world.

The results from EGEM detailed above show that the form of policy co-ordination by the developed world, and the importance of different economic factors, will be very different depending on which of these two potential targets is enforced, since they represent qualitatively different changes in economic behaviour.

Reducing the sum of G7 emissions to 1990 levels over the next thirty-five years will not stabilise fossil fuel demand growth, and imposes a welfare cost of around 0.8 per cent of G7 output, depending on what type of co-ordination instrument is used. There is also not a large amount of economic efficiency to be gained from using efficient policy instruments, such as tradable permits, compared to regional stabilisation targets or harmonised international taxes. However, imposing carbon taxes to reduce energy demand produces tax revenue at a lower macroeconomic cost than taxes on labour paid by employers. As these revenues are three times as large as the welfare costs of abatement, revenue recycling considerations are very important. Consequently, the distribution of tradable permits, and the resulting transfers, have a large effect on the equilibrium abatement distribution. Recycling tax revenues through employers' labour taxes also produces significant new employment in the short to medium term, and if this is valued – in and of itself – by governments, the optimal distribution of abatement can again change quite markedly. Given these important uses for carbon tax revenues, and the small (< 9 per cent) benefits of using a tradable permits scheme, regional stabilisation targets seem on balance to be the best policy instrument with which to reach such a modest target. This is because they are relatively 'equitable' – in that all regions lose similar proportions of their GDP in complying – their transaction costs are low, and they allow all carbon tax revenues to be retained for domestic recycling.

These arguments do not hold if agreement is reached to stabilise emissions at 25 per cent below 1995 levels. Imposing taxes large enough to meet this target almost halts the growth of fossil energy demand in EGEM, and stabilises emissions into the future with virtually no further increases in energy taxes. This is because long run energy demand in EGEM is based on past measurements of how price rises stimulate growth in energy efficiency; these improvements then become incorporated in technology, and can be applied costlessly into the future. The production structure of EGEM means that the immediate switch to energy efficient technology implies less investment in improving labour productivity, and this has the combined effect of lowering output, and increasing employment (or rather slowing down on-going reductions in labour intensity). This large investment switch reduces G7 welfare by 2.8–3.5 per cent over the next thirty-five years, depending on which policy instrument is used, but these output costs become progressively smaller in the future.

Unlike the smaller target, the distribution of these costs when imposing regional stabilisation varies considerably from 1.8 per cent in Japan to 5.7 per cent in Europe. Using an efficient policy instrument can reduce overall costs by 25 per cent, which is the welfare cost of hitting the 1990 target. The larger costs of abatement in this case mean that using tradable permits provides a good approximation to the optimal taxation scheme, however they are distributed. This is because the importance of creating new employment and recycling revenues declines as abatement costs increase, and the economy is moved nearer 'full' employment. This result becomes more robust and important if high carbon taxes cause significant migration of energy intense industries between the G7 and non-abating countries in the developing world; which is a feature of the model built in Chapter 9. These factors mean that it will be worthwhile constructing a full tradable permit infrastructure in order to reach this target efficiently, and with enough flexibility for countries to be able to respond individually to the industrial relocation of energy intense industries.

Tradable permits become even more desirable when the stability of any treaties is considered. Any efficient distribution of abatement will involve North America in strongly reducing its fossil fuel use, and unless it has relatively high damage costs this will make co-operation unprofitable. To ensure that the full gains of co-operation are reached, America must be compensated for its involvement; given the distribution of damage costs derived here, the minimum level of payments sufficient to do this amounts to over half of the benefits of its inclusion. This means that it would be possible to use side payments to build an efficient agreement, if countries can provide transfers at the same cost as their face value; that is, the deadweight loss from taxation is low. Modelling this process in EGEM suggests that, even taking into account the macroeconomic costs of providing such transfers, it is feasible to provide sufficient side payments, but this result was somewhat dependent on terms of trade effects.

Future extension of the FCCC into a successful global abatement regime will depend on whether countries which suffer high climate damages can afford to give such stabilising side payments to polluting nations with less interest in abatement. If such payments cannot be funded, the human misery caused by climate change will be far too high, considering the relatively low *welfare* cost of many abatement measures, and the local environmental benefits of reducing dependence on fossil fuels. The distribution of global income will have a non-trivial effect on negotiations to allocate abatement responsibilities, and implicit or explicit transfers, between nations. This represents possibly the largest political challenge in confronting climate change, and it seems unlikely that significant global abatement will be possible without the altruistic (in strict economic terms) involvement of the developed world, in funding abatement efforts and transferring technology to the developing world. Whether the incentive for this involvement comes from an ethical motivation, or from the recognition of the industrialised world's past responsibility for rising GHG concentrations, is not very important; what is necessary is that high income countries play a leading role in stimulating global agreement.

Therefore, there seems to be both a necessity for, and few practical obstacles to, the developed countries' agreeing to substantial co-operative abatement, as a first step to binding global GHG emissions limits. As targets become more severe some of the complications surrounding abatement issues, such as the effects on employment, seem to diminish in importance; while others, such as the effect of unilateral abatement on competitiveness, industrial relocation and emissions in uncommitted countries, seem to loom larger. To simplify negotiations at the international level it seems better to concentrate on abatement issues, while creating a co-ordination framework flexible enough to deal with the other macroeconomic issues connected to large scale abatement at the national level. The first-best option – if there was full information – would be to agree to abatement targets in each region which gave efficient abatement, and provide lump sum compensation payments between countries to ensure profitability and stability. In the absence of full information a similar role can be played by distributing tradable permits to countries which will abate most, but have the least damages. Unlike most simple models of tradable permit systems this will require significant centralised knowledge, because atomistic trading of emissions permits may provide efficiency of the agreement but there is no guarantee that it will give stability. Thus, agreeing the distribution of permits will be a non-trivial negotiating task, and there are many pitfalls to be overcome in moving from theory to practice with such far reaching and radical proposals. However, the gains to be had from using innovative policy co-ordination instruments seem too great – both financially and environmentally – to avoid taking advantage of them at the nearest opportunity.

APPENDIX 11.1: TRADABLE PERMITS EQUILIBRIA WITH REVENUE RAISING COSTS

Assume a single country is assessing whether to buy or sell its permits allocation given an international selling price (P) which it takes as being exogenous. The only abatement instrument available is a pollution tax and so full costs of abatement to this country are:

$$M(e) - G[R(e) + P.(T-e)] = \text{Total Costs} \qquad (11.1:1)$$

where M(e) are the macroeconomic and welfare costs of maintaining a certain level of CO_2 emissions e; R(e) is the revenue raised from taxing these emissions; P is the exogenous permit price; T is the initial allocation of permits, and G[. .] are the macroeconomic benefits of recycling these taxation revenues into the economy; that is, the difference in tax distortion between energy taxes and the worst revenue instrument currently available.

The characteristics of these functions are:

$$dM/de < 0,\ d^2M/de^2 < 0$$

$dR/de < 0$; that is, energy demand is inelastic.

$$dG/d\$ > 0,\ d^2G/d\$^2 < 0$$

Therefore, revenues rise as emissions are abated ($R(e) = te$, where t is the carbon tax level) and the marginal effectiveness of a recycled dollar drops as more is put through the system.

The level of emissions which minimises total costs is therefore defined by:

$$dM/de - dG/d\$.(dt/de.e + t - P) = 0 \qquad (11.1:2)$$

If the marginal effectiveness of recycling changes with the amount of revenue, that is $d^2G/d\$^2 \neq 0$, then for each distribution of permits T, dG/d$ will change and there will be a different optimum level of emissions for the country.

Simplifying the model and assuming that $dG/d\$ = \alpha$, $d^2G/d\$^2 = 0$, the optimum equation for a single country becomes:

$$dM/de - \alpha.(dR/de - P) = 0 \qquad (11.1:3)$$

Therefore, for two countries A and B, where $dG_A/d\$ = \alpha$, $dG_B/d\$ = \beta$, the equilibrium trading position is:

$$1/\alpha.dM_A/de - dR_A/de = 1/\beta.dM_B/de - dR_B/de \qquad (11.1{:}4)$$

And country B will buy/sell to country A when (11.1:5) holds:

$$dM_B/de = \beta/\alpha.dM_A/de + \beta.(dR_B/de - dR_A/de) \qquad (11.1{:}5)$$

That is, the costs of abatement are weighted by the ratio of the macro-economic distortions caused by raising, and not recycling, energy tax revenues.

Part IV
OVERVIEW

12

SUMMARY AND CONCLUSIONS

INTRODUCTION

This final chapter presents a non-technical summary of the research contained in the rest of the book; it is intended to be both a stand-alone piece, and a means of introducing the reader to the main body of the text. This book has grown from a research project, and its structure reflects the methodological logic used: starting with an outline of the environmental and policy problems; explaining the derivation of the economic models used to address these issues; and then combining these tools with theoretical analysis so as to study policy areas with a common economic theme. While this makes sense from an academic viewpoint, it is less logical for a non-technical specialist primarily interested in research results and how they link to real world concerns. Therefore, this summary is arranged around policy themes, and much of the material in the book has been re-ordered to make it fit this framework.

After introducing the main aspects of climate change as a scientific problem, we split the policy analysis into two distinct sections: achievement of collaborative effectiveness – that is, how closely any international agreement to control GHG emissions is likely to approximate to a sustainable, welfare maximising solution; and operational effectiveness, or whether the implementation of commonly agreed targets will be achieved in a way that is both economically efficient and keeps countries in the agreement (often termed treaty stability).

Empirical research into the economic impacts of climate change is still very immature, therefore most of the analysis on collaborative effectiveness is theoretical. Modelling results are only used to give illustrative examples of how different factors (for example, uncertainty and progress in energy efficiency) affect prescriptions for optimal climate policy. The operational effectiveness of an international agreement is easier to analyse, because it depends on the reaction of the global economy to changes in fossil fuel prices and energy use (taking carbon dioxide as the most important GHG). An econometric model of the global economy (EGEM – Environmental Global Econometric Model) is used to model policy options for developed countries over the next thirty-five years. In particular, the competitiveness effects of the OECD unilaterally agreeing to CO_2 emission reductions; the impact of energy taxation on employment and equity inside the developed countries; and the consequences of using different policy instruments (targets, taxes or tradable permits) to co-ordinate international climate change policy.

Through all this analysis we preserve the main theme that economic approaches to climate change policy must be explicitly placed inside a political and institutional context if they are to be useful to policy makers. This is because the international nature of the problem raises many macro-economic and policy issues which are irrelevant when discussing the normal case of national environmental policy. It is the implications of international relations in which we are most interested, and this focus defines the detail with which we address different questions. The sheer scale of climate change means that there can never be a definitive analysis; however, we hope that this work will illuminate parts of the problem which to date have not been given prominence in the economic literature.

THE CHALLENGE OF CLIMATE CHANGE

Climate change has the dubious accolade of being the most complex and difficult environmental problem which humankind has ever faced. A problem which will require drastic action to prevent, and if left unchecked has the potential to undermine the support systems upon which the vast majority of the world's population depends for subsistence.

In deciding how to respond to the threat of climate change, we are faced with a scientific dilemma: that is, while we clearly understand some of the main physical mechanisms involved – concentrations of greenhouse gases (GHGs) in the atmosphere unequivocally raise global temperatures – we also know about mechanisms which could mitigate, or remove, the impact of increased pollution. The natural variability of the global climate, and the long lags in response to increased radiative forcing, mean that unambiguous measurement of the effect of man-made pollution will take many decades. If we wait that long before controlling emissions we will have already committed ourselves to a large increase in atmospheric GHG concentrations and, if our worse fears are realised, the only possible response will be to defend ourselves against the effects of a changing climate, rather than preventing it happening in the first place. That said, recent measurement of global temperatures, the frequency of extreme climatic phenomena, and other climate sensitive features – such as the extent of polar ice coverage – has led to a consensus among the vast majority of physical scientists that some global warming is definitely taking place.

Unfortunately, the current state of the measurable evidence cannot tell us relative strength of the opposing forces governing climatic regulation. They also cannot be found by the traditional scientific method of controlled experimentation, because the earth is the only available laboratory. Therefore, we must resort to predicting the impacts of increased GHG concentrations using complex mathematical models which look centuries into the future. This causes great difficulties for policy making, which does not have a good history of dealing with long run environmental problems under a high degree of uncertainty. This is shown by the history of other environmental problems surrounded by similar uncertainties; for example, the long run health effects of low dose radiation, and potentially cancerous chemicals.

In all these situations, experimental science is only able to analyse part of the system involved, and waiting for the effects to unambiguously show themselves could lead to unacceptable loss of human life and health. Generally, the policy response has been to take a precautionary approach; aiming to limit exposure as much as possible without necessarily banning useful substances from being used in production. However, the extent of controls on substances with similar potential for dangerous side effects has been determined by many factors, including the cost of control, public awareness of the issues and the strength of industrial/military interests in the political process. Policy in many of these areas is fraught with controversy, with experts arguing that science indicates both tightening and loosening controls as the 'rational' course of action!

Current international policy on climate change is governed by the Framework Convention on Climate Change (FCCC; UNEP, 1992), which contains a precautionary principle: 'Where there are threats of serious or irreversible damage, lack of full scientific certainty should not be used as a reason for postponing such measures [to slow climatic change]' (Article 3.3). The convention also stipulates that any policy to halt climate change must be cost-effective and not hurt the global economy – while preventing dangerous interference with the global climate. The huge stakes involved mean that this implicit balancing of the costs and benefits of controlling climate change must rest on more solid decision making than has occurred in the past; this will require radical advances in the politics, economics and institutions of global environmental governance.

The work of the Inter-Governmental Panel on Climate Change (IPCC) has initiated such a change in the policy process, by bringing together experts from around the world to assess the science and economics of a changing climate. However, its two reports so far – the latest published in April 1996 – have shown that work in the physical sciences is far ahead of that on the politics and economics of decision making. This may just be a result of the relative amount of funding given to each area, but it also seems to show the reluctance of economists to invest their energies into an essentially empirical problem, which seems to contain little of theoretical interest.

The political economy of climate change

In fact climate change is a supremely challenging problem to economics, or perhaps more correctly, to the 'old-fashioned' discipline of political economy, because global politics cannot be separated from the determination of the 'correct' economic policy. There is no pure economic answer because this problem lacks an existing legal framework within which property rights, obligations and contractual enforcement allow classically defined economic interactions to take place. Inside nation states, democratic government, with its various checks and balances, is the mechanism which is supposed to balance the welfare of different groups. However, greenhouse gases emitted in any country affect all others, so climate change is a problem to be solved between nation states, and, barring war (military or economic), there is no way of enforcing majority rule between sovereign countries.

For conceptual clarity, the policy problem can be divided into two stages, collaborative effectiveness and operational effectiveness; though in actual negotiations these categories will blur and overlap. Collaborative effectiveness is concerned with what the international community agrees to do about climate change in the *aggregate*: the timing, severity and scope of reductions in GHG emissions, or actions to mitigate the impact of climate change. Operational effectiveness is about how well this agreement is actually put into practice: the success of policy instruments and institutions used to co-ordinate global action to prevent climate change.

THE COLLABORATIVE EFFECTIVENESS OF CLIMATE CHANGE NEGOTIATIONS

In economic terms, negotiations to control climate change will have achieved full collaborative effectiveness if they maximise global welfare into the future; that is, perfect co-operation, or collaboration, has been reached. This seems like a simple criterion, but the complexity arises in defining what maximising global welfare implies. A workable concept of global welfare requires a reliable way of adding up some measure of happiness (or utility) across present and future populations – a task that welfare economics has long recognised as impossible. Problems in aggregation arise from non-comparability of preference choices, the disparity of incomes between different countries, and the unknowable preferences of future generations. To help solve these problems, many commentators have urged the application of two supplementary ethical criteria for climate change policy: intra-generational equity and inter-generational equity. Intra-generational equity means there is an acceptable balance between the interests of rich and poor countries. Achieving inter-generational equity would preserve the options of future populations, and is subsumed under the more general concept of sustainable development (often defined as – 'development that meets the needs of the present without compromising the ability of future generations to meet their own needs' – WCED 1987).

This approach to climate change policy would turn the economic problem from one of full welfare optimisation, to one of constrained optimisation; with the constraints ensuring that the interests of vital groups are not lost in the calculation of costs and benefits. Many developing countries see this as a reasonable reaction to a perceived Northern bias in economic methods and market valuation; unsurprisingly, some economists see it as irrational interference in the science of economic decision making. These differences have caused serious disputes in the IPCC process, but do not seem to have advanced understanding of the problems of climate policy. However, it is more incumbent upon economists to understand why there have been violent objections to their 'optimal' prescriptions, than upon elected representatives of national governments to withdraw their opinions. Governments ideally represent the values and preferences of their populations; economists should accurately measure and process these interests when attempting to calculate 'optimal' global policy. A mismatch would suggest fault on the part of economists, rather than the often imputed irrationality of policy makers.

Intra-generational equity and global welfare

If climate change only affected material goods which could be priced in markets, policy making would be relatively straightforward, and the optimal amount of pollution prevention could be found by balancing financial costs and benefits. This of course is not the case; it is likely that most of the impact of a changing climate will be in increased human misery and ill-health as lands flood, ecosystems collapse and communities are forced to migrate away from previously fertile land. The resources needed to prevent pollution lie in the North, the majority of welfare damage is most likely to be in the South. However, the 'market value' of this huge welfare loss is capped at the very small cash incomes of those who will suffer, and will be easily outweighed by the amount industrialised countries would be prepared to pay to continue burning fossil fuels.

Such a policy choice, where the luxuries of the rich outweigh the necessities of the poor, is clearly inequitable but – lest we forget – is also the existing state of our current global economic system. Given that current levels of development aid are so small as to imply that the ethic of intra-generational equity is virtually non-existent, why should it apply more strongly in the case of climate change? Beyond pure altruism, perhaps the only reason why rich countries would be prepared to pay for the welfare of the poor at above their 'market rate', is that the achievements of these wealthy societies have been built on the pollution which has caused the problem. Therefore, the rich can be said to have a past-indebtedness to the poor for using up their share of the atmosphere's assimilative capacity for GHGs. A recognition of the rich countries responsibility for past emissions is written into the FCCC, but it remains to be seen if it will be an important factor in the negotiations.

As any global valuation of climate damage will implicitly contain assumptions as to the relative worth of people's lives and health in different countries, if these are not made explicit then the study cannot be considered good economics. Therefore, though politics of equity between North and South are far from being clear cut, they must be determined *before* any usable economic valuation of climate damage can be drawn up. At one extreme, decisions could be made by a form of global democracy – one person, one vote – which implies an equal share for everybody in global resources; at the other extreme, market values, based on the existing distribution of resources, could be used. A set of illustrative simulations in EGEM showed that the optimal level of abatement for the OECD using the 'democratic' valuation system was nearly twice that when using market valuations. This is a larger difference in policy prescription than resulted from using the extremes of the uncertain climate change damage distribution. Despite the acknowledged problems with this modelling, the results illustrate that the treatment of non-market welfare costs introduces variations of the same order of magnitude as our ignorance over climate damages, and so will be a critical part of determining optimal policy.

Assuming everybody has similar preferences – if different incomes – global welfare would be greater if the democratic system was used, despite the well-known flaws in this type of public choice system (Fishburn 1973), but this is unlikely to occur in practice. So this aspect of the collaborative effectiveness of the FCCC is likely to be determined by power politics, links to other issues (for example, trade), and the damage rich countries think they might suffer from climate change if poorer countries carry on industrialising at their present rate. Accurate economic analysis of this problem lies in the realm of game theory, and will have to wait until we have much better data on the distribution of economic damages from climate change.

Inter-generational equity and sustainability

The balancing of interests between rich and poor countries will be determined by face-to-face negotiation, but as future generations have no place at the table, their interests risk being lost in the political process. This has serious environmental implications, because effectively tackling the causes of climate change will require investments in non-polluting technologies now, but climate inertia means that the impacts of current pollution will occur decades into the future. Inter-generational equity implies that future generations must agree with our resource-use decisions, regarding the amount of man-made wealth bequeathed to them ample substitute for any irreversible loss of environmental assets. However, the dynamics of climate change encourage a short term outlook by political leaders, who will value the financial consequences of present abatement far higher than the environmental implications of future damage.

Current IPCC endorsed estimates suggest that a 50–70 per cent cut in GHG emissions from current levels would be needed to stabilise GHG concentrations in the atmosphere; the actual long run emission level depending on the stable concentration chosen. This would be a daunting target even if global GHG emissions were not projected to rise by around 300 per cent over the next thirty years. However, none of the cost-benefit analyses carried out to date have recommended an optimum level of emission reductions of more than ≈ 20 per cent from business-as-usual (BAU) levels. Adopting such a weak policy is obviously unsustainable, as it will only slow the rate of temperature rise, and will not halt it even it the long run. Long run sustainability requires a constant atmospheric concentration of GHGs, or a very slow build-up which is comparable to previous natural changes which have preserved ecosystem stability. In the long run it will make little difference postponing a particular temperature increase by a few years, if human and environmental systems will still have to face the same absolute change. Though there will, of course, be minor advantages from merely slowing the rate of temperature increase, as this will allow easier adaptation.

If environmental sustainability is an aim of global policy – and the vast majority of governments signed an agreement to that effect at the 1992 Rio Conference on Environment and Development – why is this not reflected in existing cost-benefit analyses? There are two possible reasons: firstly, that it will be cheaper to mitigate the impacts of climate change in the future, rather than preventing them in the first place; secondly, that the cost-benefit analyses have been fundamentally flawed in the way they assess the value of abatement. As the cost of adaptation to climate change has not been exhaustively studied (except for combating sea level rise), or realistically incorporated into past studies, it would seem that the problem lies in the methodology of these analyses.

It is well recognised that existing costing studies on climate damage have suffered from a staggering lack of good information and very limited scenarios of future temperature rise; we would also argue that there are fundamental methodological flaws which further data collection cannot solve. These problems surround the mechanism by which future damages and current abatement costs are calculated and compared in cost-benefit analysis (CBA) – specifically, the use of high discount rates for assessing future damage, and erroneous projections of how much future generations would be prepared to pay to avoid serious climate change. In CBA these factors – discount rates and future prices – are two sides of the same coin, and the problems in their determination stem from the same root cause: the lack of substitutability between natural and man-made goods.

One of the assumptions behind traditional CBA is that the value of all factors can be reduced to a numeraire good – money – which can then be mathematically manipulated to show if costs outweigh benefits. Costs in the present are compared to benefits in the future by discounting; this assumes that if instead of investing in GHG abatement the money was placed in a bank account, we would want to abate if the accumulated capital and interest invested in the account is less than the climate change damages resulting from the increased emissions, and vice versa. In CBA this comparison is performed by reducing the value of future damage costs at the inverse of the interest rate – the discount rate – which is conceptually identical. Theoretically, the correct value for this discount rate is the return on the next-best investment the decision maker can make which has the same risks attached.

However, as many climate change impacts are irreversible (for example, salt water flooding of farmland), money is not a perfect substitute to future generations because it cannot reproduce what has been lost (unlike man-made objects which can usually be replicated). For goods priced in markets (for example, agricultural land) irreversible losses will cause the cost of other similar goods to rise, because the supply has permanently shrunk. For example, it has been suggested that the complete destruction of agricultural productivity in the USA has a maximum economic cost of 3 per cent of GDP, because this is its current contribution to national output; however, even simple estimates of welfare loss show that in reality decreases in agricultural production can be far more damaging (see Chapter 8). Similarly, for non-market goods which have poor market substitutes (for example, personal health), future generations will need higher money payments to compensate for irreversible losses than would be calculated from the usual methodology of measuring how much *current* generations would pay to prevent these damages (Hanemann 1991). These changes in future prices, which result from the non-marginal impact of climate change on critical environmental goods, could be included in CBA either by reducing discount rates or raising future prices. Both methods have advantages, but this is a technical issue, the important policy implication is that unless the limited substitutability of man-made and natural capital is taken into account, CBA studies are likely to recommend erroneous, unsustainable polices.

The degree of substitutability between man-made and environmental goods can sometimes be measured technically (for example, the cost of coastal defences and water purification if wetlands are inundated). However, many climate costs are non-market, such as the loss of unique species or productive resources, and for non-market costs substitutability becomes an ethical and political choice of the type of world it is right to leave to our descendants. Options include: 'economic optimality', where man-made and environmental capital are assumed to be complete substitutes; 'weak' sustainability, where different environmental assets can be substituted for each other, but the total stock must be constant; and 'strong' sustainability, where all existing resources must remain unaltered between generations. The economics of inter-generational equity are based on a set of public choices which have not yet been made. Therefore, the acceptability of using CBA inside the policy process will depend on its implications for the achievement of our preferred vision of the future, rather than on the academic acceptability of the methodology used in its calculation.

Cost-benefit analysis and uncertainty

Determining climate change policy is an iterative process where positions change as more information becomes apparent and critical issues gradually emerge from the debate. While ethical and political debate will decide the framework within which economic valuation can be useful in policy decisions, economic techniques will be vital in measuring and articulating preferences, and in rigorously quantifying the consequences of decisions. In particular, economic logic can give important insights into how uncertainty over the impacts of climate change should be taken into account by policy makers.

The predictive calculations needed to translate gross climatic changes into specific, quantifiable economic impacts are orders of magnitude more complex than exist in current models. Secondary climatic changes in rainfall, weather patterns, and so on must be defined at the regional level, and then translated into tertiary impacts on specific biological and human systems (agricultural productivity, coastal inundation, disease incidence, migration, etc.). There are also several important, if only marginally understood, feedback systems between vegetation and regional climate systems which complicate this hierarchy of modelling. Finally, economic modelling and analysis must assign monetary values to all these effects through a combination of market and non-market valuation techniques, which will be heavily influenced by unpredictable changes in technology and preferences. It is clear that economic assessment is the weakest link in this analysis; not only is it dependent on the inputs from the physical modellers, but while the mechanisms underlying climate and biological models can be assumed to be constant and thus derivable from past events, the rules underlying the evolution of economic systems have no such permanence. The past can only therefore be a partial, if useful, guide to the future.

Significant uncertainty as to the costs of climatic change is therefore an endemic part of this problem, and will not be solved by future research, no matter how complex and sophisticated it seems. Policy will therefore be made on the balance of future probabilities and outcomes. In this context, it is important to remember that investing in strong abatement measures now can still be considered a 'good' decision, even if it turns out that climate change only results in minimal future damages. The essence of invoking a precautionary principle is that sometimes we will make mistakes, but they should be the right mistakes!

To formally describe uncertainty, analysts assign a probability to all the different states of the world they think will occur, thus creating a probability distribution of different outcomes. By definition, a probability distribution can only contain eventualities we know about (often termed risks or 'soft' uncertainty), and has nothing to say about things we do not know, so-called 'hard' uncertainty. Because of this, probabilistic approaches to decision making are most useful when dealing with events that have happened many times – such as airline crashes – rather than infrequent events such as nuclear power plant failures.

For climate change, the distribution of outcomes is probably bounded at no adverse impacts at one end, and an extreme climatic disaster – such as complete melting of polar ice sheets – at the other. The distribution of likelihood between these points defines what we know about the future, and in turn can be reduced to a few key parameters: for example, the mean value, or average damage, and the variance of the distribution. An increase in uncertainty implies increasing variance in the distribution – for example, a moving apart of the likely extreme outcomes – this does not imply that severe climate change is less likely to happen (as in the common usage of something becoming 'more uncertain'), but just that we are less certain which outcome will prevail because they have all become more equally probable.

Therefore, there are two distinct facets of climate policy: how to deal with 'known' uncertainty which can be expressed reliably as a probability distribution, and the process of learning more about previously hard uncertainty – learning which may increase or decrease the variance of the known damage distribution. For example, research may discover that ocean currents are far more sensitive to surface water temperatures than previously believed, thus increasing the probability of very serious impacts, and increasing the distribution variance. This is an important feature to bear in mind, as the usual debate in climate change policy is whether to pursue a preventive policy of acting now or wait-and-see on the assumption that research will reduce uncertainty over outcomes. In reality, learning usually just turns hard uncertainty into soft uncertainty, and its policy relevance will depend on how these different types of ignorance are treated inside decision making.

The correct method of dealing with the different types of uncertainty is to assess the importance of probabilistic risk using the techniques of stochastic optimisation, and to take account of unknowable hard uncertainty by adopting a risk-averse hedging strategy.

Stochastic optimisation can be used to find the strategy – in this case at global abatement level – which gives the best result over all future outcomes *included in the probability distribution*. For simple cases, this can be done by balancing cost and benefits using the mean of the probability distribution, and this has been the method usually employed in previous CBA studies. However, it is a well-known mathematical result that this approach fails when the system contains significant non-linearities (Hall and Stephenson 1990). This is unsurprising, as when trying to optimise an uncertain future it would seem natural to include in the calculation all the non-random features of the future which are known in the present (that is, the complete parameterisation of the probability distribution, including the variance and skewness), and not just the mean of the distribution.

The dangers of simply using mean damage values when calculating optimal abatement levels are best shown by an example. If, as in the EGEM model, technical progress in energy efficiency in the future depends on past energy prices – higher prices inspire more research – this introduces a major non-linearity into controlling climate change. If energy taxes are set so as to optimise abatement costs against the mean damage value, but in the future damages turn out to be much higher than the mean, we will want to increase abatement significantly over a short period of time. However, technology takes time to develop and infrastructure to change, so it will cost more to respond to high damages than if investment had already been stimulated by increased energy prices. Here the existence of a non-linearity in the system means that optimising to the mean results in too little pre-emptive abatement. Of course, a similar argument holds if abatement results in irreversible capital commitments and damage turns out to be low; optimising to the mean has again raised costs compared to the optimum, but the optimum policy is lower pre-emptive abatement. Full stochastic optimisation takes into account the dynamics of control reactions to each potential outcome, and so can handle these type of effects. Whether this results in more or less current abatement compared to an optimisation using the mean damage level will depend on the relative magnitude of different effects, not theoretical analysis.

The most important non-linearities in climate change are: changes in the marginal damage caused by GHG emissions due to previous abatement; dynamics in the evolution of technological change; irreversibility of climate change damages or abatement measures; and how dynamics in GHG emissions affect climate change damage. Using a damage function based on IPCC models of how emissions link to temperature rise, and some illustrative damage estimates, EGEM was used to investigate the first three non-linear effects. From this modelling, it seemed that while technology dynamics could greatly affect the choice of optimal policy, the effect of past GHG concentrations was relatively insignificant. The latter result stems from the insensitivity of *marginal* damage costs to atmospheric GHG concentrations, as the relationship between concentrations and damage is virtually linear. For while the temperature increase associated with each unit of GHG emissions falls as concentrations increase, damage is assumed to grow faster than temperature; the two non-linear effects cancel each other out giving a nearly linear resultant. This linearity also reduced the influence of the irreversibility of climate damage on abatement policy. However, as stated before, the immaturity of these models means this result should be treated with caution.

The damage model inside EGEM could not simulate the costs associated with different rates of climate change, which could be very large when considering ecosystems which can only move very slowly; for example, tropical rainforests. Omitting such effects will tend to lead to insufficient pre-emptive abatement, because discounting encourages decisions to be made at the last possible moment. Emissions will therefore continue rising in the near term, with abatement to stable levels being postponed to future decades. The resulting 'spike' of radiative forcing may result in serious ecosystem damage, which would be prevented if a smoother path to stabilisation was chosen.

Powerful as they are, the techniques of stochastic optimisation will give misplaced confidence in the resulting policy if used naively. The probability distribution of climate change damages is itself highly uncertain, as we only have limited and partial information about past occasions when the globe has warmed significantly. Without a rich set of frequency observations the assumptions of probability theory will not hold and so cannot be used to give quality policy advice. Dealing with non-probabilistic, or hard, uncertainty is essentially an *ad hoc* affair because we have no data with which to make sophisticated calculations; this is when action must be justified by the precautionary principle outlined in Article 3.3 of the FCCC. Policy makers must adopt a risk-averse hedging strategy, where future policies which involve uncertainty – such as continual increases in concentrations of GHGs – are valued less highly than certain outcomes which seem to have the same financial value. For example, avoiding uncertain and irreversible climate damage leaves us with the option of burning fossil fuels in the future, but taking the uncertain course commits us to an outcome we cannot change. There is therefore value in preserving options and taking the less uncertain path.

A true hedging strategy would involve a mixture of pre-emptive abatement to slow climate change, investment in mitigation measures such as coastal defences, and aggressive research and development into clean energy sources. The size of each policy component is by definition unquantifiable by objective methods – otherwise we could just use stochastic optimisation to find the best policy mix. However, it is unlikely that a true hedging strategy would involve 'no-regrets' as currently defined (that is, policies which have little net cost to the economy). Providing a true hedge against climate change has to involve a significant slowing of the growth of atmospheric GHG concentrations – and this is likely to involve real costs. However, these costs should be balanced against the 'secondary' benefits of lowering fossil fuel use which are not linked to climate change; for example, reductions in local air pollution, traffic congestion and acid rain. With the economic costs of climate damage so hard to accurately measure, it would be more policy-relevant for empirical researchers to attack the more tractable problem of accurate accounting for these secondary benefits.

Though hard uncertainty surrounding climate change damage makes deciding the correct response strategy harder, it does have some advantages. One of the largest dangers in formulating climate policy is that some countries with energy-dependent economies will only suffer minor damage (as seems to be the current impression in both the USA and Russia); these countries may therefore fail to co-operate with global abatement policies, or will demand large 'side payments' to ensure their co-operation. If most damage is non-market and concentrated in developing countries, such payments will not be affordable, and excessive climate change and human misery will result. However, if damage is uncertain enough as to be unpredictable for any specific country, it may inspire a 'Rawlsian' veil of uncertainty (Rawls 1971), where countries will agree to an equitable and effective agreement because none knows which will suffer the greatest damage from uncontrolled climate change.

Collaborative effectiveness – conclusions and issues

Collaborative effectiveness refers to how near a negotiated international agreement on climate change is to the theoretical welfare-maximising solution. This is a substantive issue because, unlike the case of national environmental policy, there are currently no effective international institutions able to compel individual countries to curtail polluting activity. Therefore, countries must decide to abate voluntarily, even though emissions of transboundary pollution are not covered by UN rules on sovereignty and so are a legitimate concern of other countries. In the future, international environmental agreements may be enforceable by trade sanctions, but though this issue is currently being looked at by the World Trade Organisation, it is not likely to be a useful measure unless directed against a very small set of countries. In the absence of enforcement, countries can fail to co-operate fully in setting targets because the benefits of abatement are unevenly distributed, and side payments are not available to make sure all countries benefit from controlling their emissions.

Even with global participation, global welfare maximisation – the benchmark for collaborative effectiveness – is impossible to define objectively. Any comparison of the costs and benefits of climate change will always contain an implicit opinion on the relative value of people in rich and poor countries, and the worth of future generations. Economic methods cannot define the split of responsibility between rich and poor countries for preventing climate change, and cannot calculate the type of world we wish to leave the next generation. These decisions depend on political processes and ethical judgments, the result of which will define the economic value of certain outcomes, and not vice versa. Until we have a firm political settlement on these issues, the main contribution of economics in deciding the level of aggregate global abatement is to improve measurement of the costs of continuing climate change, analyse the costs of reaching different targets, and calculate the secondary benefits from reducing fossil fuel use – such as lower urban pollution. Economics is an input to the decision process and a mechanism for comparison, but not the assignor of value, or arbiter of 'correct' choices.

THE OPERATIONAL EFFECTIVENESS OF CLIMATE AGREEMENTS

The analysis of operational effectiveness starts from the assumption that an absolute level of global abatement has been decided upon by some legitimate process. This policy commitment now remains to be put into practice effectively. In particular, the structure of the agreement, institutions and policy instruments used must ensure that the aims of the agreement are met; that these aims are reached efficiently; and that the agreement is stable, with none of the parties free-riding to gain unilateral advantage. While economics must take a back seat in *prescribing* climate change policy, it has an important role in *describing* the consequences of different actions; therefore, it is critical to examining the operational effectiveness of any agreement.

From the above discussion it will be clear that though useful as analytic concepts, operational and collaborative effectiveness overlap greatly in practice. The ability to reach a target, and the efficiency of carrying out abatement, will help determine the optimal size and timing of an emissions target which is acceptable to all parties. The stability of an agreement will be determined by the distribution of costs and benefits, or in the absence of reliable national damage costs the apportioning of abatement responsibilities. However, we can model most of the issues surrounding operational effectiveness without having to know detailed damage cost estimates, and so economic research in these areas will produce results which are more useful and reliable for policy makers.

Given a global abatement target, emission reductions must be apportioned among countries in a reliable way using a co-ordinating instrument. There are three possible types of policy instrument: national emission targets, international GHG taxes (collected nationally or internationally) and tradable permit schemes (where each country is given a quota of GHG emissions, but they can buy and sell parts of this to other states if that is worthwhile). The FCCC also contains Joint Implementation, where countries gain some credit for paying for abatement in another country, but in many ways this is just a preliminary form of tradable permits scheme.

All these instruments have advantages and disadvantages:

- **National targets** are cheap to administer, give flexibility of national policy and ensure global abatement is met. However, as abatement costs will vary between countries, an equal percentage reduction target in each one would be inefficient; the cost of meeting the target is uncertain; and efficient target setting involves too much centralised information.
- **International taxes** ideally lead to economically efficient abatement, if existing tax energy schemes are harmonised beforehand and energy markets are perfect. However, they erode national sovereignty, may interfere with other national policy objectives (e.g. industrial and export policy), make global abatement targets uncertain, and preclude the use of cost saving instruments (e.g. efficiency standards, eco-labeling, etc.) to overcome market failures.
- **Tradable permits** should ideally combine the advantages of international taxes and targets, albeit at a higher administration cost; the global target will be met efficiently, and without superseding national powers or restricting national policy choices. However, if the money needed to buy permits is raised from energy taxation, and so cannot be used to correct existing labour tax distortions in the economy, there is a deadweight loss in spending revenues abroad, rather than saving energy at home. As a percentage of revenues spent on permits EGEM calculates this deadweight loss to be 7 per cent in North America, 16 per cent in

Japan and 39 per cent in the European countries. These deadweight losses mean that permit schemes will not automatically tend to towards the most efficient solution, but may have a less efficient equilibrium partially determined by the initial distribution of permits to different countries.

The distribution of costs and benefits from abatement will also determine the stability of the treaty, and in an efficient treaty some of the countries abating most (e.g. USA) may require side payments to ensure their co-operation. National targets do not provide a source of international revenue to make such side payments, so it is unlikely that they can be both efficient (assuming full information) and stable. International taxes provide potential for transparent side payments, and so can support stable agreements; however, the potential for inefficiency, and clashes with national policy on industrial, transport and equity issues could erode support for keeping to international commitments. The initial distribution of tradable permits can also be used to make implicit side payments, if there is enough centralised knowledge about the lowest cost distribution of abatement. However, the distribution of permits giving stability may reduce the efficiency of abatement, if the deadweight losses from buying permits abroad are large compared to gains from increased efficiency. In this case there will be a trade-off between efficiency and stability in using tradable permits.

Therefore, in most ways, the choice of which policy instrument will give the greatest operational effectiveness is empirical, and depends on the relative costs and benefits in specific cases. Most of the costs involved are amenable to standard economic modelling techniques, though others – such as the administrative costs of the different schemes, the costs associated with not reaching environmental targets and the benefits of non-price policies – are not. To introduce EGEM's numerical results on these issues, the next section describes the rationale behind the modelling structures used, and the important assumptions and caveats to bear in mind when interpreting its conclusions.

ECONOMIC MODELLING OF CLIMATE CHANGE POLICY

In the long run, preventing climatic change will involve a drastic reduction of our current dependence on fossil fuels. Industrialised countries currently spend between 3 and 5 per cent of their incomes on fossil fuels, and the value fossil energy provides to the economy is much greater than this. Energy is the largest single input to production and consumption, after wages and capital goods. Lowering the proportion of fossil energy to a sustainable level – by developing clean energy sources and increasing energy efficiency – will require fundamental shifts in the technology of production and patterns of consumption. The task of economic modelling is to try and quantify the cost to society of such a shift, both in terms of forgone material consumption and decreased personal welfare (for example, increased inconvenience caused by using public, rather than private, transport).

Predicting the consequences of such a major economic shift up to 100 years into the future is fraught with difficulty, as forecasts of economic performance over the next 2–5 years are often wrong by as much as 50 per cent. The reasons for this inaccuracy are many, but boil down to the fact that underlying economic structures evolve unpredictably. As we only have the past (and some good guesses) to guide us, forecasts are always bound to be flawed.

The main factors determining the long run economy are: population growth and workforce participation; technological progress; investment in education and productive capital (machines, infrastructure, etc.); availability of natural resources and services; and changes in preferences for different goods (for example, increased discretional spending on health as incomes rise). All these processes are endogenous to the economy in that they will be altered by income distribution, prices and government policies. Their evolution over time as a system will determine how much material production we produce and the impact this production has on the natural environment. However, our understanding of the links between different factors is very patchy. In general, economic modellers assume that preferences are unchanged in the future, that technology and population growth are unaffected by economic behaviour, and that only investment in productive capital responds to government policies such as climate change prevention. The contribution of environmental resources and services to production is not usually modelled.

There are two main methods used to assess the cost of reductions in fossil fuel use: top down, or macroeconomic models; and bottom-up, or engineering models. Top-down modellers construct mathematical frameworks describing the whole economy, based on assumptions about the underlying trends and theoretical economic relationships. They calculate the amount of production (Gross Domestic Product – GDP) predicted in the future, and then measure the difference between this business-as-usual case and a scenario with controls on carbon dioxide emissions. In contrast, bottom-up modellers usually concentrate on the energy-using sectors of the economy, use engineering data to model the potential for new technologies which will save energy and emissions, and then calculate how much these would cost to apply.

To reach the same abatement target, top-down models usually produce higher cost estimates than bottom-up analyses. This is because top-down models assume that the initial growth path of the economy represents an economically optimal use of inputs to production, and therefore that any deviation from the path caused by reducing energy use involves real costs. Bottom-up models contain no such assumptions of optimality, and find that 10–30 per cent of carbon dioxide emissions could be reduced at no net economic cost. The differences between the modelling methods are axiomatic, not empirical, and have led to much heated but ultimately sterile debate. However, both techniques have their strengths: top-down models in assessing the reaction of the whole economy to resource use changes; bottom-up models in explicitly describing new technology – and many of the latest generation of models incorporate both approaches.

The hybrid model approach is particularly useful in developing countries, where markets are very immature and prone to large distortions, capital scarcity, and information deficits. Chapter 6 of this book gives a detailed description of such a hybrid model of India, and clearly illustrates the differences and synergies between the two approaches. However, in constructing EGEM we have not adopted this approach, but instead have constructed a purely top-down model based on statistical measurements of past economic behaviour.

Rationale behind EGEM's structure

The focus of our policy analysis is the political economy of international action to prevent climate change over the next few decades. This is in contrast to most other economic analyses, which have looked at the potential for lowering global emissions over the next 50–150 years. Over very long timescales, the dynamics and details of economic behaviour do not matter much, and the role of such models is to provide broad sensitivity analysis of different influences on policy. However, when considering immediate policy responses, the dynamics of economic behaviour, unemployment, trade, and inflation will be vital to both the political settlement which is reached, and the operational effectiveness of any agreement.

The only way to model the detailed dynamic behaviour of economies is to fit a general theoretical model to past economic data using the statistical techniques of econometrics; the validity of the relationship between economic variables is decided by their accordance with past observed behaviour. In this way, both the structure and the parameterisation of the model are guided by the data, and constructing the model involves learning about the structures of the real economy. This approach contrasts with the general equilibrium macroeconomic models usually used to study climate policy; these have an a priori theoretical structure into which miscellaneous data is incorporated in order to parameterise the equations, and the behaviour of the model is not rigorously compared to the real economy. The advantages of the econometric approach are proven consistency with past data, rich dynamic structure showing transition behaviour, and data driven structural development of the model. The main disadvantages are that the model structure may not be logically consistent in the long term; that the model's statistical validity only holds over ranges of data (for example, changes in oil prices) experienced in the past; and that it can only be rigorously applied in countries with mature markets and a long history (15–20 years minimum) of collecting accurate economic data.

The final structure chosen for EGEM shows the trade-offs inherent in model construction. Though estimated econometrically, many of the equations are also derived from economic theory, giving consistent long run solutions particular economic areas (e.g. fossil fuel substitution). However, the model as a whole is not completely theoretically consistent, and so can only be confidently used over the medium term, here defined as the next thirty-five years (1995–2030). Data limitations and the estimation workload limited detailed economic modelling to the G7 countries (USA, Canada, Japan, Germany, France, Italy, UK). The rest of the world is represented by a full regional trade matrix, and varying degrees of detail in modelling regional economies. This scope limits the policy questions that can be tackled by EGEM, but not too drastically, because the seven main countries emit 88 per cent of CO_2 emissions covered by Annex II of the FCCC; that is, countries which have agreed to quasi-binding GHG emission controls in the near future.

Modelling energy use in EGEM

In EGEM, each country's production is a function of three aggregate inputs: labour, capital and energy. Energy use is divided into fossil and non-fossil fuels; and fossil use is further disaggregated into coal, oil and gas, because these give off different quantities of CO_2 per unit of delivered energy. The main tasks in constructing EGEM were to model the determinants of energy and fuel use in the past, substitution between different fossil fuels, and then measure how changes in energy use affect aggregate production and the use of other inputs. EGEM does not have an estimated consumer welfare function; therefore, as in other models, we measure the cost of abatement as a drop in GDP, not as the more correct form of changes in consumer welfare.

For each of the main countries, a stable cointegrating relationship was established between aggregate energy use per unit of GDP (energy intensity), relative energy prices, and a dummy trend representing technology. This assumes that economies expand using the same amount of energy per unit of production, unless either technology improves or prices rise. Cointegration techniques identify reliable long run correlations between trended economic variables, avoiding the problem of spurious regressions (seemingly significant relationships between actually uncorrelated data) inherent in standard statistical methods. For all countries this estimation found that price levels had a very small effect of energy use (elasticities of ≈ 0.1), and most changes in energy intensity were explained by the dummy technology trend. This is an undesirable result because it suggests that increased taxes on energy use, a favoured policy option, will have little effect in stopping pollution. Therefore, this relationship was re-estimated using a trend term which included endogenous economic factors – cumulative energy prices, investment, industrial production – and an element to represent random changes in (exogenous) technical progress. This model assumes that investments in developing energy efficient technology are driven by current and expected prices, and, once developed, improved technology will stay in use even if prices fall. Technological development is therefore partly endogenous, with only a residual exogenous component which results from random scientific discoveries unconnected to economic forces.

In all the countries studied, cumulative energy prices proved to be a strong determinant of technological change in energy efficiency, with industrial production and investment much more secondary factors. The simulation properties of this model have important policy implications. In other top-down models, trend energy efficiency is assumed to be exogenous and unaffected by prices, and so stabilisation of fossil energy use can only be accomplished by continually rising energy taxes suppressing demand. However, the EGEM model can stabilise fossil energy use into the future with a finite tax rate (approximately US$750 (1990) per tonne of carbon), as this price increases advances in energy efficiency to the point where they outweigh the growth in energy demand caused by an expanding economy. In this aspect EGEM behaves like a bottom-up model with a non-fossil 'backstop' energy technology (for example, nuclear fusion), which comes into use once energy prices reach a certain level, stabilising carbon emissions into the future.

The advantage of EGEM's energy modelling structure is that decisions over the timing of abatement policy will affect technological progress, and thus delaying the introduction of carbon taxes has an economic cost. In models without this feature, delaying abatement measures can only have environmental costs and, as these are difficult to measure, such models have an in-built bias towards inaction in the short run.

Energy use and production in EGEM

Measuring the influence of macroeconomic factors on energy use is a reasonable, if complex task. Estimating the corresponding linkage – how changes in energy use affect production – is much harder, because even large shifts in energy use (at 3–5 per cent of input costs) are small compared to the size of random economic 'noise' and business cycles.

Rational behaviour by energy consumers implies that the marginal productivity (or welfare) of a dollar's worth of each input (capital, labour and energy) must be equal; otherwise, a costless shift in the input mix would increase output/welfare. If the price of one input rises its use will drop and that of the others will rise until this rational equilibrium is reached again. This model would therefore predict that increasing fossil energy prices with a carbon tax will always produce a fall in aggregate output and welfare. Though conceptually useful, this model assumes that technologies exist which are productive with any mix of inputs; this is probable for whole economies, if not at the industry level. However, the evolution and dynamics of technological development are not explained, nor are the intuitively plausible links between investment in capital and the diffusion of technological advances through the economy.

This production model is not easily compatible with the type of endogenous technical change embodied in the energy demand equations, because it assumes improvements in technology to be costless. If we assumed that all the changes in energy intensity modelled by the price-driven trend came from such pure technical progress (the disembodied technology assumption), then to be consistent with traditional production models the costs of abatement will be very low. However, if the price-driven trend is really modelling shifts in *investment* between labour-saving and energy-saving capital, with the new technology being embodied in these machines and equipment, we would be vastly understating the cost of abatement by incorporating the assumption of disembodied technical change into a traditional model.

The available empirical evidence seems to favour the embodied technology assumption. Engineering studies indicate that energy and pollution reduction into the foreseeable future will usually be accomplished by investing in more capital intensive equipment (wind turbines, electric cars, etc.). If this was new investment, an increase in energy prices would unambiguously raise economy-wide investment, but the econometric evidence of past behaviour shows that this link is statistically insignificant. However, higher energy prices can be reliably linked to increases in labour use; this is consistent with companies switching investment from saving labour to energy efficiency, inside an overall investment constraint. Lower investment in labour saving increases employment, but lowers output and wages, while leaving capital unchanged apart from accumulation effects. In the long run, technological change advances through learning-by-doing, and replacing installed energy efficient devices at the end of their useful life is assumed to involve neither extra cost nor further diversion from labour productivity enhancing investments.

EGEM was estimated using this framework of embodied technical change, and also a more traditional production structure where it was assumed that there were *no* increases in technical efficiency in the energy sector due to higher prices. The models were then compared to check comparative simulation properties. Using an embodied technology approach reduced the cost of achieving any particular abatement target by over 50 per cent in some cases. The relative cost of abatement in each country was also different in the two models. In the traditional model, low fossil energy use in – for example, Japan – implies high energy productivity, and so reducing energy use has high costs. However, this simple relationship is complicated in the embodied technical change model because the impact of investing in energy saving will depend on the contribution of capital to labour productivity, and the underlying rate and turnover of investment. Abatement costs were much lower for Japan (by an order of magnitude) in the embodied model, because even substantial investments in energy saving are relatively small in Japan's investment and capital rich economy.

Policy implications of EGEM's modelling structure

With all complex models the devil truly lies in the detail, and this is especially true when considering climate change. Choosing a particular modelling methodology will inevitably bias policy results one way or another, and it is important these deficiencies are mentioned *inside* the resulting policy analysis. Theoretically-based econometric models such as EGEM will tend to *overestimate* the cost of abatement efforts because they ignore no-regrets options by assuming inputs are being used to maximum efficiency in the business-as-usual case. Also, basing parameter values on a relatively short estimation period implies that the economy is more inflexible when faced with price changes than it really is in the long run. In EGEM these problems have been partially overcome by incorporating price-driven technical change, and restricting simulations to thirty-five years – though this is still longer than the estimation period. However, one advantage of an econometric model over a general equilibrium approach is that it allows for imperfections in the labour market and taxation policy. Therefore, recycling carbon tax revenues through labour taxation can reduce economic distortions, lessening abatement costs; and these effects are very significant in EGEM.

Perhaps more important for general policy implications, given that the absolute size of model predictions are always rather suspect, are assumptions underlying the modelling of technology. Two main arguments in climate policy are what constitutes a good hedging strategy in the face of uncertainty, and how this strategy can be most effectively implemented. If technology is modelled as being exogenous to the economic system, early abatement only prevents climate damage caused by rapid temperature rise, but incurs large costs by not waiting for more efficient technology to be available. However, if research and development is a dynamic economic activity driven by energy prices, significant early abatement will be a good hedging strategy against both climate damage and the future costs of abatement. The results from EGEM mentioned above show that the policy differences between the two modelling methods are real and substantive, and so models with exogenous technical change *cannot* be used to assess the dynamics of climate change policy over the next few decades.

ANALYSIS OF CLIMATE CHANGE POLICY ISSUES

The Framework Convention on Climate Change provides an basic structure of international law, within which countries can negotiate to their environmental and economic advantage. At present the aims of the treaty are relatively weak, but it is to be expected that they will be strengthened in the future. To fulfill its aims of preventing dangerous interference with the global climate, aggregate abatement targets will have to be a significant proportion of GHG emissions in the medium term; otherwise the marginal benefit of abatement – in terms of risk mitigation – will be very small.

Significant medium to long term abatement requires that the developing world reduce its emissions in the future, but this is unlikely to happen if the developed countries cannot agree to limit their own emissions, or to provide the money and technology for abatement in other regions. Therefore, successful abatement in the developed world is a necessary precursor for wider global efforts. This implies that when setting abatement targets the OECD countries will have to take into account damage in the developing world as well as in their own regions, even though the developing countries bear no binding commitments. If these damage estimates are based on purely financial measures, then it is likely that the majority of global damages, which seem to be non-market costs in tropical countries, will not influence OECD abatement levels. This is consistent with the OECD expecting compensation for abating at a higher level than their own damage costs would justify, but not with co-operation based around mitigating *global* welfare losses.

If they did not consider global welfare costs, illustrative scenarios run in EGEM suggest the OECD countries could plausibly agree to maintain emissions at 1990 levels over the next thirty-five years; in EGEM this represents a cumulative reduction of 23 per cent from BAU levels over this period. If the developed countries considered the welfare costs of climate change in the developing world this could push the target up to ≈ 25 per cent reduction from 1990 levels over the same period; a 42 per cent cumulative reduction from BAU. EGEM calculates that this more stringent target would cost 3–4 times as much to reach, even using the most optimal distribution of abatement. However, it seems unlikely that the developing world will accept future restrictions on their emissions unless such an major effort was forthcoming, especially as the vast majority of current increases in GHG concentrations have come from economic activity which has benefited the developed world.

Commitment by developed countries to stabilising GHGs at 1990 emission levels over the next 10–20 years is probably the weakest target being currently considered inside the FCCC process; a reduction of 25 per cent below 1990 levels is the largest target likely in the near term. Therefore, these are useful scenarios with which to examine how economic interactions, and the choice of co-ordinating policy instruments, are likely to affect the operational effectiveness of any agreement.

Achieving global abatement targets

No global agreement could be considered successful unless its timetable for global emission reductions is kept. Accurate achievement of a treaty's aims is not only environmentally important: institutional mechanisms must be seen to be working for countries to have confidence in the process. If it seems that some countries are not abating stringently enough, or that somehow targets are not being met, this could seriously weaken trust between parties, and thus lower co-operation – with consequent adverse effects on the collaborative effectiveness of negotiations.

The most basic requirement for achieving aims is a comprehensive monitoring of GHG emissions (and enhanced sinks) from different countries, and this is already mandated under the FCCC. However, the type of co-ordinating instrument used to enforce the FCCC mandate will also affect the accuracy of reaching targets. The complex nature of international energy markets – only oil has a true international spot trading market, and even this is rather thin – and the many distortions present mean that setting an international tax rate (as in the European Union carbon/energy tax proposal) will lead to an unpredictable level of emission reductions. Given the environmental importance of ensuring timely abatement, this argues in favor of using national targets or tradable permits to co-ordinate abatement. National governments can then use their far superior information to implement policies – which can include national carbon taxes or regulations – that will ensure targets are hit.

It has been suggested that under a tradable permits scheme, countries should be able to shift their emission reductions in time as well as in space (i.e. permit trading between countries). This could be organised by giving each country a cumulative quota of emissions over several decades, which could be bought and sold between countries like any property right (this is sometimes termed the 'nationalisation' of the atmosphere). Such a scheme could reduce abatement costs by avoiding premature scrapping of capital equipment – for example, coal power stations – though these reductions must be weighed against counter-arguments that early abatement stimulates technological development (as discussed above); however, such a scheme will not ensure the treaty's aims are met. Allowing countries to choose when they abate will probably result in abatement measures being delayed, if only because this will be politically expedient. This reorientation of emissions reductions could significantly alter the rate-of-change of temperature rise, if large emission reductions are being rescheduled over 30–50 years. No individual country has an incentive to consider the global environmental impacts of this shift; there will therefore be an external climate damage cost which does not enter decision makers' calculations. As with any externality, the only way to ensure optimality is to co-ordinate activity at the international level. Thus, the consequences of allowing countries discretion to shift abatement measures in time must be very carefully modelled before such increased flexibility is given, if it is given at all.

Carbon leakage

Even when abatement targets and scheduling are strictly defined, the fact that only the developed countries are likely to face binding targets in the next decades introduces further complications in achieving global emission reductions. In an incomplete treaty, Joint Implementation with uncommitted countries cannot ensure accurate abatement, because there is no emission baseline to refer to in the abating country. More importantly, market behaviour may undermine the effect of emission reductions in one region by increasing pollution from uncontrolled countries – so-called 'carbon leakage'. There are two main avenues by which this may take place: reductions in global fossil fuel use could lower international fuel prices, causing higher consumption in uncontrolled countries; and energy dependent industries could be made uncompetitive by imposition of a large carbon tax, causing them to lose market share, or migrate, to uncontrolled areas. In contrast, energy use in uncontrolled areas could be lowered if there are significant income reductions in the developed world from imposing CO_2 reductions, as exports of manufactured goods would fall (the income effect).

Significant carbon leakage would increase the cost of developed country abatement, reducing the size of abatement target likely to be agreed, and potentially undermining progress on stricter targets. Of course, if the developed countries refuse to implement abatement because of leakage, developing countries are unlikely ever to join the treaty and non-cooperation will result, to the detriment of all parties.

The importance of carbon leakage can only be found by modelling; however, the long run values of critical parameters needed, such as trade elasticities and oil market supply elasticities, are not well-known and almost certainly vary over time. Therefore, in EGEM we have attempted to use the best available estimates from the literature, along with the econometrically estimated energy demand model for the OECD, to model the magnitude of the various avenues for leakage. Simulations were run over a range of abatement levels up to 25 per cent below 1990 levels, though analysis concentrated around stabilisation at 1990 levels.

It appears that oil price leakage effects are small, which – despite the obvious uncertainties in modelling world oil markets – is a fairly robust conclusion. The main proviso on these results is the assumption that gas and coal markets do not become sufficiently internationalised to contribute their own effects. As global abatement increased towards a 25 per cent reduction from 1990 levels, the oil price effect in EGEM increased more than proportionately, indicating its potential importance in future negotiations. A major limitation of the model is that fuel markets outside the OECD are simplistic, including no interfuel substitution. Statistical work when building EGEM showed that coal and gas prices inside the OECD are often partially determined by oil prices, as well as extraction and delivery costs; thus a fall in oil prices could also increase consumption of higher carbon content coal and lower carbon content gas. The immaturity of energy markets in the important coal consuming areas, especially China, India and the former Soviet Union, makes the type of econometric analysis of price reactions we performed in the G9 countries inappropriate here, so the size of these effects can only be 'guesstimated' at the present time.

EGEM models both aggregate changes in trade patterns from price increases and explicit leakage through industrial relocation. Relocation is limited to the most energy intensive industries of steel, concrete, and bulk chemicals; in which energy accounts for 10–15 per cent of input costs, rather than the 3–6 per cent common in other industrial sectors. Intuitively, we might expect low carbon taxation levels to produce little industrial relocation, as transaction costs would outweigh cost savings; however, with a very large tax, all of the most energy intensive industries might move away. Therefore, proportionately, the influence of relocation will decline compared to oil price and income effects as abatement targets increase. Under these circumstances, the highest rates of leakage from relocation will occur at medium taxation levels, which means incremental movement to higher taxes may encounter strong, but temporary, opposition from the affected industries. It should also be remembered that 50–70 per cent of carbon emissions in developed countries come from non-traded sectors (domestic use, services, etc.); this gives an upper bound to carbon leakage which has often been exceeded in past studies!

When stabilising at 1990 levels, trade effects seem to outweigh fuel price effects, accounting for most of the 9.5 per cent carbon leakage. Total leakage increases to up to 37 per cent of OECD abatement for the 25 per cent reduction target; though neither of these figures include negative leakage from changes in OECD incomes. The model attempts to give a probable upper bound for leakage, by using high values for trade elasticities but limiting the effect to the most energy intensive industries. However, it should be noted that if process efficiency of the relocated industry in non-OECD countries remained much lower than OECD, leakage would be higher than modelled here. The baseline projection is also uncertain, as energy intense industries are likely to relocate to the developing world over the simulation period even in the absence of carbon taxes – demand for energy rich materials, such as steel and bulk chemicals is growing fastest there.

Reductions in OECD incomes due to carbon taxes have very significant effects on leakage and can completely remove it at some tax levels. OECD income losses may be small (≈ 1 per cent) for stabilisation at 1990 emissions, but they have larger than proportional effects on trade and thus emissions outside the OECD. Leakage including income effects is virtually zero when emissions are stabilised at 1990 levels, and increases to 19–25 per cent (depending on policy instrument used) with reductions of 25 per cent below 1990 emissions. However, the results at the higher abatement level should be treated with some caution, given the problems with modelling such large trading shifts using a fixed real exchange rate regime.

Another weakness of our modelling is that it does not reflect how changing patterns of worldwide consumption will affect trade, making non-OECD countries less dependent on exports for growth. Changes in the pattern of OECD consumption due to energy taxes will also affect energy use outside the OECD, by stimulating technological innovation, and possibly causing a shift away from energy intensive processes. New theories of how countries innovate and imitate at different stages of development suggest that, even with no worldwide agreements to limit emissions, any new low energy technologies will be disseminated worldwide as they become cheaper, and as 'standard' products incorporate previously state-of-the-art technology (Grossman and Helpman 1992). In fifty years time the OECD may be importing renewable energy technologies rather than cars from South-East Asia! To model such shifts numerically is impossible, but the potential for such synergistic effects, which have been repeatedly observed in the past, should be borne in mind when interpreting the likely biases in any quantitative results.

Nonetheless, in the short term it appears that while carbon leakage may reduce the effectiveness of carbon taxes, a tax throughout the OECD would lose a relatively small proportion of its abatement through leakage, of the order of 10–20 per cent for the most stringent target. This is certainly not large enough to justify abandoning abatement efforts in the developed countries. However, the actual size of leakage is heavily dependent on income effects, and relying on this avenue is not a sustainable way of reaching international emissions targets.

In the long term, to address carbon leakage or global stabilisation effectively, non-OECD countries must face some emission restrictions, and so must be aided by technology transfer to ensure their energy intensity is kept low. This may include anything from efficient power generation and consumption technologies, to effective biomass and renewable energy plant. Implementing efficient market incentives for carbon abatement, via some form of joint implementation, or 'producer taxes' on energy (particularly oil) that are then partially redistributed to developing countries, could also play their part in stabilising developing country emissions without restricting their economic growth.

Efficiency in international agreements

The standard economic meaning of 'efficiency' involves balancing the cost and benefits of any action; the efficient solution being defined as the point at which the marginal cost of doing an action equals marginal benefit resulting from it. Here, this type of efficiency is covered by considerations of collaborative effectiveness, and is far more complex than a simple piece of marginal analysis. Therefore, this section considers the simpler case of achieving a previously defined target at least cost.

The FCCC explicitly states that achievement of any international target should be done cost-effectively; this may seem a rather obvious statement, but it actually contains many important political and institutional issues. The most general condition for the *aggregate* efficiency of a policy instrument for controlling emissions, is that the marginal cost of emission reductions (GHG abatement or sink enhancement) be equal in all complying countries. Therefore, there are no opportunities for reducing total costs by shifting a unit of emission reduction from one country to another. The co-ordination problem is how to ensure this condition holds, when internationally the cost of abatement in each country is unknown.

Flexibility in the location of emission reductions is given by tradable permits, and because equilibrium is reached by individual countries trading with each other there is no need for centralised information on abatement costs. Efficiency will not be given by imposing equal per centage reduction targets in each country, because heterogeneous national energy needs and existing fuel prices will lead to different marginal abatement costs. If all markets are perfect, an equal international carbon tax will also produce emission reductions at least cost, as long as existing taxes on fuels are harmonised beforehand (that is, equalising existing national energy taxes before adding additional ones). This is because with harmonised taxation all consumers will face the same cost of energy, and so will conserve until the marginal cost of abatement equals the energy price. Of course, if energy taxes are not harmonised consumers will face differing energy prices, and a flat-rate international tax will be inefficient. Even harmonised taxes will be inefficient if energy markets are not perfect, as an equalisation of marginal costs will not take place. Within the OECD existing energy taxes are highly heterogeneous, especially on oil products, and analysis of the energy demand equations in EGEM showed that these taxes accounted for the majority of past differences in energy market responses to price rises. A harmonised international carbon tax evens out these differences; therefore the cost of inefficiencies from existing imperfections in energy markets should be relatively small.

The advantages of flexibility in locating emission reductions are relatively clear inside national boundaries, but there are some dangers in using flexible instruments internationally. First, the allocation of emission reductions means that some countries will incur more costs than others, and so some type of compensation will be needed to keep countries in the treaty; this issue of treaty stability is considered in detail in the next section. Second, the cost-effectiveness of such flexibility will depend on the domestic political economy of emission reductions, and how different priorities are balanced by national governments.

Domestic political economy of carbon taxes

The domestic economic impact of a carbon tax is composed of direct welfare effects, macroeconomic effects on output, and benefits from recycling taxes through the labour market in a way that reduces unemployment. Each of these factors is distributed differently: direct welfare losses from reductions in energy consumption proportionately damage the poor more, macroeconomic effects are relatively evenly spread (except for employees in energy intense industries which lose competitiveness), and tax recycling options differentiate between employed and unemployed workers. The timing of any taxation scheme is also important, since different types of taxation and recycling can shift the cost of mitigation further into the future, penalising future generations, but benefiting current voters! Balancing the concerns of different groups over efficiency, employment and equity is the job of national governments, and will determine the domestic acceptability of abatement commitments agreed internationally.

The most contentious domestic economic issue is whether imposing a carbon tax, and recycling the revenues through employers' labour taxes (employers' national insurance in the UK), will reduce unemployment and raise output – the so-called triple dividend (environment, employment and output) of ecological tax reform. This result is possible because taxes on inputs to production produce distortions which lower output; for example, a tax on investment earnings reduces the amount of capital in the economy by lowering returns to potential investors. In a democracy these distortions are considered a price worth paying for the public goods (social services, policing, defence, etc.) which can only be funded by this revenue; though it should be noted that public services such as education, training, health care and infra-structure investment also have a strong beneficial effect on economic production. However, it is reasonable to assume that marginal taxation levels suppress material production.

The efficient way to raise revenue, analogous to the distribution of abatement, requires that the decrease in output per unit of revenue raised be equal for all taxation sources. If this is not true, then redistributing the tax burden would raise output at no cost. It is argued that carbon/energy taxes provide a ideal source of revenue with which to perform such a tax shift, because reducing their use has environmental benefits and reducing the tax burden on employment will create jobs by lowering costs to employers. In reply, economists often argue that if there were such an opportunity to increase output and employment, any rational government would have done it already, and so energy taxes must incur real costs to produce environmental benefits.

Assumptions of governmental rationality are always contentious, but the real point of the economists' argument is that benefits from re-optimising the tax system cannot be 'credited' to climate change policy if they could also be found in other ways; for example, through increased taxation on general consumption. However, carbon taxes may indeed be the only feasible instrument given current world-wide aversion to income- and capital-based taxes, and the regressivity of consumption taxes. The unique feature of carbon/energy taxation is that the associated environmental benefits (related both to climate change and local issues) tend to mitigate the distributional and efficiency effects, thus making it, on net, less distortionary than other available instruments. EGEM estimates that agreement to stabilise G7 CO_2 emissions at 1990 levels over the next thirty-five years would give income streams of $\approx$ 1–4 per cent of GDP with which to perform such tax reform in each country. The resulting tax shift would reduce unemployment by 35 per cent, and lower output losses by 40 per cent compared to recycling funds through income taxes. However, these gains will reduce proportionately as the economy grows and the revenue stream shrinks relative to the national wage bill; virtually disappearing after thirty-five years.

Carbon taxes impact employment, environmental quality and economic output, but EGEM's results suggest there are no *permanent* win–win–win situations where all three are improved. These will exist if the costs of climate change are mainly impacts on production, not welfare, but we cannot model these costs at the moment. Even though they do not completely outweigh the costs of abatement, the benefits of recycling energy taxes through labour taxes, rather than income taxes, are highly significant. EGEM calculates the extra output produced, as a percentage of revenue collected, to be 7 per cent in North America, 16 per cent in Japan and 39 per cent in the European countries. However, carbon taxes are quite regressive; this is counteracted by the fact that most new jobs are likely to be created at low income levels, and so be poverty alleviating. The regressivity is perhaps less than would be expected, because car ownership attracts a large proportion of carbon taxes, which by-pass the poor. We calculate that a redistribution of up to 20 per cent of carbon tax revenues would give generous compensation for increased energy prices in the G7 countries, if at the cost of fewer new jobs and higher aggregate output losses.

Balancing the positive side of revenue recycling, the results from EGEM also show that the dynamic demand-side impacts of taxation and recycling in the initial 10–20 years are of a similar size to the supply-side effects (the focus of more long run studies). A significant reorientation of the tax system could lead to an inflationary spiral as workers bid up wages to take account of higher prices; this would not occur with income tax recycling because take-home pay would instantly adjust to increased prices. If these inflationary effects are assumed to be severe, EGEM estimates that the consequent higher interest rates needed to maintain macroeconomic stability will wipe out all the output gains from labour tax recycling, even though the employment gains would largely remain.

However, probably the largest problems associated with ecological tax reform in general, and carbon taxes in particular, are in the political dynamic of implementation. The majority of the electorate will immediately lose from imposition of a tax – though reduced prices would eventually balance most of these losses – and in energy intensive industries there will be large job losses. These industry and labour constituencies will be far more vocal than the unemployed who find work due to lower producer wages, or the present and future population who benefit from cleaner air, less traffic congestion and reduced risk of adverse climatic change. Even if the environmental benefits become clear and garner mass support, the temptation to channel revenues through income tax reductions will be great, because EGEM shows that in many countries this will provide more short term benefits to the majority of the electorate than reducing employers' labour taxes.

Effect of national policy on international agreement efficiency

The choice of taxation levels, instruments and revenue recycling policy is far more complex than finding the distribution of abatement which produces the lowest aggregate cost. Institutions for co-ordinating international action to reduce emissions will have to recognise the constraints imposed by domestic politics and economics, and these will form a large component of any substantive negotiations on sharing the abatement burden. For while North-South interactions on this issue are currently driven by ideas of equity and opportunity for development, negotiations between the industrialised countries will be based on relative macroeconomic costs and the impact of taxes on domestic welfare in all its facets.

In modelling these effects, the various G7 countries are aggregated for convenience into three regional blocks: North America (USA and Canada), Japan and Europe (Germany, France, Italy and UK). EGEM has no explicit technologies, so non-price abatement policies cannot be modelled, and the distribution of abatement between these regions involves setting appropriate carbon tax rates in each one. For each stabilisation target – 1990 emissions and 25 per cent below 1990 levels – five different policy instruments were modelled: regional stabilisation targets; a flat-rate international tax; harmonised international taxes; tradable permits; and an optimal international tax. Optimal international taxation (that is, collectively minimising costs by setting different taxes in each region) is practically unfeasible because of information constraints; however, it gives a benchmark against which to measure the efficiency of tradable permits. In the absence of revenue raising costs, tradable permits should be as efficient as optimal taxation. Three initial distributions of tradable permits were modelled: equal permits per caput; equal permits per unit of output; and permits based on baseline energy consumption.

Stabilisation of cumulative G7 emissions at 1990 levels over the next thirty-five years is not a very ambitious target, and this is reflected in the low cost of abatement for reaching it (0.75–0.87 per cent of G7 GDP). Unit abatement costs do not differ vastly between regions or with abatement level over the range in question, and so the ability to lower costs by using efficient policy instruments is limited. Switching from a system of regional stabilisation targets to an optimally differentiated tax gives only a 14 per cent drop in output losses. Dead-weight taxation losses from buying and selling permits proved to significantly affect the distribution of abatement using tradable permits, reducing their efficiency compared to an optimal tax. Distributing permits based on population or carbon intensity results in a 7 per cent efficiency loss compared to optimal taxation, which – though small – is half the saving gained by not using regional stabilisation targets. Distributing permits based on baseline energy use in each country results in an equilibrium very close to the optimum, and very little buying and selling of permits. Therefore, for this abatement target tradable permits only guarantee efficiency if the optimal distribution of abatement can be imposed from the start, reducing their attraction compared to international taxes. Surprisingly, harmonised taxes produced *larger* output losses than flat-rate taxes (though smaller welfare losses), because they resulted in fewer new jobs in Europe (Europe having high initial energy taxes). Higher initial unemployment in Europe means that these jobs have a large output premium (39 per cent of revenue recycled compared to 7 per cent in the USA), and this materially affected aggregate abatement costs.

Achieving these small efficiency increases involves a large shifting of abatement responsibilities, reducing abatement in Japan and increasing output losses in North America by 60 per cent compared to hitting a national stabilisation target. Efficient abatement also reduces the amount of employment created from carbon tax recycling and labour/energy substitution by over 30 per cent. Therefore, the side effects of efficiency from both the shift in abatement, and lower employment creation, could undermine domestic political support for carbon abatement. If regions use the flexibility of tradable permits to balance increasing employment against abatement efficiency, the results seem to be highly sensitive to the relative valuation of jobs created. Using the *ad hoc* assumption that governments value a new job at 1.5 times the unemployment benefits it saves, overall G7 employment in EGEM actually falls, but much more is created in Japan where benefits are relatively high. The efficiency cost of targeting employment means that for some permit distributions G7 abatement costs are nearly the same as when using regional stabilisation targets.

Tradable permits allow individual governments more flexibility in balancing the conflicting goals of efficiency and employment, without having to discuss these issues explicitly in an international forum. In contrast, if an optimal taxation regime was to be negotiated under the FCCC all these factors would complicate the division of abatement effort and the setting of overall abatement targets. The interaction of all these issues suggests that for such a modest abatement target, the expense of constructing a full tradable permit regime will not be compensated for by either a marked increase in abatement efficiency, or a greater ability to balance international commitments with domestic agendas on employment and equality. Therefore, in the medium term an extension of regional stabilisation targets beyond the year 2000 will be almost as efficient (especially if some JI between developed countries is used), easier to agree on, and cheaper to co-ordinate than a tradable permits system.

These arguments do not hold if agreement is reached to stabilise emissions at 25 per cent below 1990 levels. Imposing taxes large enough to meet this target almost halts the growth of fossil energy demand in EGEM, and stabilises emissions into the future with virtually no further increases in energy taxes. This is because long run energy demand in EGEM is based on past measurements of how price rises stimulate growth in energy efficiency; these improvements then become incorporated in technology, and can be applied costlessly into the future. The production structure of EGEM means that the immediate switch to energy efficient technology implies less investment in improving labour productivity, and this has the combined effect of lowering output, and increasing employment (or rather slowing down ongoing reductions in labour intensity). This large investment switch reduces G7 output by 2.8–3.5 per cent over the next thirty-five years, depending on which policy instrument is used, but these output costs become progressively smaller in the future.

Unlike the smaller target, the distribution of these costs when imposing regional stabilisation varies considerably from 1.8 per cent in Japan to 5.7 per cent in Europe (undiscounted cumulative GDP loss). Compared to imposing regional stabilisation targets, using any other policy instrument reduces overall costs by around 25 per cent, which is equal to the welfare cost of hitting the 1990 target. In this case, using tradable permits provides a good approximation to the optimal taxation scheme, no matter how they are initially distributed. This is because the relative importance of creating new employment and recycling revenues declines as abatement costs increase, the economy moves towards 'full' employment, and the marginal productivity of new jobs falls. This increase in costs means that even if governments value new jobs for their own sake, as modelled above, the distribution of abatement is changed by under 1 per cent.

Carbon leakage also becomes more important at this abatement level. When the impact of energy intense industries migrating between the G7 and non-abating countries is taken into account, output savings from using tradable permits increase from 25 per cent to 32 per cent. These figures should give a reasonable approximation to the difference in efficiency between regional targets and tradable permits, because they both give the same freedom to pursue no-regrets non-price policies. Unlike the lower target, regional stabilisation imposes much higher costs when a significant amount of abatement is agreed to, therefore it will be worthwhile constructing a full tradable permit infrastructure. Tradable permits will allow each country to reach this target efficiently, and with enough flexibility to be able to respond individually to the industrial relocation of energy intense industries.

The efficiency gains from using tradable permits and optimal taxes in EGEM are lower than some other models have found. This is surprising given that EGEM is based on individual measurements of each G7 country's economy, and not the more usual method of regional assumptions. Some of the differences arise because we are only moving abatement between the developed countries, which have similar levels of energy use per unit output; the largest efficiency savings will come from moving abatement to the developing world which is still using out-of-date technology. Such an inclusive system will depend on the developing world agreeing to emission restrictions; otherwise they cannot be allocated permit distributions. This would result in large payments from North to South for the right to pollute, and there are many governance problems to be solved over this process. For example, given the immaturity of democratic political mechanisms in many countries, would elites use the selling of permits to gain hard currency, while reducing the ability of their countries to develop in the interests of the mass of people? Questions of governmental legitimacy and representation are very difficult to discuss in an international forum, such as the FCCC, as they depend on the fiction of the representative nation state. Therefore, taking advantage of the flexibility of a global tradable permit scheme will probably take decades to achieve, and be accompanied by much political argument and dispute.

The stability of international agreements

EGEM's results show that for a significant cut in emissions, the total cost of abatement is highly dependent on the type of policy instrument used. The use of tradable permits produces savings over a system of regional targets, but the costs of raising revenue reduces these gains compared to the hypothetical equilibrium found by an optimally differentiated international tax. Such an international tax is unlikely to be used because of sovereignty issues. The corollary of the sovereignty issue is that using tradable permits allows governments to construct their own objective functions, and so weigh the benefits of increasing employment and reducing tax regressivity against abatement cost reductions. This increased flexibility is likely to make domestic political agreement with international policy easier to achieve, as it allows political trade-offs between different internal constituents.

However, even with the flexibility to use different domestic policy instruments, and to optimise the political acceptability of abatement commitments, there is no a priori reason why countries should co-operate together to reduce GHGs emissions. In fact, countries always have an incentive to free-ride, because by defecting they save their abatement costs but still benefit from the abatement of other countries. Incentives to free-ride are greatest if a country has low damage costs and low abatement costs, as it will abate substantially in an efficient agreement, but will gain little from reduced emissions, so making a net loss from co-operation.

The stability of coalitions made up of actors who are working together to achieve common benefits, but have an incentive to unilaterally cheat on any agreement, has been extensively studied in the economics literature; especially in the context of cartel formation. However, empirically-based analysis of the potential for forming lasting coalitions is rare and difficult to transfer to the real world. Even when cost and benefits are well-defined, the stable agreements found by modelling are highly sensitive to assumptions about the order in which countries act, the information they possess, and the institutional arrangements surrounding agreement. Usually, no one equilibrium is more likely than any other; thus, great care must be taken in interpreting modelling results.

North America is both the largest source of GHGs and, from EGEM's results, the cheapest place to abate; therefore, any cost-effective agreement will require significant changes in behaviour in both the USA and Canada.

In climate negotiations, North America also seems to assume its damages from climate change will be low, implying that major abatement obligations would not be politically acceptable. America could be encouraged to participate if side payments from other countries compensated for its abatement effort. However, these transfers cannot outweigh the benefits North America brings to any existing agreement, since all other regions must also see a improvement in their welfare from its accession. Such transfers can be split into two types (from Botteon and Carraro 1995):

- Profitability payments, which ensure that all countries benefit from agreement.
- Stability payments, which ensure that no country can gain by defection.

By definition, a stable treaty must be profitable, but the reverse does not hold. It is possible that there is no way of distributing the benefits from co-operation which leaves all parties better off, and simultaneously dissuades them from free-riding.

Ensuring the profitability of a treaty involves having a rule with which to share the surplus produced by co-operation. The simplest case – an example of a Nash Bargain – is defined between two equal sized parties with unlimited access to transfers, and has an equilibrium where the gains from co-operation are split evenly. However, EGEM shows that providing transfers has a significant macroeconomic cost, and so the equilibrium bargaining position will be affected by the form of revenue recycling and by the distribution of any tradable permits between regions.

While Nash Bargains tend to provide an equitable sharing of the benefits of co-operation, they do not necessarily reflect the full negotiating power of the different parties. For example, if the gains from co-operation are redistributed based on the added value each participant brings to a coalition by their accession, benefits will be more unequally distributed between the co-operating regions than the Nash case, favouring those that contribute most. This is more likely to give a stable agreement than distributing transfer payments based around a Nash Bargain, because a country will be less likely to gain from defection; however, there is no theoretical reason why this distribution will be stable. In fact, there may be no way of forming a stable agreement using side payments which will produce an improvement to all parties compared to a less extensive agreement. If this is the case, then it is optimal for some countries to free-ride, and for the remaining parties to co-operate just among themselves.

The difficulty in modelling these types of interactions inside EGEM is that one side of the incentives for agreement or dispute – the damage costs seen by each country – is unknown. To look at stability using side payments, we hypothesized that the distribution of perceived climate change damages would be proportional to the cost of meeting the national abatement targets set under the FCCC; this implies that damage costs per unit of CO_2 emitted are lowest in North America and highest in Europe.

For aggregate abatement at 1990 levels EGEM calculates that co-operation will be stable for full co-operation even without transfers, except when North America joins an existing partnership of Japan and Europe. However, Japan and Europe gain so much from America's participation that they can easily afford to fund side payments to ensure profitability and stability. However, when OECD emissions are reduced 25 per cent below 1990 levels, North America loses every time the coalition expands beyond two countries. To guarantee America's participation in the full co-operative scheme, around 50 per cent of the benefits from co-operation have to be transferred to it. Of the other regions, Europe's high damage function means that it always values full co-operation and needs no extra incentive to join; however, Japan would prefer to free-ride on the abatement efforts of Europe and America unless side payments were available.

Therefore, in order to ensure the participation of either Japan or North America, Europe will have to fund the side payments as it gains most from co-operation, even when it has been free-riding beforehand. Thus Europe plays a pivotal role in ensuring co-operation, because of its high damages from climate change. In their work on global co-operation, Botteon and Carraro (1995) suggested that India and China could also perform such a role, as they had the highest damage costs. However, these damage costs were mainly derived from estimates of increased mortality, and so are subject to the problems of valuation mentioned before. Hence, though regions with high damage costs can commit to support co-operative agreements unilaterally, the result will only be stable if they have the hard currency funds and political will to give side payments.

Stability in the absence of side payments

Even if side payments are not available (due to funding constraints, or the type of policy instrument used), or not sufficient to deter free-riding, there are other mechanisms which can stabilise coalitions of countries. Free-riding may not be rational strategy for countries contributing a non-trivial proportion of abatement (that is, they are very large and/or have very low abatement costs relative to other regions), because their defection will lead to rise in co-operative abatement costs which may radically reduce abatement in the remaining co-operating countries. This decrease in abatement raises damage costs in the defecting country, and if this increase outweighs the benefits from free-riding there is no incentive to leave the agreement in the first place.

The basic intuition behind stability with no side payments is that defection by one region causes the whole co-operative structure to break down, or at least reduces co-operative abatement to a non-substantive level. If this holds, there exists a credible threat to potential defectors that if they leave the treaty it will unravel; under these circumstances, rationally they will not defect as they will be no better off than in the non-cooperative case. The credibility of the threat lies in the remaining countries not being able to negotiate a feasible and substantive equilibrium after the defection which would leave them better off than the non-cooperative case. For this equilibrium to hold there must also be institutional mechanisms (ratification levels, monitoring of abatement, etc.) to ensure potential defectors know the consequences of their actions, and understand the credible nature of the threat.

Researchers who have investigated this type of interaction suggest the maximum size of coalition it can support is three, if co-operation significantly raises abatement above the non-cooperative case. If the co-operative case does not bring very large gains, then an agreement without side payments can be stable when it includes all potential members, but of course stability in this case is unimportant environmentally as it does not significantly increase abatement.

However, if other costs follow from a country's defection – such as a fall in industrial competitiveness among the remaining co-operators – an agreement can be stable with both a large number of countries, and much larger abatement than the non-cooperative case. Modelling in EGEM showed that such an equilibrium could be enforced by competitiveness effects between the OECD regions, but only if North America defects. That is, if North America leaves any agreement, energy intensive industries relocate from co-operating nations in such quantities that these extra costs outweigh the benefits they garner from co-operation, and remaining abatement drops to very low levels. Because Japan and Europe drastically reduce their abatement in response to a North American defection, North America has no incentive to leave the treaty in the first place because it gains neither abatement benefits, nor competitiveness. However, this argument does not seem to hold if either Japan or Europe are first to leave an agreement.

These scenarios have shown that, given this particular distribution of damage costs, it is possible to construct a stable agreement between the G7 regions if side payments are available. If a high target is to be agreed, it is not obvious whether Japan is necessarily part of an efficient agreement, and it is unlikely that it would be offered side payments to induce joining. Co-operation tends to rest on Europe providing side payments, because it suffers the highest damages and so gains the most from global abatement. The agreement could also be stable because of competitiveness effects, and in this case North America has no incentive to defect, but Europe and Japan do.

Stability rests on the complex interplay of national damage costs, abatement costs, competitiveness effects, and the ability to use side payments to support agreements. Currently, the distribution of damage costs is not well known, and this uncertainty may promote stability as countries will have more equal bargaining power inside any agreement. The importance of transfers for ensuring stability, especially when reaching a high abatement target, again argues in favour of using a tradable permit system to achieve global GHG abatement targets; as this is the only policy instrument that can accommodate side payments, achieve efficiency and is acceptable to governments on sovereignty grounds.

CONCLUSIONS

The research summarised above spans theoretical analysis, econometric estimation and scenario modelling. Different policy questions are illuminated by different techniques: theory structures inquiry and suggests empirical questions; econometrics measures how economies have behaved in the past; and modelling allows the impacts future policies to be investigated. However, results can only really be understood from a knowledge of their assumptions, but these are hard to communicate to policy makers. Given the structure of this study – and the derivation of the EGEM model – the following conclusions can be drawn about climate policy; some of which are general points, while others are results specific to this study.

Defining global policy on climate change

Economics argues that the uncertain damages from climate change, the potential for adaptation, and the cost of controlling GHGs must be balanced to derive correct policy targets. However, the FCCC argues for precautionary action, and has rejected previous attempts to use cost-benefit analysis to define policy. The following issues are probably the most important contributors to dispute in this area:

- Political choices over what constitutes equitable measurement of non-market damage in developing countries (the largest category of climate damage costs) can alter the optimal policy defined by CBA by the same degree as scientific uncertainty over climate change.
- The failure of previous analysis of optimal climate change policy to account for the irreversibility and unique nature of climate damage has undervalued the cost of climate damage to future generations, and resulted in the recommendation of environmentally unsustainable abatement policies.
- Economic models which do not measure how different rates of climate change impact ecosystems, or how the development of GHG abatement technology is stimulated by abatement policy commitments, cannot be used to form GHG abatement policy over the next 10–50 years, but only in the very long term (100–200 years).
- Climate change is still dominated both by probabilistic uncertainty and large areas of complete ignorance. As abatement policies are more reversible than future climate damage, learning about *greater* uncertainty over future outcomes should lead to a policy of more GHG abatement in the near term.
- A risk-averse, precautionary approach to climate change implies *greater* current abatement than that defined by 'no-regrets' policies which would be cost-effective either directly, or taking into account alleviation of other environmental problems (e.g. local pollution). The Precautionary Principle implies that a decrease in future risks is valued by policy

makers in and of itself; this value requires abatement over and above no-regrets levels.

Modelling the costs of climate change policy

Reducing GHG emissions, especially CO_2, implies enormous changes in economic systems. Discounting the direct costs caused by climate change, these shifts will result in lower economic growth as resources are diverted from consumption to abatement and mitigation. The assumptions underlying the modelling of these costs are a critical input to climate policy, but often remain buried in technical reports. Our research shows this is very dangerous because:

- Econometric study of how economies have responded to changes in energy prices in the past does not support many of the crucial assumptions used in the most common economic models of GHG abatement costs.
- Long run changes in energy use appear to be more dependent on how technology responds to price signals, than on traditional models of direct substitution between different factors of production.
- Future modelling of the costs of climate change must explicitly include an endogenous description of technological development in different sectors, and its interaction with investment, prices and global markets. The current assumption that technology will be unaffected by climate policy is untenable given past economic behaviour.

Putting climate policy into action

Agreeing on a global target for international GHG abatement is the first step in any international process. However, to be effective, the institutions implementing any international agreement must also ensure: that the aims of the agreement are met; that abatement is carried out efficiently; and that stable co-operation is maintained between countries, even without coercive instruments such as trade sanctions.

- Achieving the environmental aims of the climate convention is best done by using quantitative global targets rather than international emission taxes. A treaty must also specify global timetables for abatement, because determining the pathway of emission reductions at the national level will lead to dangerous increases in climate damage.
- Unilateral stabilisation of GHGs at 1990 levels by the OECD, as a precursor to a larger agreement, will lead to negligible increases in CO_2 emissions in uncommitted countries. OECD abatement to 25 per cent below 1990 levels will probably increase emissions outside the agreement by 20 per cent of the total emissions reduction. Carbon leakage is therefore a problem, but not one large enough to undermine the aims of an OECD-wide agreement.

- OECD stabilisation of GHGs at 1990 levels will cost ≈ 0.8 per cent of GDP and shifting abatement between countries can save 7 per cent of these costs. Reductions of 25 per cent below 1990 levels will cost up to 3 per cent of OECD GDP, but 25 per cent savings can be achieved by concentrating abatement in the lowest cost countries.
- Recycling the revenues from carbon taxes by reducing employer's labour taxes cuts unemployment by 35 per cent for abatement at 1990 levels. These extra jobs lower the cost of abatement by up to 40 per cent compared to recycling revenues through income taxes. However, these gains die out over time (thirty-five years), and are vulnerable to inflationary and political pressures if implemented rashly.
- When using carbon taxes to lower CO_2 emissions, national governments must balance difficult issues of employment, efficiency, equity and trade. These trade-offs are best facilitated by using targets or tradable permits to co-ordinate international abatement, not internationally set carbon taxes. Restrictions on national policy options may remove the domestic political consensus for implementing international abatement obligations.
- There is no easy way to force countries to join international agreements to abate GHGs, and any country can gain from free-riding on the efforts of others. The stability of an agreement will critically depend on the distribution of climate damage, the size of competitiveness losses from unilateral abatement, and the ability to make international side payments. Abatement targets in each region do not allow financial transfers between countries and therefore will probably not support a stable agreement at high abatement levels.

For significant abatement, an international tradable permits scheme seems to be the only way to ensure abatement targets are met, efficiency is achieved and a mechanism (permit distribution) is available for ensuring all countries gain from remaining in a treaty. Permits also allow countries to craft domestic policies which are appropriate to their national policy agendas, and exploit all available cost-effective abatement measures.

Negotiating a substantial global abatement target and effectively implementing it, is essentially a political problem; however, careful design of the economics and incentives provided by different institutions will also facilitate agreement and co-operation. The biggest threat to co-operation comes from heavily polluting countries who think they will suffer few damages, as they have little incentive to abate GHGs. As altruism is seldom seen in international politics, we must hope that the regional distribution of damages remains uncertain in the near future, or that countries can agree to fund side-payments which ensure co-operation. However, if most damages occur in the developing world which is likely such stabilising payments will not be fundable, and climatic change will be far more extreme than future generations would wish. Future generations will be the arbiter of whether current decisions trading off pollution and abatement were correct. The verdict always lies far in the future: to predict it will require complex techniques of policy analysis, but more importantly, a leap of imagination to ensure short term political issues do not cloud long run policy choices.

NOTES

1 AN INTRODUCTION TO CLIMATE CHANGE

1 Montreal Protocol on Substances that Deplete the Ozone Layer, UNEP 1987.
2 Global Warming Potentials are calculated as:

$$GWP(t) = \int_{t_0}^{t+t_0} a.c(u)du \Big/ \int_{t_0}^{t+t_0} a_{CO2}.c_{CO2}(u)du$$

where a is the radiative forcing per unit mass, c(u) denotes concentration at time u after release and subscripts CO_2 denote the values for CO_2 and the index is over time t.

2 INTERNATIONAL CO-ORDINATION OF CLIMATE CHANGE PREVENTION

1 The European Union proposal for a carbon/energy tax has not been imposed because of a condition requiring all other OECD countries to enforce similar taxes.
2 It is a principle of the convention that countries bear 'common but differentiated responsibilities' in responding to climate change. The split of responsibilities is defined by past and present per capita emissions ('Noting that the largest share of historical and current global emissions of greenhouse gases has originated in developed countries', Preamble).
3 The thirty-five countries in Annex I are: Australia, Austria, Belarus,* Belgium, Bulgaria,* Canada, Czechoslovakia,* Denmark, Estonia,* Finland, France, Germany, Greece, Hungary,* Iceland, Ireland, Italy, Japan, Latvia,* Lithuania,* Luxembourg, Netherlands, New Zealand, Norway, Poland,* Portugal, Romania,* Russian Federation,* Spain, Sweden, Switzerland, Turkey, Ukraine,* United Kingdom of Great Britain and Northern Ireland and the United States of America. A * denotes an 'economy in transition', none of which is in Annex II.
4 'Parties may implement such policies and measures jointly with other Parties and may assist other Parties in contributing to the achievement of the objective of the Convention' (Article 4.2.a).
5 Though theoretically the same in a simple analysis it should be noted that auctioning permits does create some serious practical problems. Given different energy elasticities in different countries and different energy needs due to climate and geography, the amount of money which must be redistributed to ensure poor countries can buy enough permits is very difficult to determine. As the penalty price if the quota is exceeded is very high and is denominated in hard currency, auctioning permits increases the expected costs of compliance to poor countries. This is because it is 'costless' for them to hold given permits as

insurance as they can always sell at market price if permits are unnecessary, but it is very costly to buy permits if they are in danger of exceeding their limit, because they have to draw on their reserves of foreign currency. Of course the precise dynamics of this will be determined by the slack in the market, the compliance rules and deadlines and the existence of rigidities and market power.

6 The collection of a carbon tax internationally encounters severe political resistance, and, while the initial distribution of tradable permits can be based on transparent allocative principles, calculating appropriate side payments from international taxes relies on centrally modelling global macroeconomic impacts. This is bound to be more difficult and the results would be disputed by national governments using their own calculations of macroeconomic impacts.

7 FCCC Article (14.1): 'In the event of a dispute between any two or more Parties concerning the interpretation or application of the Convention, the Parties concerned shall seek a settlement of the dispute through negotiation or any other peaceful means of their own choice.' If agreement is not forthcoming the issue can be referred by one of the parties to conciliation by the Council of Parties or by *mutual* agreement of both the Parties involved to the International Court of Justice.

8 The possibility for such sanctions are written into the Montreal Protocol on Substances that Deplete the Ozone Layer, which was signed in 1987.

3 A REVIEW OF MODELLING ISSUES AND PAST WORK

1 If a model, derived from the social sciences, is used to inform the policy process its results have to be explainable in the medium of that process, i.e. verbal debate and written argument. This contrasts with the majority of natural science modelling, where the results can be checked by controlled experiments, and the details of the process do not need to be understood by decision makers to ensure the results are accepted. Modelling of the global climate provides an interesting (partial) exception to this rule, because controlled experiments cannot be carried out on the whole system; therefore, we see climate modellers going through the same process as economic modellers have done for many years, and legitimising their models rather than their results.

2 Cline (1992) reports the following divergences in fossil fuel prices in GREEN using USA = 100 as an index: North America = 100, Soviet Union = 22, China = 49, Europe = 132, OECD–Pacific = 144.

3 A few national level studies assess the costs of different types of taxation to achieve emission reductions. Jorgenson and Wilcoxen (1990) use a general equilibrium model for the USA and find that a carbon tax is the cheapest alternative with an annual loss in GDP of 0.50 per cent for a 20 per cent reduction. Next comes the energy tax with an annual loss in GDP of 0.6 per cent. An *ad valorem* tax is the most expensive with the cost of abatement at 1 per cent loss in GDP. The revenue generated is the largest in the case of the *ad valorem* tax. However, in all cases the revenue is returned to the consumers by reducing labour taxes so that the differences in GDP losses are the net distortionary effects of changing the taxation system.

4 A more recent dynamic version of the model incorporates backstop technologies, increases regional disaggregation to twelve regions, endogenises oil prices, revises the production structure and extends the model to 2050 (Burniaux *et al.* 1991a). However, it is expected that results will not change by much and costs of reducing carbon emissions will be within 2–3 per cent of annual GNP.

5 MODELLING THE MACROECONOMIC IMPACTS OF CARBON ABATEMENT

1 These are USA, Japan, Germany, France, Italy, UK, Canada – the G7.

2 Designing a stable exchange rate model between many countries is very difficult if they all experience a simultaneous price shock, and have uncoordinated interest rate regimes. As prices rise initially, interest rates will rise, strengthening the currency. This 'exports' inflation to other countries which must import at a higher domestic cost, thus exacerbating their own inflation. With seven countries in this system it can quickly become unstable under simple uncovered interest rate parity modelling, and this results in runaway nominal exchange rate growth in one country. A more robust method of modelling would be to derive optimal control rules based around learning equations (Hall 1994); however, it was considered that this was overly complex and would add little information to the questions being studied here.

3 The Griffen *et al.* study estimated a trans-log production function on pooled time series data from the G9 industrialised countries and concluded that the own price elasticity of energy is $\approx 0.8 \pm 0.07$ in all cases. This result is very different from the elasticities found in our own estimation which showed a far greater degree of heterogeneity, reflecting, as we believe, the large differences in energy markets and existing energy prices between these countries. Therefore, following Peseran and Smith's work we interpret the homogeneity of elasticities in the Griffen work as being an artefact of the panel data estimation approach, rather than a reflection of the close similarities between different countries.

4 Investment was only significant in explaining energy use trends in the USA and Belgium. In general the most important explanatory variables were relative prices and industrial production, which was included as a proxy for long run sectoral shifts from manufacturing into services.

5 Forty years can be taken as a plausible maximum upper limit as this is the longest lifetime of mechanical equipment associated with energy conversion, i.e. thermal power stations. However, it should be remembered that at least 60 per cent of energy use in most industrialised countries also depends on infrastructure investment in transport and housing so the true long run for turnover of all energy related capital is nearer 100 years.

6 Non-fossil fuel energy such as hydroelectric power and nuclear power is essentially produced by capital and labour with no other significant inputs. The value added of these sectors can therefore be treated in the same way as any other manufactured good with no reference to 'energy'.

7 This structure did not give good results in France so exogenous technical progress was applied equally to Labour and Capital (Hicks Neutral) in this case.

8 Despite the simplicity of this formulation empirical work has indicated that profit maximising labour demand equations derived from a CES production structure perform at least as well as more data intense models such as the trans-log (Madsen 1991).

9 Augmented Dickey–Fuller Unit Root tests were used to determine the order of integration of each series; an I(1) series being defined as one whose natural logarithm exhibits a unit root, but which is stationary when first differences are taken of the logged series.

10 The bias in labour productivity could be partly deterministic and partly endogenous, but in estimating the derived demand for labour the deterministic part will not affect the outcome. This is because the CES function can always be divided through by the deterministic trend, leaving total factor productivity on the outside and an extra factor with the capital stock; as the capital stock is then substituted out of the expression, along with its multipliers, it is irrelevant

where the deterministic trend is included in the expression. This would not be true if the capital demand equation was also going to be estimated using this structure.

7 OPTIMAL CLIMATE CHANGE POLICY

1 Precautionary Principle: 'Where there are threats of serious or irreversible damage, lack of full scientific certainty should not be used as a reason for postponing such measures [i.e. emissions control measures].'
2 A rule which prescribes the preserving of the natural environment from generation to generation *unchanged* in composition is often described as a 'strong' sustainability criterion; while one that allows environmental resources to be substituted for each other, while the total productive stock remains constant, is termed 'weak' sustainability. Obviously the grounds for these criteria are primarily ethical, not economic, and there are many different coherent and logical conditions which could be imposed.
3 Examples of irreversible impacts include species loss, inundation of agricultural land by sea water, and destruction of ecosystems such as marshes, mangrove swamps and coral reefs which can only regenerate over very long time periods.
4 An example of the differences in underlying assumptions embodied in the FCCC is given by the following passage from the Preamble: 'Recognizing that States should enact effective environmental legislation, that environmental standards, management objectives and priorities should reflect the environmental and developmental context to which they apply, and that standards applied by some countries may be inappropriate and of unwarranted economic and social cost to other countries, in particular developing countries.'
5 'Taking into full account the legitimate priority needs of developing countries for the achievement of sustained economic growth and the eradication of poverty,' FCCC Preamble.
6 'Noting that the largest share of historical and current global emissions of greenhouse gases has originated in developed countries, that per capita emissions in developing countries are still relatively low and that the share of global emissions originating in developing countries will grow to meet their social and development needs,' FCCC Preamble.
7 In the general case, characterising the output of a non-linear system as $f[Y,X,\varepsilon]$, where vector Y contains endogenous variables, vector X exogenous variables and vector ε gives the stochastic error terms surrounding the exogenous parameters. Solving this model for any solution (here set to equal zero) gives:

$$E\{f\ [Y^*,X,\varepsilon]\} = 0$$

Where: E is the expectations operator, Y^* is the mathematical expectation of Y. If however the stochastic nature of the exogenous variables is removed by taking the expected value of the errors (here set to zero for convenience) the solution is:

$$f\ [Y^{`},X,0] = 0$$

If the model is non-linear, the reaction of the endogenous terms to ε will be non-linear, and so $Y^* \neq Y^{`}$ and the solution using expected values will be incorrect. In the case of climate change Y would represent the level of abatement measures and X a vector of technological options and uncertain damage functions.
8 Even if energy efficient investments were large the timescale for switching back to a carbon intense economy would be much smaller than the time needed for the atmosphere to return to the natural greenhouse equilibrium. Any changes in

technology, as are modelled in Chapters 4 and 5, will only be used in the future if they are cheaper than the original processes; therefore, the permanent costs of adjustment will be the opportunity costs of developing these particular resource saving technologies and not other (i.e. labour saving) ones.

9 That is, the evolution of the distribution of marginal damage costs does not conform to a Markov Process or a transformation (i.e. logged) of a Markov Process.

10 For example, hard uncertainty characterises the probability of a nuclear reactor failing because we have few observation points; while the probability of an aeroplane crash is much better understood because of the frequency and history of observations. In economic theory hard uncertainty is often used to describe a situation where the actions of economic agents generate endogenous uncertainty in the economy, and so assumptions of rational expectations cannot be used to close a general equilibrium model.

11 Strictly rational, as opposed to psychological, risk aversion can be seen as part of the classic disagreement between frequency-based and Baysian probabilities in decision theory: because anthropogenic climate change can only happen once in our lifetime assigning Baysian probabilities to outcomes is relatively meaningless unless these probabilities have been calculated from frequency data of similar occurrences. The problem with using geological data to provide this information is that we cannot control for all the climatic differences between the past and present, and so are left with a question of their relevance to anthropogenic climate change. Therefore, for 'one-off' occurrences there will always be a large amount of uncertainty surrounding any modelling results and predictions, even if the underlying system is deterministic.

8 QUANTITATIVE MODELLING OF OPTIMAL INTERNATIONAL ABATEMENT POLICIES

1 For any time period t the discounted marginal cost of an extra unit of emissions E is:

$$D_{NPV} = \int_{t}^{t+200} e^{-\rho t}.(GDP)^{\eta}.\frac{dD_t}{dE}\,dt$$

2 Assuming equal percenatge (β) financial losses in all regions the global costs are:

$$\text{Global costs GC} = \beta.(Y_N + Y_S)$$

where $Y_{N,S}$ are the material output of North and South respectively; and $P_{N,S}$ are their respective populations. Aggregating utility in North and South means that global costs must equal a constant propportion of regional per capita income ($\beta = \gamma$):

$$\text{Global costs GC} = B.P_N.Y_T/P_T + \gamma.P_S.Y_T/P_T$$

Now assuming that the utility loss to the South is twice the per capita value to the North ($2\beta = \gamma$), and the North is prepared to internalise this, we can substitute out for β to give the financial value of global utility losses as a proportion of the original global financial costs:

$$\text{Welfare costs} = GC.(P_N + 2.P_S) \,/\, (P_N + P_S)$$

In our base-case, population in the North increases from 0.8 to 1.3 billion over the simulation period 1995–2030, while in the South population nearly doubles from 4.5 to 8.8 billion. This gives a multiplier on damage costs of 1.87 over the financial value case.

9 CARBON ABATEMENT IN INCOMPLETE INTERNATIONAL AGREEMENTS

1 France, Germany, Italy, UK, Belgium, Netherlands, USA, Canada and Japan.
2 A good example of the failure of empty threats is given by Barrett (1994d) discussing the dispute between the USA and Canada over seal hunting in the late nineteenth century. In response to Canada's refusal to agree to quota arrangements in the face of diminishing stocks, the USA claimed it would wipe out the whole population of seals once and for all. The Canadians successfully called the Americans' bluff because they knew it was not in the USA's interest to destroy the seal stocks, and their value was too large for it to be done out of pique.

10 THE DOMESTIC POLITICAL ECONOMY OF CARBON TAXES

1 In solving this model we consider the steady state growth case where all private inputs and output are growing at the same rate; it is assumed that public goods cost a constant proportion of income and the change in marginal utility of income and leisure as the economy grows is assumed to reflect this fact.
2 By keeping the amount of public funds raised constant we are assuming that none of the energy tax revenue is spent by the government on direct energy efficiency or technology forcing projects, but it is all recycled through the existing taxation system. If public spending did increase it is possible that aggregate welfare could fall due to the increased distortion from added taxation. If the government is behaving optimally this would imply that the energy tax should not be levied, or should be levied at a lower rate.
3 When the environment, which is a non-growing good and input to production, is considered in a steady state growth optimum several special conditions must hold. Firstly, the value share of the environment in utility must remain constant. This means that the *unit* utility of the environment must be increasing at the same rate as the *total* utility of income and energy. Similar arguments also hold for other public goods, though some of these such as public amenities and social support may grow in absolute (as well as value) terms over time. On the production side, this condition would mean that the environment retains a constant value share in output. Given the observed drop in primary resource based GDP as countries develop this seems rather implausible in the long run, but changing this assumption introduces unnecessary complications into the analysis.
4 The only case where this will not hold is when the energy tax is higher than the optimal environmental tax in the base-case; that is, on purely revenue raising grounds. Obviously the externality is irrelevant here as pollution is already below its optimum level for environmental protection.
5 The difficulty the British government had in introducing VAT on fuel, nominally a global warming measure but really a revenue raising tax, may be seen as symptomatic of this problem even though redistributive payments to some of the poorest groups were made. On the other hand the annual 5 per cent escalator on petrol duty was relatively unopposed, probably because people understand the environmental impacts of vehicle use more easily than global warming, travel is less 'essential' than heating and the income distribution of those affected is narrower.

6 There are obviously many problems associated with an employment subsidy and many different ways in which energy taxes could be recycled through other areas such as marginal tax rates and tax allowances for the low paid, and training funds and job subsidies targeted at the long term unemployed. However, these presuppose different sources of disequilibria in the labour market which we do not consider here, and by raising public spending could also increase tax distortions in the economy.

7 North America – USA and Canada; European Union – Germany, France, Italy and the UK.

8 Social security payments to the unemployed are calculated as a proportion of real wages – the replacement ratio. The figures used in model simulations were derived from OECD Country Surveys and Layard *et al.* (1991): USA – 0.19, JP – 0.37, GE – 0.39, FR – 0.33, IT – 0.15, UK – 0.18, CN – 0.29.

11 OECD CO-OPERATION UNDER THE FCCC

1 Using a linear approximation to the demand function, and the results from EGEM, in response to a rise in energy prices from p1 to p2 and a subsequent drop in direct energy consumption from q1 to q2, the loss in consumer surplus is:

$$\Delta CS = (p2 - p1) \times q2 + 0.5 \times (p2 - p1) \times (q2 - q1).$$

REFERENCES

Adelman, M. A. (1989) 'Mideast governments and the oil price prospect', *Energy Journal*, 10(2): 15–24.

Al-Sahlawi, M. A. (1989) 'Oil price changes and non-OPEC oil supply: an empirical analysis', *OPEC Review*, Spring, pp. 11–19.

Allen, C., Srinivasan, T. G. and Vines, D. (1994) 'Analysing external adjustment in developing countries: a macroeconomic framework', *Centre for Economic Forecasting Discussion Paper* 21–94, London Business School, London.

Allen, C. and Urga, G. (1995) 'Derivation and estimation of interrelated factor demands from dynamic cost function', *Centre for Economic Forecasting Discussion Paper* 10–95, London Business School, University of London.

Allen, C. and Nixon, J. (1995) 'Two concepts of the NAIRU', *Centre for Economic Forecasting Discussion Paper* 28–95, London Business School, London.

Alsmiller, R. G. Jr, Horwedel, J. E., Marshalla, R. A., Nesbitt, D. M. and Haas, S. M. (1985) 'A model of the world oil market with an OPEC cartel', *Energy* 10(10): 1089–1102.

Anderson, D. and R. Williams (1993) 'Cost effectiveness and the investments of the global environment facility', *Background Paper for the Scientific and Technical Advisory Panel of the GEF*, Global Environment Facility, Washington.

Anderson, G. and Blundell, R. (1984) 'Consumer non-durables in the UK: a dynamic demand system', *Economic Journal* 94 (Supplement): 35–44.

Association of Small Island States, (1994) *Draft Protocol to the United Nations Framework Convention on Climate Change on Greenhouse Gas Emission Reduction*, Association of Small Island States FIELD, SOAS, London.

Atkinson, J. and Manning, N. (1995) 'A survey of international energy elasticities', in T. Barker and P. Ekins (eds) *Global Warming and Energy Elasticities*, Routledge, London.

Austwick, O. G. (1992) 'Limits to oil pricing: scenario planning as a device to understand oil price developments', *Energy Policy*, November, pp. 1097–1105.

Bacon, R. (1991) 'Modelling the price of oil', *Oxford Review of Economic Policy* 7(2): 17–34.

Baldwin, N. and Prosser, R. (1988) 'World oil market simulation', *Energy Economics*, July, pp. 185–98.

Baranzini, A. and Bourguignon, F. (1994) 'Is sustainable growth optimal?', *Cahiers du Département D'Economie Politique* 94–06, Faculté des Sciences Economiques et Sociales, Université de Genève.

Barker, T. (1993) 'Secondary benefits of greenhouse gas abatement: the effects of a UK carbon/energy tax on air pollution', *Energy Environment Economy Modelling Discussion Paper* 4, Department of Applied Economics, University of Cambridge.

Barker, T., Baylis, S. and Madsen, P. (1992) 'The carbon tax: economic and policy issues.', *Energy-Environment-Economy Modelling Discussion Paper*, No. 3, Dept. of Applied Economics, University of Cambridge.

Barker, T., Baylis, S. and Madsen, P. (1993) 'A UK carbon/energy tax: the macroeconomic effects', *Energy Policy*, March, pp. 296–308.

Barker, T. (1994) 'Taxing pollution instead of employment: greenhouse gas abatement through fiscal policy in the UK', *Department of Economics Discussion Paper* 94–10, University of Birmingham.

Barker, T. and Gardiner, B. (1994) 'Employment, wage formation and pricing in the European Union: empirical modelling of environmental tax reform', paper presented at the International Workshop on Environmental Taxation, Revenue Recycling and Unemployment, Fondazione ENI Enrico Mattei, Milan, Italy.

Barr, D. G. and Cuthbertson, K. (1991) 'Neo-classical consumer demand theory and the demand for money', *Economic Journal* 101 (July): 855–76.

Barrett, S. (1993) 'Self-enforcing international environmental agreements', mimeo, London Business School and Centre for Social and Economic Research on the Global Environment, University of London.

Barrett, S. (1994a) 'Heterogeneous international environmental agreements', mimeo, London Business School and Centre for Social and Economic Research on the Global Environment, University of London.

Barrett, S. (1994b) 'Trade restrictions in international environmental agreements', mimeo, London Business School and Centre for Social and Economic Research on the Global Environment, University of London.

Barrett, S. (1994c) 'Self-enforcing international environmental agreements', *Oxford Economic Papers* 46: 878–94.

Barrett, S. (1994d) 'Managing the international commons', in G. Brown and V. K. Smith (eds) *Resource and Environmental Economics*, Cincinnati: South-Western.

Barrett, S. (1995) 'Towards a theory of international environmental co-operation', mimeo, London Business School and Centre for Social and Economic Research on the Global Environment, University of London.

Beaver, R. D. and Huntington, H. G. (1991) 'A comparison of aggregate energy demand models for global warming policy analyses', *EMF OP 33 Energy Modelling Forum*, Terman Engineering Center, Stanford University, California.

Bergman, (1990) 'The development of computable general equilibrium modelling', in L. Bergman, D. W. Jorgenson and E. Zalai (eds) *General Equilibrium Modeling and Economic Policy Analysis.*

Black, J., Levi, M. D. and de Meza, D. (1992) 'Creating a good atmospere: minimum participation for tackling the "Greenhouse Effect"', *Economica* 60: 281–93.

Boero, G., Clarke, R. and Winters, L. A. (1991) *The Macroeconomic Consequences of Controlling Greenhouse Gases: A Survey*, DOE Environmental Economics Research Series, HMSO, London.

Bohm, P. (1993) 'Incomplete international carbon emissions quota agreements: efficiency and distributional implications of joint implementation and quota tradability', paper presented at the International Institute of Applied Systems Analysis (IIASA), Laxenberg.

Bohm, P. (1994) 'On the feasibility of joint implementation of carbon emissions reductions', *University of Birmingham Discussion Paper* 94–05, University of Birmingham.

Bohm, P. (1995) 'Environmental taxes, carbon taxes, tax recycling and tax distortions', mimeo, Department of Economics, University of Stockholm.

Boone, L., Hall, S. G., Kemball-Cook, D. and Smith, C. (1995) 'Endogenous technological progress in fossil fuel demand', in T. Barker, P. Ekins and N. Johnstone (eds) *Global Warming and Energy Demand*, Routledge, London and New York.

Botteon, M. and Carraro, C. (1993) 'Is the European carbon tax really effective?', *Fondazione ENI Enrico Mattei Working Paper* 9–93, Milan.
Botteon, M. and Carraro, C. (1995) 'Burden-sharing and coalition stability in environmental negotiations with asymmetric countries', *Fondazione ENI Enrico Mattei Working Paper* 78–95, Milan.
Bovenberg, L. and Smulders, S. A. (1994) 'Environmental policy in a two sector endogenous growth model', in *Quantitative Economics for Environmental Policy*, OCfEB – Research Centre for Economic Policy, Erasmus University, Rotterdam.
Bovenberg, L. and Van der Ploeg, F. (1994) 'Tax reform, structural unemployment and the environment', paper presented at the International Workshop on Environmental Taxation, Revenue Recycling and Unemployment, Fondazione ENI Enrico Mattei, Milan, Italy.
Bovenberg, L. and Goulder, L. (1994) 'Costs of environmentally motivated taxes in the presence of other taxes: general equilibrium analysis', paper presented at the International Workshop on Environmental Taxation, Revenue Recycling and Unemployment, Fondazione ENI Enrico Mattei, Milan, Italy.
Bovenberg, L. and de Mooij, R. (1994) 'Environmental levies and distortionary taxation', *American Economic Review* 84(4): 1085–9.
Boyd, G. A. and Karlson, S. H. (1993) 'The impact of energy prices on technology choice in the United States steel industry', *Energy Journal* 14(2): 47–56.
British Petroleum (1994) *Statistical Review of World Energy*.
Burniaux, J.-M., Martin, J. P., Nicoletti, G. and Martins, J. O. (1991a) 'The costs of policies to reduce CO_2 emissions: initial simulation results with GREEN', *OECD Working Paper* 103, OECD, Paris.
Burniaux, J.-M., Martin, J. P., Nicoletti, G. and Martins, J. O. (1991b) 'GREEN – a multi-region dynamic general equilibrium model for quantifying the costs of curbing CO_2 emissions: a technical manual', *OECD Working Paper* 104, OECD, Paris.
Burniaux, J.-M., Martin, J. P., Nicoletti, G. and Oliveira-Martins, J. (1992) 'The costs of reducing CO_2 emissions: evidence from GREEN', *OECD Economics and Statistics Department Working Paper* 115, OECD, Paris.
Capros, P., Georgakopolous, P., Zografakis, S., Proost, S., Van Regemorter, D., Conrad, C., Schmidt, T., Smeers, Y. and Michiels, E. (1994) 'Double dividend analysis: first results of a general equilibrium model (GEM–E3) linking the EU-12 countries', paper presented at the International Workshop on Environmental Taxation, Revenue Recycling and Unemployment, Fondazione ENI Enrico Mattei, Milan, Italy.
Carraro, C. and Siniscalco, D. (1993) 'Strategies for the international protection of the environment', *Journal of Public Economics* 52: 309–28.
Carraro, C. and Soubeyran, A. (1994) 'Environmental taxation and employment in a multi-sector general equilibrium model', paper presented at the International Workshop on Environmental Taxation, Revenue Recycling and Unemployment, Fondazione ENI Enrico Mattei, Milan, Italy.
Commission of European Communities (1992) *Proposal for a Council Directive Introducing a Tax on Carbon Dioxide Emissions and Energy*, COM (92) 226 final, CEC, Brussels.
Chang W. W. (1970) 'The neoclassical theory of technical progress', *American Economic Review*, 60(5): 912–23.
Chemical Marketing Reporter (1994) 'DSM mulls big move for melamine unit', 14 March, p. 9.
Chung, J. W. (1994) *Utility and Production Functions*, Blackwells, Cambridge, Mass, USA.
Chung, J. W. (1987) 'On the estimation of factor substitution in the translog model', *Review of Economics and Statistics*, 64: 409–17.
Clarke, R., Boero, G. and Winters, L. A. (1993) 'Controlling greenhouse gases: a

survey of global macroeconomic studies', *University of Birmingham Discussion Paper* 93–18, University of Birmingham.
Clifford, M. (1993) 'Scottish mill for Malaysia: Indonesian firm plans steel plant in Trengganu', *Far Eastern Economic Review*, 156(10): 58.
Cline, W. R. (1991) 'Economic models of carbon reduction costs: an analytical survey', mimeo, Institute for International Economics, Washington DC.
Cline, W. R. (1992) *The Economics of Global Warming*, Institute for International Economics, Washington DC.
Collard, D. (1988) 'Catastrophic risk: or the economics of being scared', in D. Collard, D. Pearce and D. Ulph (eds) *Economics, Growth and Sustainable Environments*, Macmillan, London.
Crouch, E. and Wilson, R. (1987) 'Risk assessment and comparisons: an introduction', *Science* 236: 267–70.
Cuthbertson, K., Hall, S. G. and Taylor, M. P. T. (1992) *Applied Econometric Techniques*, Phillip Allen, London.
Daniel J. S., Solomon, S. and Albritton, D. L. (1995) 'On the evaluation of halocarbon radiative forcing and global warming potentials', *Journal of Geophysical Research: Atmospheres* 100: 1271–85.
Dean, A. and Hoeller, P. (1992) 'Costs of reducing CO_2 emissions: evidence from six global models', *OECD Economics Department Working Paper* 122, OECD, Paris.
Deaton, A. S. and Muellbauer, J. (1980) *Economics and Consumer Behaviour*, Cambridge University Press, Cambridge.
Dixit, A. and Pindyck, R. (1993) *Investment Under Uncertainty*, Princeton University Press, Princeton, New Jersey.
Donsimoni, M. P., Economides, N. S. and Polemarchakis, H. M. (1986) 'Stable cartels', *International Economic Review* 27: 317–27.
Eckhaus, R. (1993) 'Laissez faire, collective control, or nationalisation of the global commons?', *Fondazione ENI Enrico Mattei Working Paper* 31–93, Milan.
Edmonds, J. and Barnes, D. W. (1990a) *Estimating the Marginal Cost of Reducing Global Fossil Fuel CO_2 Emissions*, Pacific Northwest Laboratory, Washington DC.
Edmonds, J. and Barnes, D. W. (1990b) *Factors affecting the Long-Term Cost of Global Fossil Fuel CO_2 Emissions*, Global Environmental Change Programme, Pacific Northwest Laboratory, Washington DC.
Edmonds, J. and Reilly, D. W. (1983a) 'Global energy and CO_2 to the year 2050', *Energy Journal* 4(3): 21–47.
Edmonds, J. and Reilly, D. W. (1983b) 'A long-term global energy-economic model of CO_2 release from fossil fuel use', *Energy Economics* 5(2): 74–88.
Ehrenfeld, J. (1992) 'Risk assessment and managment: a critique of current practices and policy implications', unpublished mimeo, Programme in Technology and Policy, MIT, Boston.
Ekins, P. (1994) 'Rethinking the costs related to global warming', *Department of Economics Discussion Paper* 94–07, University of Birmingham.
Ekins, P. and Jacobs, M. (1994) 'Are environmental sustainability and economic growth compatible?', *Energy–Environment–Economy Modelling Discussion Paper* 7.
Engle, R. F. and Granger, C. W. J. (1987) 'Co-integration and error correction: representation, estimation and testing', *Econometrica*, 55(2): 251–76.
European Roundtable of Industrialists (1994) *The Climate Change Debate: Seven Principles for Practical Policies*, ERT, Brussels.
Falk, I. and Mendelsohn, R. (1993) 'The economics of controlling stock pollutants: an efficient strategy for greenhouse gases', *Journal of Environmental Economics and Management* 25(1): 76–88.
Fankhauser, S. (1993) 'Global warming damage costs: some monetary estimates', *CSERGE GEC Working Paper* 92–29, University College, London.
Fankhauser, S. (1994a) 'The social costs of greenhouse gas emissions: an expected value approach', *Energy Journal* 15(2): 157–84.

Fankhauser, S. (1994b) 'The economic costs of global warming damage: a survey', *Department of Economics Discussion Paper* 94–04, University of Birmingham.
Fankhauser, S. (1995) *Valuing Climate Change: The Economics of the Greenhouse*, Earthscan, London.
Felder, S. and Rutherford, T. F. (1993) 'Unilateral CO_2 reductions and carbon leakage: the consequences of international trade in basic materials', *Journal of Environmental Economics and Management* 25: 162–76.
Foundation for International Environmental Law and Development (1994) *Improving compliance with international environmental regimes for protection of the global commons, with particular emphasis on the international negotiations in respect of the UN Framework Convention on Climate Change*, draft report for the Economic and Social Research Council (ESRC), FIELD, SOAS, London University.
Filar, J. and Zapert, R. (1995) 'Uncertainty analysis of a greenhouse effect model', *Nota di Lavoro* 27–95, Fondazione ENI Enrico Mattei, Milan.
Fishburn, P. (1973) *The Theory of Social Choice*, Princeton University Press, Princeton, New Jersey.
Fisher, F. M. (1965) 'Embodied technical change and the existence of an aggregate capital stock', *Review of Economic Studies* 32: 263–88.
Gately, D. (1984) 'A ten-year retrospective: OPEC and the world oil market', *Journal of Economic Literature* 22 (Sept.): 1100–14.
Golombek, R., Hagem, C. and Hoel, M. (1993) 'The design of a carbon tax in an incomplete international climate agreement', *Fondazione ENI Enrico Mattei Working Paper* 22–93, Milan.
Government of India (1991) *Sectoral Energy Demand in India*, Planning Commission, New Delhi.
Greene, D. L. (1991) 'A note on OPEC market power and oil prices', *Energy Economics*, April, pp. 123–9.
Greene, O. (1993) 'International environmental regimes: verification and implementation review', *Environmental Politics* 2(4): 156–73.
Griffen, J. M. and Gregory, P. R. (1976) 'An international translog model of energy substitution responses', *American Economic Review* 66 (December): 845–57.
Griliches, Z. (1988) *Technology, Education and Productivity: Early Papers with Notes to Subsequent Literature*, Blackwells, Cambridge, Mass.
Grossman, G. M. and Helpman, E. (1992) *Innovation and Growth in the Global Economy*, MIT Press, Cambridge, Mass.
Gupta, S. (1995) *Carbon abatement costs in an empirical model of the Indian economy: an integration of micro and macro approach*, PhD dissertation, London Business School, London.
Haas, P. and Sundgren, J. (1993) 'Evolving international environmental law and changing practices of national sovereignty', in N. Choucri (ed.) *Global Accord: Environmental Challenges and International Responses*, MIT Press, Cambridge, Mass.
Hall, D. O. (1991) 'Biomass energy', *Energy Policy*, October, pp. 711–37.
Hall, R. E. (1968) 'Technical change and capital from the point of view of the dual', *Review of Economic Studies* 35: 35–46.
Hall, S., Kemball-Cook, D. and Smith, C. (1993) Inter-fuel Substitution and CO_2 Emissions in Nine Countries, *Centre for Economic Forecasting Discussion Paper* 22–93, London Business School, London.
Hall, S. and Stephenson, M. (1990) 'Optimal control of stochastic non-linear models', in S. G. B. Henry and K. D. Patterson (eds) *Economic Modelling at the Bank of England*, Chapman and Hall, London.
Hall, S., Kemball-Cook, D. and Smith, C. (1993) 'Inter-fuel substitution and CO_2 emissions in nine countries', *Centre for Economic Forecasting Discussion Paper* 22–93, London Business School, London.
Hall, S. (ed.) (1994) *Applied Economic Forecasting Techniques*, Harvester Wheatsheaf, London.

Hanemann, W. (1991) 'Willingness-to-pay and willingness-to-accept: how much can they differ?', *American Economic Review* 81(3): 635–47.

Harcourt, G. C. (1972) *Some Cambridge Controversies in the Theory of Capital*, Cambridge University Press, Cambridge.

Hardin, G. (1968) 'The tragedy of the commons', *Science* 162: 1243–8.

Helliwell, J. F. (ed.) (1976) *Aggregate Investment: Selected Readings*, Penguin, Harmondsworth.

Hoel, M. (1992) 'International environment conventions: the case of uniform reductions of emissions', *Environment and Resource Economics* 2: 141–59.

Hoel, M. (1993) 'Stabilising CO_2 emissions in Europe: individual stabilisation versus harmonisation of carbon taxes', *Fondazione ENI Enrico Mattei Working Paper* 8–93, Milan.

Hoeller, P., Dean, A. and Nicolaisen, J. (1991) 'Macroeconomic implications of reducing greenhouse gas emissions: a survey of empirical studies', *OECD Economic Studies* 16, OECD, Paris.

Hoeller, P. and Coppel, J. (1992) 'Energy taxation and price distortions in fossil fuel markets: some implications for climate change policy', *OECD Economics Department Working Paper* 110, OECD, Paris.

Hoeller, P., Dean, A. and Hayafuji, M. (1992) 'New issues, new results: the OECD's second survey of the macroeconomic costs of reducing CO_2 emissions', *OECD Economics Department Working Paper* 123, OECD, Paris.

Hogan, W. W. and Jorgenson, D. W. (1991) 'Productivity trends and reducing CO_2 emissions', *Energy Journal* 12(1): 67–86.

Horton, G. K., Rollo, J. M. C. and Ulph, A. (1992) 'The implications for trade of greenhouse gas emission control strategies', *Environmental Economics Research Series*, Department of Trade and Industry/Department of Education, HMSO, London.

Hourcade, J.-C. (1993) 'Economic issues and negotiation on the global environment: some lessons from recent experience of the greenhouse effect', *Fondazione ENI Enrico Mattei Working Paper* 24–93, Milan.

Howarth, R. B. and Norgaard, R. B. (1993) 'Intergenerational transfers and the social discount rate', *Environmental and Resource Economics* 3(4): 337–58.

Howarth, R. B., Schipper, L. and Andersson, B. (1993) 'The structure and intensity of energy use: trends in five OECD nations', *Energy Journal* 14(2): 27–45.

Huntington, H. G. (1994) 'Oil price forecasting in the 1980s: what went wrong?', *Energy Journal* 15(2): 1–22.

International Monetary Fund (1994) *World Economic Outlook*, IMF, Washington DC.

Ingham, A., Ulph, A. and Ulph, D. (1993) 'Carbon taxes and energy markets', *Fondazione ENI Enrico Mattei Working Paper* 2–93, Milan.

International Energy Agency (1993) *Energy database for non-OECD countries*, Organisation of Economic Co-operation and Development, Paris.

Inter-governmental Panel on Climate Change (1990) *Climate Change: the IPCC Scientific Assessment*, Cambridge University Press, Cambridge.

Inter-governmental Panel on Climate Change (1992) *Climate Change: the 1990 and 1992 IPCC Assessments*, United Nations Environment Program and World Meteorological Organisation, Cambridge University Press, Cambridge.

Jaffe, A. B., Peterson, S. R., Portney, P. R. and Stavins, R. N. (1994) 'Environmental regulation and the competitiveness of US manufacturing: what does the evidence tell us?', *Journal of Economic Literature*, forthcoming.

Johansen, S. (1991) 'Estimation and hypothesis testing of cointegration vectors', *Econometrica* 59: 1551–80.

Johansson, P.-O. (1993) *Cost-Benefit Analysis of Environmental Change*, Cambridge University Press, Cambridge.

Jorgenson, D. W., Gollop, F. M. and Fraumeni, B. (1987) *Productivity and US Economic Growth*, North-Holland, Oxford.

Jorgenson, D. W. and Wilcoxen, P. J. (1990) 'The costs of controlling US carbon dioxide emissions', paper presented at Workshop on Economic/Energy/Environmental Modeling for Climate Policy Analysis, Washington DC, October.

Jorgenson, D. W. and Wilcoxen, P. J. (1992) 'Energy, the environment and economic growth', *Harvard Institute of Economic Review Discussion Paper* 1604, Harvard University, Cambridge, Mass.

Jorgenson, D. W. and Wilcoxen, P. J. (1993) 'Reducing US carbon emissions: an econometric general equilibrium assessment', *Resource and Energy Economics* 15(1): 7–25.

Keeling, B. (1992) 'Structural change in the world steel industry: a North–South perspective', in G. van Liemt (ed.) *Industry on the Move: causes and consequences of international relocation in the manufacturing industry*, International Labour Organisation, Geneva, pp. 149–78.

Kennedy, P. (1994) 'Rethinking sustainability', paper presented at Fifth Annual Meeting of the European Association of Environmental and Resource Economists, Dublin, June 1994.

Kolstad, C. (1993) 'Mitigating climate change impacts: the conflicting effects of irreversibility in CO_2 accumulation and emission control investment', paper presented to the International Workshop on the Integrative Assessment of Mitigation, Impacts and Adaption to Climate Change, International Institute of Applied Systems Analysis, Laxenburg, October 1993.

Kristrom, B. and Riera, P. (1994) 'Is the income elasticity of environmental improvements less than one?', paper presented at the Second Conference on Environmental Economics, Ulvon, Sweden, 2–5 June.

Layard, R., Nickell, S. and Jackman, R. (1991) *Unemployment: Macroeconomic Performance and the Labour Market*, Oxford University Press, Oxford.

Lazarus, M., Greber, L., Hall, J., Bartels, C., Bernow, S., Hansen, E., Raskin, P. and Von Hippel, D. (1992) 'Towards global energy security: the next energy transition', draft report for Greenpeace International, Stockholm Environment Institute – Boston Centre, Boston, Mass.

London Business School (1993) 'GEM Model Manual', unpublished working document, London Business School/National Institute.

Ligthart, J. and Van der Ploeg, F. (1994) 'Environmental policy, tax incidence and the cost of public funds', paper presented at the International Workshop on Environmental Taxation, Revenue Recycling and Unemployment, Fondazione ENI Enrico Mattei, Milan, Italy.

Lovins, A. B. and Lovins, L. H. (1991) *Least Cost Climatic Stabilisation*, Rocky Mountain Institute, Old Snowmass, Colo.

Lucas, R. (1988) 'On the mechanics of economic development', *Journal of Monetary Economics* 22: 3–42.

Luckert, M. and Adamowicz, W. (1993) 'Empirical measures of factors affecting social rates of discount', *Environmental and Resource Economics* 3(1): 1–21.

Mabey, N. (1995a) 'International CO_2 abatement agreements with competitiveness externalities', *Unpublished Working Paper*, London Business School, London.

Mabey, N. (1995b) 'The Macroeconomic modelling of carbon taxes', *Centre for Economic Forecasting Discussion Paper* 11–95, London Business School, London.

Madsen, J. (1991) 'Labour demand under different technology assumptions', *Working Paper* 223, Department of Economics, Australian National University.

Manne, A. S. and Richels, R. G. (1992) *Buying Greenhouse Insurance: The Economic Costs of CO_2 Emission Limits*, MIT Press, Cambridge, Mass.

Manne, A. S. and Richels, R. G. (1989) *CO_2 Emission Limits: An Economic Cost Analysis for the USA*, Electric Power Research Institute, Palo Alto, Calif.

Manne, A. S. and Richels, R. G. (1990) *Global CO_2 Emission Reductions: the Impacts of Rising Energy Costs*, Electric Power Research Institute, Palo Alto, Calif.

Manne, A. and Olsen, T. (1994) 'Greenhouse gas abatement: toward pareto-optimal decisions under uncertainty', *University of Birmingham Discussion Paper* 94–06, University of Birmingham.

Manne, A. and Rutherford, T. F. (1994) 'International trade in Oil, Gas and carbon emission rights: an intertemporal general equilibrium model', *Energy Journal* 15(1): 57–76.

Mayes, D. and Young, G. (1994) 'Improving estimates of the UK capital stock', *National Institute Economic Review*, National Institute for Economic and Social Research, No. 147, London.

MERGE (1992) *Macroeuropean Energy Resource Global Environmental Model*, Enrico Mattei Foundation and Greta Associates, Venice.

Michel, P. (1993) 'Pollution and growth towards the ecological paradise', *Nota di Lavoro* 80–93, Fondazione ENI Enrico Mattei, Milan.

Mintzer, I. (1993) 'Implementing the framework convention on climate change: incremental costs and the role of the GEF', *GEF Working Paper* 4, Global Environment Facility, Washington DC.

Morgan, T. (1987) 'Oil prices: a long term perspective', *OPEC Review* 11(4): 325–47.

Naqvi, N. and Schneider, K. (1994) 'Unilateral environmental policy and international competition: a dual approach', paper presented at Fifth Annual Meeting of the European Association of Environmental and Resource Economists, Dublin, June.

Neuburger, H. (1992) 'Energy use in an era of rapidly changing oil price: how OPEC did not save the world from the greenhouse effect', *Environment and Planning A* 24: 1039–50.

Nordhaus, W. D. and Yohe, G. W. (1983) 'Future carbon dioxide emissions from fossil fuels', in *Changing Climate*, pp. 87–153. National Academy: Washington DC.

Nordhaus, W. D. (1990b) 'An intertemporal general equilibrium model of economic growth and climate change', paper presented at Workshop on Economic/Energy/Environmental Modeling for Climate Policy Analysis, Washington DC, October.

Nordhaus, W. D. (1991a) 'To slow or not to slow: the economics of the greenhouse effect', *Economic Journal* 101: 920–37.

Nordhaus, W. D. (1991b) 'A survey of cstimates of the costs of reduction of greenhousc gases', *Energy Journal* 12(1).

Nordhaus W. D. (1993) 'Rolling the "DICE": an optimal transition path for controlling greenhouse gases', *Resource and Energy Economics* 15: 27–50.

OECD (1992) *Financial and Economic Statistics*, OECD, Paris.

O'Mahoney, M. (1993) 'Capital stock and productivity in industrial nations', *National Institute Economic Review*, National Institute for Economic and Social Research, No. 145, London.

Oliveira-Martins J., Burniaux, J.-M. and Martin, J. (1993) 'Trade and the effectiveness of unilateral CO_2 abatement policies: evidence from GREEN', *Fondazione ENI Enrico Mattei Working Paper* 47–93, Milan.

Olsen, M. (1971) *The Logic of Collective Action*, Harvard University Press, Cambridge, Mass.

Parra, F. R. (1994) 'OPEC and the price of oil in 1993', *Energy Journal* 15(1): 17–30.

Pearce, D. and Turner, K. (1990) *Economics of Natural Resources and the Environment*, Harvester Wheatsheaf, London.

Perroni, C. and Rutherford, T. F. (1993) 'International trade in carbon emission rights and basic materials: general equilibrium calculations for 2020', *Scandinavian Journal of Economics* 95(3): 257–78.

Peseran, M. H. and Smith, R. (1995) 'Estimation of long-run relationships from dynamic heterogeneous panels' *Journal of Econometrics* 68(1): 79–114.

Pezzey, J. (1991) *Impacts of Greenhouse Gas Control Strategies on UK Competitiveness*, Department of Trade and Industry, HMSO, London.

Pezzey, J. (1992) 'Analysis of unilateral CO_2 control in the European Community and OECD', *Energy Journal* 13(3): 159–71.

Pindyck, R. S. (1979) 'Inter-fuel substitution and the industrial demand for energy: an international comparison', *Review of Economics and Statistics* 61 (May): 169–79.

Poterba, J. M. (1991) 'Tax policy to combat global warming: on designing a carbon tax', in R. Dornbusch and J. M. Poterba (eds) *Global Warming: Economic Policy Responses*, MIT Press, Cambridge, Mass.

Putman, R. D. (1988) 'Diplomacy and domestic politics: the logic of two-level games', *International Organisation* 42(3): 428–60.

Rabin, M. (1993) 'Incorporating fairness into game theory and economics', *American Economic Review*, December, pp. 1281–1302.

Rasmusen, E. (1989) *Games and Information, An Introduction to Game Theory*, Basil Blackwell, Oxford and New York.

Rawls, J. (1971) *A Theory of Justice*, Harvard University Press, Cambridge, Mass.

Rothenberg, J. (1993a) 'Economic perspectives on time comparisons: an evaluation of time discounting', in N. Choucri (ed.) *Global Accord: Environmental Challenges and International Responses*, MIT Press, Cambridge, Mass.

Rothenberg, J. (1993b) 'Economic perspectives on time comparisons: alternative approaches to time comparisons', in N. Choucri (ed.) *Global Accord: Environmental Challenges and International Responses*, MIT Press, Cambridge, Mass.

Rutherford, T. (1992) 'The welfare effects of fossil carbon restrictions: results from a dynamic trade model', *OECD Economics Department Working Paper* 112, OECD, Paris.

Sinclair, P. (1992) 'The macroeconomics of oil', unpublished D.Phil. thesis, Brasenose College, Oxford.

Sjoberg, H. (1993) 'First time around: creating the global environmental facility', draft mimeo, University College, London.

Smith, C. (1994) 'Carbon leakage in incomplete CO_2 abatement treaties', *Centre for Economic Forecasting Discussion Paper* 26–94, London Business School, London.

Smith, C., Hall, S. and Mabey, N. (1995) 'Econometric modelling of international carbon tax regimes', *Energy Economics* 17(2): 133–46.

Smith, S. (1992) 'The distributional consequences of taxes on energy and the carbon content of fuels', *European Economics* 1, European Community Publications, Luxembourg.

Solow, A. R. (1991) 'Is there a global warming problem?', in R. Dornbusch and J. M. Poterba (eds) *Global Warming: Economic Policy Responses*, MIT Press, Cambridge, Mass.

Science and Technology Advisory Panel (1993) *Analytical Frameworks for Global Warming, Biodiversity and International Waters*, STAP, Global Environment Facility, Washington DC.

Starrett, D. A. and Zeckhauser, R. (1992) 'Treating external diseconomies: markets or taxes?', in A. Markandya and J. Richardson (eds) *Environmental Economics*, Earthscan, London.

Steininger, K. (1994) 'Environmental policy in a trade-interconnected world: is there scope for unilateral action?', paper presented at Fifth Annual Meeting of the European Association of Environmental and Resource Economists, Dublin, June.

Storck, W. (1993) 'CMA says energy tax will cut jobs, increase costs', *Chemical and Engineering News* 71(14): 21.

Symons, E. J., Proops, J. R. L. and Gay, P. W. (1994) 'Carbon taxes, consumer demand and carbon dioxide emissions: a simulation analysis for the UK', *Fiscal Studies* 15(2): 19–43.

Tata Energy Research Institute (1992) *TERI Energy Data Directory and Year Book 1992*, Tata Energy Research Institute, New Delhi.

Tata Energy Research Institute (1993) *Energy Scene in India with Special Reference to Natural Gas*, Draft Report, Tata Energy Research Institute, New Delhi.

Titus, J. (1992) 'The cost of climate change to the United States', in S. K. Majundas, L. S. Kalkstein, B. Yarnal, E. W. Miller, and L. M. Rosenfeld (eds) *Global Climate Change: Implications, Challenges and Mitigation Measures*, Pennysylvania Academy of Science, Pennysylvania.

Toman, M., Pezzey, J. and Krautkraemer, J. (1994) 'Neoclassical growth theory and sustainability', *Discussion Paper* ENR93-14-REV, Resources for the Future, Washington DC.

Tucci, M. (1995) 'Stochastic sustainability in the presence of unknown parameters', *Nota di Lavoro* 64–95, Fondazione ENI Enrico Mattei, Milan.

Ulph, A. (1994) 'Environmental policy and international trade: a survey of recent economic analysis', *Fondazione ENI Enrico Mattei Working Paper* 53–94, Milan.

Ulph, A. and Ulph, D. (1994a) 'Global warming: why irreversibility may not require lower current emissions of greenhouse gases', *Discussion Paper in Economics and Econometrics* 9402, University of Southampton.

Ulph, A. and Ulph, D. (1994b) 'Who gains from learning about global warming?', *Department of Economics Discussion Paper* 94–11, University of Birmingham.

United Nations Conference on Trade and Development (1994) *Combating Global Warming: Possible Rules, Regulations and Administrative Arrangements for a Global Market in CO_2 Emission Entitlements*, United Nations, New York.

United Nations Environment Programme (1992) *United Nations Framework Convention on Climate Change*, United Nations, Geneva.

United Nations Environment Programme Collaborating Centre on Energy and Environment (1992) *UNEP Greenhouse Gas Abatement Costing Study*, UNEP–CCEE, Riso National Laboratory, Denmark.

United Nations (1990) *Annual Bulletin of General Energy Statistics for Europe*, 23.

United Nations (1992) *The United Nations Energy Database*, UN Statistical Division, New York, June 1992.

Vercelli, A. (1994) 'Hard uncertainty and the environment', *Fondazione ENI Enrico Mattei Working Paper* 46–94, Milan.

Verleger, P. K. (1993) *Adjusting to Volatile Energy Prices*, Institute for International Economics, Washington DC.

Vogler, J. (1995) *The Global Commons: A Regime Analysis*, Wiley, London.

Weitzman, M. (1974) 'Prices vs. quantities', *Review of Economic Studies* 41(4): 477–91.

Weiss, E. B. (1993) 'Intergenerational equity: toward an international legal framework', in N. Choucri (ed.) *Global Accord: Environmental Challenges and International Responses*, MIT Press, Cambridge, Mass.

Whalley, J. and Wigle, R. (1991) 'The international incidence of carbon taxes', in R. Dornbusch and J. M. Poterba (eds) *Global Warming: Economic Policy Responses*, MIT Press, Cambridge, Mass.

Whitely, J. (1994) *A Course in Macroeconomic Modelling and Forecasting*, Harvester Wheatsheaf, London.

Williams, R. H. (1990) 'Low-cost strategy for coping with CO_2 emission limits (a critique of "CO_2 emission limits: an economic cost analysis for the USA" by Manne and Richels)', *Energy Journal* 11(4).

Winters, A. L. (1992) 'The trade and welfare effects of greenhouse gas abatement: A survey of empirical estimates', in K. Anderson and R. Blackhurst (eds) *The Greening of World Trade Issues*, Harvester Wheatsheaf, London.

Wirl, F. (1994) 'Pigouvian taxation of energy for flow and stock externalities and strategic, non-competitive energy pricing', *Journal of Environmental Economics and Management* 26(1): 1–18.

World Bank (1992) *World Development Report 1992: Development and the Environment*, Oxford University Press, New York.

World Bank (1993) *World Development Report*, Oxford University Press, New York.

World Commission on Environment and Development (1987) *Our Common Future* (the Brundtland Report), Oxford University Press, Oxford.

World Energy Council Commission (1992) *Energy for tomorrow's world: the realities, the real options and the agenda for achievement. Global report*, WECC, Madrid.

World Resource Institute (1990) *World Resources: A Guide to the Global Environment*, Oxford University Press, Oxford.

World Resource Institute (1992) *World Resources: A Guide to the Global Environment*, Oxford University Press, Oxford.

Young, O. (1993) 'Negotiating an international climate regime: the institutional bargaining for environmental governance', in N. Choucri (ed.) *Global Change: Environmental Challenges and Institutional Response*, MIT Press, Cambridge.

INDEX

Note: references in italics refer to figures and tables.